D1707518

Illegal Logging

The Earthscan Forestry Library
Jeffrey A. Sayer, Series Editor

Illegal Logging: Law Enforcement, Livelihood and the Timber Trade
Luca Tacconi (ed)

Logjam:
Deforestation and the Crisis of Global Governance
David Humphreys

The Forest Landscape Restoration Handbook
Jennifer Rietbergen-McCracken, Stewart Maginnis and Alastair Sarre (eds)

Forest Quality:
Assessing Forests at a Landscape Scale
Nigel Dudley, Rodolphe Schlaepfer, Jean-Paul Jeanrenaud, William Jackson and Sue Stolton

Forests in Landscapes:
Ecosystem Approaches to Sustainability
Jeffrey A. Sayer and Stewart Maginnis (eds)

The Politics of Decentralization:
Forests, Power and People
Carol J. Pierce Colfer and Doris Capistrano

Plantations, Privatization, Poverty and Power:
Changing Ownership and Management of State Forests
Mike Garforth and James Mayers (eds)

The Sustainable Forestry Handbook 2nd edition
Sophie Higman, James Mayers, Stephen Bass, Neil Judd and Ruth Nussbaum

The Forest Certification Handbook 2nd edition
Ruth Nussbaum and Markku Simula

Illegal Logging

Law Enforcement, Livelihoods and the Timber Trade

Edited by Luca Tacconi

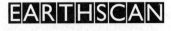

London • Sterling, VA

First published by Earthscan in the UK and USA in 2007

Copyright © Luca Tacconi and CIFOR, 2007

All rights reserved

ISBN-13:   978-1-84407-348-1
Typeset by JS Typesetting Ltd, Porthcawl, Mid Glamorgan
Printed and bound in the UK by Cromwell Press, Trowbridge
Cover design by Susanne Harris

For a full list of publications please contact:

Earthscan
8–12 Camden High Street
London, NW1 0JH, UK
Tel: +44 (0)20 7387 8558
Fax: +44 (0)20 7387 8998
Email: earthinfo@earthscan.co.uk
Web: **www.earthscan.co.uk**

22883 Quicksilver Drive, Sterling, VA 20166-2012, USA

Earthscan is an imprint of James and James (Science Publishers) Ltd and publishes in association with the International Institute for Environment and Development

A catalogue record for this book is available from the British Library

Library of Congress Cataloging-in-Publication Data
Illegal logging : law enforcement, livelihoods and the timber trade / edited by Luca Tacconi.
    p. cm.
  Include bibliographical references.
  ISBN-13: 978-1-84407-348-1 (hardback)
  ISBN-10: 1-84407-348-3 (hardback)
 1. Logging – Economic aspects – Developing countries. 2. Logging – Corrupt practices – Developing countries. I. Tacconi, Luca.
  HD9768.D44I45 2007
  363.25'92–dc22
                                                                            2006039464

This publication is printed on FSC-certified and totally chlorine-free paper. FSC (the Forest Stewardship Council) is an international network to promote responsible management of the world's forests.

Contents

List of Figures, Tables and Boxes vii
List of Contributors x
Foreword by Frances Seymour xii
List of Acronyms and Abbreviations xiv

1 The Problem of Illegal Logging 1
   Luca Tacconi

2 The Social Basis of Illegal Logging and Forestry Law Enforcement 17
   in North America
   Michael R. Pendleton

3 From New Order to Regional Autonomy: Shifting Dynamics of
   Illegal Logging in Kalimantan, Indonesia 43
   Anne Casson and Krystof Obidzinski

4 Turning in Circles: District Governance, Illegal Logging and
   Environmental Decline in Sumatra, Indonesia 69
   John F. McCarthy

5 Illegal Logging, Collusive Corruption and Fragmented Governments
   in Kalimantan, Indonesia 91
   *Joyotee Smith, Krystof Obidzinski, Sumirta Subarudi and
   Iman Suramenggala*

6 Forest Law Enforcement and Rural Livelihoods 110
   David Kaimowitz

7 Rural Livelihoods, Forest Law and the Illegal Timber Trade in
   Honduras and Nicaragua 139
   *Adrian Wells, Filippo del Gatto, Michael Richards, Denis Pommier and
   Arnoldo Contreras-Hermosilla*

8 Livelihoods and the Adaptive Application of the Law in the Forests
   of Cameroon   167
   Guillaume Lescuyer

9 Forest Law Enforcement and Rural Livelihoods in Bolivia   191
   Marco Boscolo and Maria Teresa Vargas Rios

10 Sustainable Forest Management and Law Enforcement:
   A Comparison between Brazil and Finland   218
   Sofia R. Hirakuri

11 Verification and Certification of Forest Products and Illegal
   Logging in Indonesia   251
   Luca Tacconi

12 Illegal Logging and the Future of the Forest   275
   Luca Tacconi

Index   *291*

List of Figures, Tables and Boxes

Figures

2.1	Tree theft spectrum	25
3.1	The district of Berau, East Kalimantan	52
3.2	The district of Kotawaringin Timur, Central Kalimantan	56
7.1	Legality and illegality in forest production in Central America	142
7.2	Power relations governing forestry exploitation in the Sico-Paulaya Valley, Honduras	155
8.1	Evolution of the annual area fee over the last decade	178
10.1	Deforestation rate in the Amazon from 1988 to 2004	225
10.2	Required and actual logging permit procurement process, inspection and enforcement actions in Brazil	234
11.1	Destination of timber product exports from Indonesia in 2003 (volume)	263

Tables

3.1	Estimated legal and illegal log production in East Kalimantan, 2000	50
3.2	Estimated legal and illegal production in Central Kalimantan, 2000	51
3.3	Total log production in Kotawaringin Timur	58
4.1	District government incomes, South Aceh	73
4.2	Total district government budget, South Aceh	73
5.1	Tax revenues and estimated informal payments from local logging permits: Bulungan, Malinau and Nunukan districts, North-East Kalimantan, Indonesia, August 2001	97
5.2	Estimates of unaccounted log exports from North-East Kalimantan to Sabah, Malaysia, 2000 and 2001	100
6.1	Threats to rural livelihoods from illegal forestry activities and from forest law enforcement	119

6.2	Options to address threats to rural livelihoods from illegal forestry activities and from forest law enforcement	130
7.1	Gross and net income and its use by the Collective Society of Romero Barahona and Associates, Sico-Paulaya, Honduras, 2000 and 2001	159
8.1	Study areas	169
8.2	One common tool to assess law contributions to livelihoods	170
8.3	Forest classification in Cameroon	171
8.4	Classification procedure: Contribution to rural livelihoods	173
8.5	Assessment of impact of hunting and NTFP gathering regulations on livelihoods	175
8.6	Main features of the exploitation concession	176
8.7	Contributions of commercial timber exploitation to livelihoods	180
8.8	Costs and revenues from small-scale timber production	182
8.9	Contribution of small-scale timber exploitation to rural livelihoods	183
8.10	The hidden costs of community forests	186
8.11	Contribution of community forests to livelihoods	186
9.1	Forest access by right (hectares managed according to authorized plans)	194
9.2	Attributes of forest-dependent social groups and contribution of the forest to livelihood	200
9.3	Impact of the new forest regime on the livelihood of four forest user groups	207
9.4	Authorized versus actual harvesting in Bolivia (cubic metres)	209
10.1	Comparison of the main features of enforcement approaches between Finland and Brazil	230
10.2	Legal requirements and limitations of logging permits in the Amazon	235
11.1	Estimated use of roundwood by sector, illegal harvest and export of timber products from Indonesia	262
11.2	Conversion factors used to calculate roundwood equivalent	274

Boxes

1.1	Types of illegal forest activities	4
7.1	The Social Forestry System and illegal mahogany exploitation in the Sico-Paulaya Valley, Honduras	145
7.2	The Social Forestry System: Benefits but no secure rights	148
7.3	Grassroots protests against logging in Olancho, Honduras	149
7.4	Volume constraints in the approval process in Honduras	151
7.5	Complex management plans as a regulatory barrier to legality for community-based forest producers in Honduras	152
7.6	Economic and social problems of the Cooperativa Agroforestal Colón, Atlántida Honduras Limitada (COATLAHL)	153

7.7	Crack-downs on illegal logging in the Rio Platano Biosphere Reserve	156
7.8	Law enforcement in a Nicaraguan frontier region	157
9.1	The National Institute for Agrarian Reform (INRA) law	193
9.2	Forest access in Bolivia	195
9.3	Key functions of the Superintendencia Forestal	196

List of Contributors

Marco Boscolo is a Boston-based consultant working on international forest policy, with experience in Latin America, Eastern Europe, Central Africa, South-East Asia and the US.

Anne Casson is a consultant based in Bali. She is also a research associate with the Resource Management in Asia-Pacific Programme, Australian National University, Canberra, Australia.

Arnoldo Contreras-Hermosilla is a fellow at Forest Trends in Washington, DC, US.

Filippo del Gatto is associate researcher at the Tropical Agricultural Research and Higher Education Centre (CATIE) in Turrialba, Costa Rica.

Sofia R. Hirakuri is a researcher at the United Nations University Institute of Advanced Studies in Tokyo, Japan. She is currently researching the interface between international trade policies and sustainable forest management.

David Kaimowitz is programme officer for Environment and Development at the Ford Foundation in Mexico City. At the time that his chapter was written, he was director general of the Centre for International Forestry Research (CIFOR) in Bogor, Indonesia.

Guillaume Lescuyer is an economist at the forestry department of the Centre de Coopération Internationale en Recherche Agronomique pour le Développement (CIRAD) in France. He is based at the office for Central Africa of CIFOR in Yaoundé, Cameroon.

John F. McCarthy is senior lecturer at the Crawford School of Economics and Government, Australian National University, Canberra, Australia.

Krystof Obidzinski is a research fellow at CIFOR in Bogor, Indonesia.

Michael R. Pendleton is senior consultant and owner of Pendleton Consulting LLC, which specializes in natural setting crime and law enforcement. He was associate director and professor of the Society and Justice Program at the University of Washington in Seattle, US, where he occasionally teaches.

Denis Pommier is an agronomist with Améliorer la Gouvernance de la Terre, de l'Eau et des Resources Naturelles (aGter) in Nogent sur Marne, France.

Michael Richards is consultancies manager at the International Non-governmental Organization Training and Research Centre (INTRAC) in Oxford, UK.

Joyotee Smith has worked in various centres of the Cooperative Group on International Agricultural Research (CGIAR). At the time that her chapter was written, she was a consultant at CIFOR in Bogor, Indonesia.

Sumirta Subarudi is a scientist at the Forest Research and Development Agency of the Ministry of Forestry in Bogor, Indonesia.

Iman Suramenggala was executive director of Yayasan Pioner Bulungan in East Kalimantan, Indonesia, at the time that his chapter was written.

Luca Tacconi is associate professor at the Crawford School of Economics and Government, Australian National University, Canberra, Australia.

Maria Teresa Vargas Rios is executive director of Fundacion Natura Boliva in Santa Cruz, Bolivia.

Adrian Wells is research fellow at the Rural Poverty and Governance Group, Overseas Development Institute (ODI), in London, UK.

Foreword

Illegal logging is an issue of profound importance for forests and the communities that depend upon them for their livelihoods. It is often associated with deforestation, in both production forests and protected areas around the world. Deforestation has led to significant losses of biodiversity and is responsible for around 20 per cent of greenhouse gas emissions. Deforestation and illegal logging also have a significant – and often negative – impact on the livelihoods and welfare of forest-dwelling communities.

Despite the recent increase in attention given to the issue, there is considerable uncertainty about the extent of illegal logging, its causes and how to tackle it. This book takes us on a tour of the forested world and explores the sociology, politics, economics, legal and livelihood aspects of illegal logging and law enforcement. In doing so, the book sheds new light on the tangled web of factors that encourage illegal logging.

Illegal logging is a manifestation of a range of underlying economic, political and social causes. Those highlighted in the case studies include community perceptions that illegal logging is not a criminal and harmful activity; biased, inconsistent or over-complex regulatory frameworks; an unwillingness or inability on the part of governments to enforce the law; conflicts between central and lower levels of governments; corruption; limited markets for legal and sustainable timber; and the profitability of illegal logging.

Unless these underlying causes are addressed, illegal logging will continue. The nature of these causes is difficult to change in the short term, given that change will depend, in part, upon structural reform. This means that illegal logging will be with us for some time.

This book also shows that law enforcement can have significant negative impacts on rural livelihoods. In Cameroon, Honduras and Nicaragua, for example, forestry regulatory frameworks are biased against rural people. The same is true of many other countries, where the strict enforcement of forest laws harms, rather than enhances, local livelihoods. This book suggests that consideration should be given to reviewing and, eventually, reforming forestry regulatory frameworks, and especially those laws that discriminate against the rural poor.

Does this mean that nothing can be done about illegal logging? No, that is not the message of this book. Given that government policies on forest management are influenced by the economic benefits that flow from forests and alternative land uses, we need to consider how forests can generate sufficient economic benefits so that governments will increasingly support sustainable forest management. With deforestation being the second largest contributor to greenhouse gas emissions, the obvious place to look for funding for sustainable forest management is the carbon market. Illegal logging may be seen by some as increasing the risks associated with investments in carbon-related forestry projects. But the fact that funding from carbon projects could make sustainable forest management economically viable would significantly reduce the perceived risks. Illegal logging is linked to climate change, and anyone interested in either of these problems needs to give serious consideration to the other.

Policy-makers should also consider encouraging the market for legal and certified forest products. Efforts in this area are useful and should continue since they help to improve the financial and economic viability of sustainable forest management. However, this book points out that large developing country markets are not environmentally sensitive, and they may not change considerably in the short to medium term. This observation implies that marginal expansions of the markets for legal and certified forest products are unlikely to result in a significant reduction in illegal logging. However, the authors point out that even if these markets were to expand markedly, the illegal clearance of forests to make way for agriculture could continue if the economics of land use did not change in favour of forestry.

Equitable and just regulatory frameworks would directly benefit rural people and indirectly foster the establishment of forestry carbon projects and sustainable forest management. Inequitable and unjust regulations exclude rural people from benefiting from natural resources. Exclusion leads to increased conflict, which in turn contributes to an increased risk for carbon forestry projects and for sustainable forest management.

The case studies presented here address a considerable range of issues and countries. However, the book rightly points out that there is still a degree of uncertainty about the extent of the various types of illegal logging and their causes. New and more detailed studies are needed to refine our understanding of the causes of illegal logging and policy options to address it. This book highlights the key issues that should be considered in further studies.

I would like to acknowledge the support provided for this book by the UK Department for International Development (DFID) and the Programme on Forests (PROFOR). The book has also benefited from the Centre for International Forestry Research's (CIFOR's) collaboration with The Nature Conservancy (TNC) and the World Wide Fund for Nature (WWF), with funding from the US Agency for International Development (USAID). I would also like to thank all the authors who have contributed their knowledge of illegal logging and have made possible this useful addition to the forestry literature.

Frances Seymour
Director General, CIFOR
November 2006

List of Acronyms and Abbreviations

AAC	annual allowable cut
AAF	annual area fee
AFE–COHDEFOR	State Forestry Administration–Honduran Forestry Development Corporation
aGter	Améliorer la Gouvernance de la Terre, de l'Eau et des Resources Naturelles
ANU	Australian National University
ASL	Asociación Social del Lugar (local community association)
ATPF	Forest Product Transport Authorization (Brazil)
CAF	Corporación Andina de Fomento (Bolivia)
CAFTA	Central American Free Trade Agreement
CATIE	Tropical Agricultural Research and Higher Education Centre
CBD	Convention on Biological Diversity
CCF	community conservation forest
CF	community forest
CFAF	Communauté Financière Africaine franc (Cameroon)
CFP	certified forest product
CGIAR	Cooperative Group on International Agricultural Research
CHT	community hunting territory
CIFOR	Centre for International Forestry Research
CIG	common initiative group
CINTROP	Centre for International Cooperation in Sustainable Management of Tropical Peatland, CINTROP
CIRAD	Coopération Internationale en Recherche Agronomique pour le Développement
CITES	Convention on International Trade in Endangered Species of Wild Flora and Fauna
cm	centimetre
COATLAHL	Cooperativa Agroforestal Colón, Atlántida Honduras Limitada
CONAMA	Conselho Nacional do Meio Ambiente, Brazil (National Council for the Environment)

DFID	UK Department for International Development
EIA	environmental impact assessment
EMBRAPA	Brazilian Agricultural Research Corporation
ENGO	environmental non-governmental organization
EU	European Union
FAO	United Nations Food and Agricultural Organization
FDC	Forestry Development Centre (Finland)
FES	*función economica y social* (socio-economic role, Bolivia)
FFMC	forestry fee management committee
FLEG	forest law enforcement and governance
FLEGT	forest law enforcement, governance and trade
FMP	forest management plan
FMU	forest management unit
FSC	Forest Stewardship Council
FUNAI	Fundação Nacional do Índio (Brazil)
G8	Group of 8 industrialized nations: Canada, France, Germany, Italy, Japan, Russia, the UK and the US
GDP	gross domestic product
GFTN	Global Forest and Trade Network
GLNP	Gunung Leuser National Park (Indonesia)
GTZ	Gesellschaft für Technische Zusammenarbeit
ha	hectare
HIPC	heavily indebted poor country
HPH	*Hak Pengusahaan Hutan* (forest concession contractor, Indonesia)
HPHH	*Hak Pemungutan Hasil Hutan* (forest products harvesting right, Indonesia)
IBAMA	Brazilian Institute of the Environment and Renewable Natural Resources
ICDP	Integrated Conservation and Development Project
IMF	International Monetary Fund
INAFOR	National Forestry Institute (Nicaragua)
INPE	National Institute of Spatial Research (Brazil)
INRA	Instituto Nacional de Reforma Agraria (National Institute for Agrarian Reform, Bolivia)
INTRAC	International Non-governmental Organization Training and Research Centre
IPPK	*Izin Pemungutan dan Pemanfataan Kayu* (timber extraction and utilization permit, Indonesia)
ITTO	International Tropical Timber Organization
IUCN	World Conservation Union
km	kilometre
LDP	Leuser Development Programme
LEI	Indonesian Eco-labelling Institute
LEO	law enforcement officer
LMU	Leuser Management Unit

m³	cubic metres
MAO	Movimiento Ambientalista de Olancho (Honduras)
METLA	Finnish Forest Research Institute
MINFOF	Ministry of Forests and Fauna (Cameroon)
MMA	Ministério do Meio Ambiente (Ministry of the Environment and Legal Amazonia, Brazil)
MSDP	Ministry for Sustainable Development and Planning (Bolivia)
NGO	non-governmental organization
NPFE	non-permanent forest estate
NTFP	non-timber forest product
ODI	Overseas Development Institute
PFE	permanent forest estate
POAF	*planes operativos de aprovechamiento forestal* (annual harvesting plans)
PROFOR	Programme on Forests
RWE	roundwood equivalent
SAKO	permit for transporting wood
SF	Superintendencia Forestal
SISNAMA	National Environmental Management System (Brazil)
SMP	Simple Management Plan
SSF	Social Forestry System (Honduras)
SSV	sales of standing volume
TCO	*Territorios Comunitarios de Origen* (Indigenous territories)
TNC	The Nature Conservancy
UK	United Kingdom
UN	United Nations
UNDP	United Nations Development Programme
UNFF	United Nations Forum on Forests
US	United States
USAID	US Agency for International Development
WWF	World Wide Fund for Nature (*formerly* the World Wildlife Fund)
YLF	Yayasan Leuser International
YLI	Leuser International Foundation

1
The Problem of Illegal Logging

Luca Tacconi

Introduction

Illegal logging has become one the most prominent global forest policy issues. The *G8 Illegal Logging Dialogue* was launched at the annual meeting of the World Bank and the International Monetary Fund (IMF) in Singapore in September 2006. This initiative will bring together legislators from the Group of 8 (G8), China, India and other key timber producer countries, such as Brazil, Cameroon, the Democratic Republic of Congo, Gabon, Ghana, Indonesia, Malaysia, the Republic of Congo and Peru. The *Dialogue* is one of several international initiatives on illegal logging, which include regional ministerial conferences on forest law enforcement and governance (FLEG) in Africa, Asia and Eastern Europe. The Convention on Biological Diversity (CBD), the Food and Agricultural Organization of the United Nations (FAO), the International Tropical Timber Organization (ITTO), the United Nations Forum on Forests (UNFF) and the World Conservation Union (IUCN) have issued statements on illegal logging, have organized meetings and prepared reports on the issue, and have included illegal logging in their work programmes. The European Commission has adopted a European forest law enforcement, governance and trade (FLEGT) action plan. Japan and Indonesia have launched the Asian Forest Partnership, which has illegal logging as one of three focal areas.

Global awareness about the need to address weak governance to support economic development spilled over into the environmental field during the 1990s and contributed to the rise of illegal logging as a major policy issue. Awareness and concern about illegal logging was raised particularly by non-government organizations (NGOs) such as the Environmental Investigation Agency (1996) and Global Witness (1999). Other environmental NGOs such as Friends of the Earth (Newell et al, 2000), Forests Monitor (2001), Greenpeace (2000a), the World Resources Institute (Brunner et al, 1998) and the World Wide Fund for Nature (WWF, 2002) have further raised public awareness about the problem and contributed to propelling illegal logging centre stage in the global forest policy arena. Illegal logging and trade is not, however, a new issue, as demonstrated by a closer look at the history of human use of forests, wonderfully told by Williams (2003, p171) in 1700s France:

> An underpaid bureaucracy of foresters accepted bribes to cut and sell wood illegally; Louis XV wanted quick revenues and alienated 800,000 acres of royal forest; a growing population needed more grain and therefore cleared agricultural land; the peasants claimed their rights to use the forest to a degree greater than good silviculture could withstand; and mining and industry stripped some areas.

And in New England:

> The years between 1722 and 1776 are muddled with illegal cutting on the part of the woodsmen, duplicity on the part of the Crown agents, and conniving on the part of the contractors. New England merchants carried on a thriving trade in masts and timber with Spain and Portugal even though these countries were officially at war with Britain, because better prices were to be had from those countries. (Williams, 2003, p231)

These examples point to the probability that illegal logging is not a result of the confluence of recent events that have led to a possible increase in its frequency. The latter statement is somewhat tentative for good reason. Despite the global interest in illegal logging, there has been limited rigorous work on the types, extent and causes of illegal logging. Therefore, this book devotes particular attention to clarifying the causes of illegal logging and the implications for policy-making, and it also considers issues and gaps in knowledge concerning the types and extent of the illegal logging. Let us consider the meaning of illegal logging, possible extent and some of its assumed impacts and causes before discussing in detail the scope of the book.

Illegal logging: The extent, its impacts and causes

What is illegal logging?

The term illegal logging is commonly used in policy forums and in the literature to refer to a range of illegal activities related to forest ecosystems, forest industries, and timber and non-timber forest products (NTFPs). The international debate on illegal logging has focused particularly on the illegal harvest of logs, possibly as a result of at least two factors. First, environmental NGOs are concerned about the ecological impacts of illegal logging. They regard illegal harvesting and illegal deforestation as having significant negative environmental impacts. Second, statistics on harvest volumes are more widely available than other information on forest management – for example, the number of infringements of harvesting regulations. Estimating illegal harvest rates is therefore easier than assessing other types of illegalities. This focus on the illegal harvest has led to the widespread use of the term illegal logging to denote the whole problem of the existence of illegal forest activities. This common use is maintained in this book given that the term is familiar to many readers and decision-makers. Nevertheless, it is important to keep sight of the fact that there are various types of illegal activities encompassed by the term. They need to be clearly recognized because they may have different impacts, causes and implications for policy-making.

These illegal activities range from acts related to the establishment of rights to land, to corrupt activities to acquire forest concessions, and to unlawful activities at all stages of forest management and the forest goods production chain, from the planning stage, to harvesting and transport of raw material and finished products, to financial management.[1] The different illegal activities summarized in Box 1.1 may be linked to each other in different ways; but two of the most significant links are worth stressing. First, violations of indigenous peoples' rights may result in the establishment of forest operations that have a legal appearance. Timber extracted by these operations may seem legal to unaware traders and consumers unless schemes aimed at verifying legality (discussed in Chapter 11) also assess that due process is followed in the allocation of forest concessions. Second, all violations can occur as the result, or at the prompting, of public officials' corruption. Corruption can affect the allocation of forestland, monitoring of forest operations and law enforcement.

It is also worth noting that the occurrence of violations does not necessarily imply that the focus should necessarily be on their prevention and repression. It is plausible that in some instances a revision of the legislation may be warranted. An example of this point is a situation in which the legislation favours large-scale industrial harvesting operations and, as a result, small-scale rural operations find themselves operating illegally. This aspect receives particular attention in this book.

Finally, the typology of illegal activities presented in Box 1.1 highlights the fact that the various infringements tend to have different impacts on the environmental, economic and social spheres. For example, infringements of forest harvesting regulations have environmental impacts,[2] whereas infringements of timber transport regulations do not normally impact on the environment, and infringements of land laws that dispossess communities of land have greater social impacts than infringements of forest harvesting regulations.

The extent and impacts of illegal logging

Illegal logging appears to occur in many countries around the world, and estimates point to some 70 countries that may be affected (WWF, 2002). Most country-level estimates of illegal logging focus on the rate of illegal harvest, and it has been reported that these rates are above 50 per cent of the total harvest in many countries (Contreras-Hermosilla, 2002; SGS Trade Assurance Services,

> **Box 1.1 Types of illegal forest activities**
>
> - Violations of indigenous peoples' rights and public or private ownership rights may involve acts against constitutional, civil, criminal or administrative law.
> - Violations of forest management regulations and other contractual agreements in either public or private forestlands are acts against forest legislation; this is the category that includes most of the acts that may be most appropriately referred to as 'illegal logging'.
> - Violations of transport and trade regulations include acts that violate forest legislation; but they may be related to legally or illegally harvested forest products. This category is referred to as illegal forest trade.
> - Timber processing activities may be regulated by industry and trade-related legislation, as well as forest legislation. In this category, a violation directly linked to illegal logging is the use of illegally harvested logs.
> - Violation of financial, accounting and tax regulations may involve acts related to legally and/or illegally harvested and traded timber. This category is referred to in this book as illegal financial activities.
>
> *Source:* adapted from Tacconi et al (2003)

2002; WWF, 2002; Tacconi et al, 2003; Seneca Creek Associates and Wood Resources International, 2004). Reported statistics appear to be, however, rather uncertain and show a large degree of variation, partly because different definitions are often used and confusion arises. For instance, Brack and Hayman (2001) reported that in the mid 1990s illegal trade irregularities were estimated to be 15 per cent of the total trade. More recently, Seneca Creek Associates and Wood Resources International (2004) reported that about 5 to 10 per cent of the value of global wood products trade can be traced to roundwood of 'suspicious' origin. That there may be significant problems with the statistics reported so far became evident as the writing of this book was coming to a conclusion, and parallel work indicated that the illegal harvest in Cameroon may not be as significant as previously thought (Cerutti and Tacconi, 2006). Rather than reporting statistics from various countries to show how significant the problem is, which in any case are available from the sources reported above, we note that current knowledge indicates that the size of the illegal harvest may be significant in many countries, but that there may be (considerable) problems with available estimates.

There is also a considerable lack of knowledge about the impacts of illegal logging. The debate has focused mostly on the perceived negative impacts. According to Global Witness (2001), Contreras-Hermosilla (2002) and Seneca Creek Associates and Wood Resources International (2004), illegal logging may:

- cause deforestation and loss of biological diversity;
- result in government revenue losses of billions of dollars;
- foster a vicious cycle of bad governance (corrupt individuals gain power through illegal revenues and then may support bad governance to maintain revenues and acquire more power);
- contribute directly to increased poverty when people lose their resources, and indirectly as a result of a reduction in government revenues, which could in turn be made available for poverty reduction programmes;
- contribute to funding national and regional conflicts, thereby exacerbating them;
- distort forest product markets.

It needs to be recognized, however, that illegal logging has a positive side for some of the stakeholders:

- The establishment of alternative land uses on illegally deforested land may provide benefits to those involved.
- National or local governments may receive higher revenues as a result of illegal or legalized land conversion and increased timber production from illegal logging.[3]
- Military and police forces derive income from illegal logging and may be more willing to support the government.
- Many people, including the poor and unemployed, derive an income from illegal logging.
- Lower timber prices increase the competitiveness of national industries.
- Consumers may benefit as a result of lower prices (Tacconi et al, 2003).

The negative and positive aspects of illegal logging receive particular attention in this book because they are fundamental to explaining why and how illegal logging persists despite so many governments and international organizations having declared their commitment to combating it.

The causes of illegal logging

Contreras-Hermosilla (2002) recognizes that understanding the causes of illegal logging is a precondition to designing effective counter-initiatives. He also notes that the causes have not been studied in detail. This situation persists despite a number of initiatives aimed at stopping illegal logging. Because of this situation, this book devotes significant attention to the causes of illegal logging. Let us summarize the main causes of illegal logging highlighted in the literature. The organization of causes under the various headings is not necessarily attributable to the sources cited under them. It is my characterization of the main causes reported in the literature.

Institutional problems

There is a range of institutional problems that may affect illegal logging. The key ones relate to the political and the property rights spheres.

Tacconi et al (2003) noted that whether a state is weak is relevant to illegal logging because such a state has, by definition, limited capacity to develop appropriate governance processes, to develop legislation, to enforce the law and to guarantee fairness in the exercise of power. A weak state may also find it difficult to control or to fund its security forces, which are then able to profit from the legal and illegal extraction of timber or simply need to rely on it to fund their operational needs. It is worth noting that apparently strong states, too, may allow security forces to profit and fund operational needs from timber harvesting, such as in the case of Indonesia during the Suharto regime (Barber and Talbott, 2003; Obidzinski, 2003). It could be argued that this is, indeed, a weakness of the state because once the regime changes from dictatorship to democracy, the state may find it difficult to stop illegal logging because the security forces still rely on it. Illegal logging in weak states needs to receive further attention, as noted later.

In relation to property rights, there is often an imbalance between government claims on forest resources and their capacity to administer it. In fact, states claim control, at least *de jure*, over most of the forested land (White and Martin, 2002). In many countries, forestland ownership rights are unclear or inexistent and, in some cases, the boundaries of public forestlands are not demarcated either on paper or on the ground (Contreras-Hermosilla, 2002). As a result, Contreras-Hermosilla (2002) remarks, those who trespass on public lands may not even know they are actually on public lands. In this respect, it is important to note that in some countries, for example Indonesia (Fay and Sirait, 2002), public ownership of land is actually disputed by local communities, who claim that their traditional land rights have been violated by the state.

Lack of government capacity

Lack of government capacity to stop illegal logging is perhaps the cause that has received most attention among practitioners involved in the design of initiatives to address illegal logging. This is demonstrated, for example, by the fact that the need to build capacity is one of the objectives of the European Union action plan on FLEGT, and that the UK Department for International Development has provided technical assistance to the Indonesian government to fight illegal logging. Let me emphasize that the lack of capacity discussed here may apply to both strong and weak states. The fundamental difference is that technical assistance projects seeking to build capacity in weak states are less likely to succeed.

Contreras-Hermosilla (2002) notes several factors that are an expression of lack of government capacity and contribute to illegal logging. First, forestry operations take place in large and remote areas removed from public scrutiny, the press and monitoring agencies. In other words, government agencies lack the capacity to monitor those areas, and the probability of detection is therefore low. Second, government forest agencies lack the capacity to carry out accurate forest inventories. This results in a weak or absent legal baseline to impose realistic forest management practices and to measure compliance. Third, forestry regulatory frameworks are unclear, incomplete and constantly changing because of lack of regulatory capacity. As a result, they are open to individual interpretation and easier to bend, therefore providing fertile ground for corruption. Fourth, and linked to the previous point, penalties for illegal forest acts are not high enough to constitute a significant deterrent to forest crime. Limited deterrence results in companies opting to carry out extensive illegal activities.

Corruption

A number of sources have noted the relationship between corruption, an illegal activity in itself, and illegal logging (Callister, 1999; Newell et al, 2000; Scotland, 2000; Lawson, 2001; Palmer, 2001; Siebert, 2001; Environmental Investigation Agency and Telapak Indonesia, 2002; Global Witness, 2004).

Contreras-Hermosilla (2002) notes several possible factors that contribute to widespread corruption in the forest sector. First, governments adopt many regulations to achieve a better use of forest resources. This proliferation of regulations generates opportunities for corruption. Second, forests are remote from decision-making centres. Therefore, field officers have broad discretionary powers and a great deal of opportunities for corrupt behaviour. Third, government officials in developing countries have relatively low salaries while they control high-value products. The incentives for corruption are therefore considerable.

The role of business

NGOs have scrutinized the behaviour of forestry companies in several countries (World Rainforest Movement and Forests Monitor, 1998; Global Witness, 1999, 2002; Greenpeace, 2000b, 2002). The NGOs have essentially pointed out that many companies violate forestry and other regulations to increase their profits. The above discussion of other causes highlights the three main reasons that explain how companies may manage to harvest illegally:

1 Companies may share profits with the military and they are therefore allowed to operate.
2 Lack of law enforcement capacity leads to non-detection or non-prosecution of companies' violations.
3 Companies bribe public officials.

The timber trade

The role of the timber trade in supporting illegal logging looms large, particularly in the work of NGOs. There have been many reports pointing out how illegally harvested timber was exported to countries in America, Asia and Europe (Glastra, 1999; Greenpeace, 2000b; Global Witness, 2001, 2005; Brack et al, 2002; WWF, 2005). The logic is that not only forestry companies harvest illegally, as noted above, but they are fundamentally supported by non-environmentally sensitive markets that demand timber products without considering whether the timber was harvested illegally. This concern about the timber trade is at the origin of a number of initiatives seeking to promote verification and certification of timber products, such as the regulatory framework for legal timber trade included in the European action plan on FLEGT, which is discussed in detail in Chapter 11.

The economics of forest management
Economic incentives and disincentives that have received limited attention within the policy circles debating illegal logging are those already noted above – that is, the low penalties faced by companies, the incentives faced in a corrupt system and the benefits generated by the trade in illegal timber. Economic incentives and disincentives do need to receive more attention because illegal logging, like logging and any other economic activity, is essentially driven by the financial benefits that it generates. It is worth pointing out that states also consider economic benefits and costs in their decision-making, in general, and on the environment in relation to both national and international issues (Hempel, 1996). How economic incentives affect the problem of illegal logging, at the company as well as the state level, therefore deserves more attention.

The potential multiple causes of illegal logging summarized above present a formidable challenge to the development of policies aimed at addressing illegal logging. Contreras-Hermosilla (2002, p10) articulates this problem well by noting:

> ... [the] causes of forest crime are rooted in the culture and governance systems of societies. Underlying causes are complex, vary from country to country, and include structural problems such as economic and political power inequalities. Their roots are linked to bureaucratic institutions and various political systems. They are always contextual and [depend] on aspects such as the country's policies, traditions, democratic levels, and so on. Thus, it is difficult to identify general causes of forest crime.

One question that arises from this quote is whether there are general causes of illegal logging, rather than just context-specific ones. Finding general causes of illegal logging would obviously facilitate the development of appropriate policies. This issue is therefore at the core of the book.

Scope of this book

The global debate on illegal logging gathered momentum in the early 2000s and lead to the first ministerial FLEG conference held in Bali in 2001. The debate at the conference and the successive meetings associated with the FLEG process particularly focused on how to 'combat' illegal logging, which initially had been considered simply a problem of law enforcement.[4] This focus is also shared by discussions on illegal logging in other regions – for example, in the process that led to the ministerial FLEG conference for Africa held in Cameroon in 2003.

The programme of research on illegal logging carried out at the Centre for International Forestry Research (CIFOR) and by researchers associated with CIFOR was increasingly indicating, however, that illegal logging had complex causes that needed to be better understood in order to assess if and what policies could address the problem effectively. Reflection on the types of illegalities also led to increased concern that lack of understanding of the nature of the problem could lead to the development and implementation of policies with negative and inequitable impacts on rural livelihoods. A third strand of the work carried out by CIFOR involved assessing the outcomes of an attempt to promote verification and certification of timber products in Indonesia, China and Japan. This work allowed us to start considering the implications of the various causes of illegal logging for practical initiatives aimed at addressing the timber trade, which is one of the apparent causes of illegal logging. This book originates in the learning generated by these three strands of work.

The complex nature of illegal logging and the limited knowledge available on it did not allow us to start from clear and testable hypotheses about the causes and impacts of illegal logging. The book is therefore based on an exploratory research approach that seeks to generate new understanding about a poorly understood problem and, hopefully, to develop new hypotheses to be tested (Babbie, 2004). The chapters included in the book were selected on the basis of our knowledge of illegal logging because they appeared to highlight key aspects of the causes and the implications for policies. The idea is to look at them together and to outline the implications for key questions that need to be answered about illegal logging:

- Are there general causes of illegal logging?
- What are the implications of the nature of illegal logging for rural livelihoods?
- What are the general implications for policies of the answers to the above questions?
- What are gaps in knowledge that need be filled to further improve our understanding of the extent, impacts, causes and policy options to address illegal logging?

Each chapter considers, to different degrees, the extent, causes and implications for policies, and, in some cases, the extent of illegal logging in the case study considered. Some chapters provide detailed policy recommendations that hopefully will contribute to policy development in the specific countries considered. These detailed recommendations are not summarized in the final chapter, which, instead, draws out the key implications derived from bringing together the arguments presented in the various chapters and from considering other relevant literature.

We do acknowledge that the book has shortcomings. The first one is that it does not question sufficiently the extent of illegal logging in the countries considered. While the work on this book was progressing, it was accepted that the available statistics on illegal logging (e.g. illegal harvest) were reasonably correct, even if approximate. As noted above, recent work indicates, however, that there may be major problems with the statistics taken for granted so far (Cerutti and Tacconi, 2006). We do consider, however, the types and known extent of illegal logging in some of the countries covered in the case studies. This analysis is aimed at showing some of the key issues that need to be addressed in estimating the extent and the impacts of the problem, as well as the approximate nature of existing assessments and the need for further research. The book would have also been strengthened by considering illegal logging in the context of weak states and the influence of corporate interests on policy-making and law enforcement. Some of its implications will be noted in the final chapter.

Outline of the book

The ongoing international debate has put significant emphasis on the need to strengthen law enforcement to control illegal logging. Indeed, it would appear logical that an illegal act be countered by tougher law enforcement. While in certain cases this approach may be warranted, putting the emphasis on repression has obscured the fact that illegal logging is not always an illegal act perpetrated by criminals. Illegal logging may not necessarily be perceived as a criminal act or condemned by the local community. There are cases in which illegal logging is conducted by (respected) members of a community, and illegal logging is sanctioned both by community members and by government authorities, including enforcement agents. In Chapter 2, Michael Pendleton discusses the importance of considering illegal logging in the context of social relations. The chapter draws on a long-term research programme conducted in Canada and the US. It shows how the logging community accommodates tree theft when it contributes to community cohesion and stability, and how accommodation of tree theft is part of the forest services' organizational culture of supporting resource-dependent communities and, more fundamentally, a forest-based rural culture. The chapter also highlights the importance of considering the factors that affect the threshold that determines when (illegal) logging becomes an unacceptable practice.

The perception of illegal logging as a criminal act is further considered in Chapter 3. Anne Casson and Krystof Obidzinski discuss illegal logging in Indonesia at the national and district levels by presenting the case of two districts in Kalimantan, the Indonesian part of the island of Borneo. They show that illegal logging is not a simple case of criminal behaviour, but a complex economic and political system involving multiple stakeholders and government authorities. Furthermore, illegal logging is not a stationary condition that can be dealt with effectively through coercive or repressive measures alone. Rather, it should be viewed as a dynamic and changing system deeply engrained in the realities of rural life in Indonesia. They also note how decentralization of government has created a supportive environment for illegal logging and allowed it to gain resilience. The findings of the study imply that the needs and the interests of the various stakeholders, including local communities, have to be understood to address illegal logging, and that the national and local governments should address the issue of illegal logging in the broader context of natural resource management and development.

Chapter 4 discusses the complex nature of illegal logging at the local level. Presenting a case study of a district on the island of Sumatra, Indonesia, John McCarthy shows, again, the complex nature of illegal logging and how district governments, businesses, local (corrupt) government officials, local leaders and rural people all benefit from illegal logging and, therefore, support it to a certain extent. He suggests that several conditions would need to change for illegal logging to decrease. These conditions include a reduction in the demand for timber; the availability of livelihood options alternative to illegal logging; an increase in district revenues from sources other than logging; the development of greater awareness of the impacts of illegal logging; and the establishment of a more accountable political system.

The chapters previously mentioned discuss how in certain situations illegal logging is accommodated by stakeholders. Certainly this does not imply that this is always the case and that illegal logging provides appropriate benefits to the stakeholders. Corruption is recognized as a means by which some agents get access to resources, including timber, to which they do not have rights. Corruption may also result in inequitable and inefficient allocation of resources. In Chapter 5, Joyotee Smith and co-authors discuss illegal logging and corruption in three districts in Kalimantan, Indonesia. They find that collusive corruption (a type of corruption in which individual government officials and businesses collude to rob the government of revenues) became widespread in Indonesia after the fall of the Suharto regime and led to increased illegal logging. They note that government-wide sustained reforms and institutional strengthening are necessary to decrease collusive corruption. They find that greater accountability of government and improved law enforcement (including legal and judicial reform) are needed to reduce corruption and illegal logging.

From considering how illegal logging is not always just a law enforcement problem because it may be entrenched in social, economic, political and institutional systems, we have come almost full circle to stressing the need for law enforcement. Obviously, law enforcement is required, and, indeed, Chapter 10 addresses in detail further issues concerning the law and forests. Previous chapters have already noted that to stop illegal logging requires more than just enforcing the law. There are other aspects of the problem that need consideration. One of these aspects is the impact of law enforcement on livelihoods. In Chapter 6, David Kaimowitz discusses how law enforcement can negatively affect livelihoods if the law or its enforcement discriminates against rural people. He highlights some of the potential negative impacts of illegal logging on livelihoods. The fact that both law enforcement and illegal logging may negatively affect livelihoods points out that to address illegal logging may present complex and difficult policy choices. Kaimowitz proposes that priority be given to reforming both forestry laws that discriminate against low-income households and the law enforcement organizations. Law enforcers, he suggests, should focus on the biggest violators, enforce the laws that favour rural livelihoods and involve local communities in law enforcement.

Laws that discriminate against the forestry activities carried out by local communities may force them into illegality. In Chapter 7, Adrian Wells and colleagues analyse the scale and dynamics of the illegal timber trade, focusing on the involvement of local communities, the legal, institutional and economic factors that drive it, and the livelihood and poverty implications (positive and negative), both directly and in terms of impacts on state revenue. They examine how the existing legal and institutional framework, within and beyond the forest sector, presents community-based forest producers with significant barriers to legal compliance. These span denial of secure tenure rights, over-complex regulation and associated corruption. They recommend the following policy changes: strengthening of community rights to land and forest resources; removing timber volume constraints (where they cap harvesting below the annual allowable cut) and restrictions on farmers' rights to plant and harvest trees on their land; and the simplification of administrative procedures to increase the returns to legal forest management by small-scale producers.

The complex relationships between the law, law enforcement, illegal forest activities and livelihoods are further explored in Chapter 8 in the context of Cameroon. Guillaume Lescuyer documents some of the contributions of forests to livelihoods, the forestry regulatory framework and how some rural livelihood options, such as hunting, gathering and small-scale timber extraction, have been outlawed. He also notes that commercial forestry could provide positive contributions to livelihoods through the devolution of a share of tax revenues to local councils. This is the formal legal picture at least. The informal aspects of forest use are somewhat different. The law that excludes communities from the direct use of forest resources is not strictly enforced (in a similar way as the case treated in Chapter 2), and potential negative effects are being avoided. On the other hand, the 'adaptive' implementation of the law may not result in significant livelihood benefits from commercial forestry given the apparent leakage or ineffective use of revenues at the district level. Lescuyer suggests revising the forestry regulatory framework with the aim of allowing rural people access to forest resources. This revision should be supplemented by simplifying administrative procedures and making them more accessible to local people and enforceable by government officials, and by providing additional means for the decentralized forestry services and better training of forestry officials in order to combine their current technical and enforcement skills with new participatory management ones.

After several chapters that discuss the various aspects of illegal logging and law enforcement, and that provide policy recommendations to address them, it is useful and refreshing to consider the case of Bolivia, a country that has already carried out extensive forest sector reforms with positive effects on livelihoods and forest management. In Chapter 9, Marco Boscolo and Maria Teresa Vargas Rios present the forestry regulatory framework and its (mainly) positive implications for rural livelihoods. They note that some problems remain, such as high direct costs of compliance, some unrealistic prohibitions (such as using chainsaws to process timber in the forest) and shortage of resources to develop forest management plans. They stress, however, that the reform process devoted particular attention to improving the conditions of disenfranchised groups, including indigenous peoples, and that key elements of the forest service system now include commitment to honesty, transparency, accountability and sustainable forest management; responsiveness to users' needs; and openness to self-criticism. The government has recognized that some problems remain, such as a certain degree of illegal timber harvest and illegal deforestation. Therefore, it has embarked on a further review of its operations with an increased focus on incentives rather than on rigid regulations. Boscolo and Vargas Rios recommend increased coordination among various government organizations, the reduction of open-access forests by transferring rights more efficiently, a reconsideration of the purpose of forest management plans, and the development of a system to monitor and evaluate the impacts of law enforcement on livelihoods.

I noted earlier in this chapter that commercial logging activities are thought to contribute significantly to the volumes of timber harvested illegally, although the relative contribution of livelihood-based and commercially based activities is unknown in many countries. Chapters 10 and 11 consider commercial illegal logging and the role of regulatory and market instruments in controlling it. Sofia Hirakuri considers forest law compliance in Brazil and Finland in Chapter 10. She notes that many countries have appropriate forestry regulatory frameworks; the problem is low compliance. Hirakuri finds that Brazil, indeed, has low forestry law compliance, which she ascribes to several factors, such as complicated administrative procedures for procuring logging permits, deficient processes for monitoring forest management, ineffective systems to impose and collect penalties, scarce financial resources for enforcing activities, and financial disincentives for sustainable forest management. Finland, conversely, is one of the most successful countries in terms of compliance. Private forest owners produce some 75 per cent of the wood used by the industry. Finland has, therefore, developed a law compliance system that relies upon a mix of financial incentives and regulatory measures. Financial incentives aimed at supporting sustainable forestry include direct financial assistance, low-interest loans and forest tax deductions. Owners who do not comply with the forestry law stand to lose financial incentives. The regulatory framework does set penalties; but it is primarily designed to foster compliance through incentives, as just noted, and through a consensus-oriented approach that uses a forest extension and monitoring system aimed at providing advice to forest owners and carrying out inspections of forest areas at low-intensity sampling of between 3 and 5 per cent. By making a comparison of forestry practices in the two countries, Hirakuri suggests ways of improving law compliance in Brazil.

In Chapter 11, Luca Tacconi considers the potential role of verification and certification of timber products in controlling illegal logging. He considers the experience of a programme carried out by The Nature Conservancy and the WorldWide Fund for Nature, which seeks to reduce illegal logging in Indonesia and the implications for other initiatives that aim to promote verification of legality and certification of timber products, such as the Voluntary Partnership Agreement scheme of the European Union. Tacconi shows that companies in Indonesia do not face sufficient incentives to adopt verification and certification given that the cost of illegally produced logs is significantly lower than that of legal logs, and there is no significant market premium for verified and certified timber products. He also finds that even if demand for verified and certified timber from export markets increased substantially, the considerable size of the domestic market for timber products implies that illegal logging would continue to supply the domestic market unless law enforcement stopped it. This suggests that a market-based approach is unlikely on its own to solve the problem of illegal logging.

In the final chapter, Luca Tacconi derives the key findings of the book by bringing together the arguments presented in the preceding chapters and in other relevant literature. First he notes key gaps in knowledge about both the extent of illegal logging and its environmental, social and economic impacts. Then he considers the major causes of illegal logging that emerge from the chapters and other studies, which include community perceptions that illegal logging is not a criminal and/or harmful activity, inconsistent or over-complex regulatory frameworks that lead forestry companies and rural people to infringe them, lack of government willingness and/or capacity to enforce the law, conflicting interests over forest management between central and lower-level governments, corruption, and the economic and financial benefits of illegal logging. While not discounting the role of other variables, particularly community views and corruption, he suggests that government objectives and the economic and financial benefits of illegal logging are the two main driving forces behind illegal logging. Changing the economics of land use in favour of forest management may change the attitude of both governments and communities towards the forest and illegal logging. This outcome is most likely to occur as a result of changes in carbon markets that allow payments for avoided deforestation.

Notes

1 This discussion of types of illegal activities and Box 1.1 are based on Tacconi et al (2003).
2 So long as these regulations are scientifically designed to maintain environmental functions, as discussed in Chapter 6.
3 That national or local governments might receive higher revenues as a result of illegal or legalized land conversion and increased timber production is perhaps counterintuitive. Illegally harvested timber may, however, be taxed by governments, as discussed in later chapters.

4 This point is exemplified by the fact that the initial proposed title of the Bali Conference was not supposed to include the term 'governance'.

References

Babbie, E. (2004) *The Practice of Social Research*, 10th edition, Belmont, CA, Thomson Wadsworth

Barber, C. V. and Talbott, K. (2003) 'The chainsaw and the gun: The role of the military in deforesting Indonesia', *Journal of Sustainable Forestry*, vol 16, no 3/4, pp137–166

Brack, D., Gray, K. and Hayman, G. (2002) *Controlling the International Trade in Illegally Logged Timber and Wood Products*, London, Royal Institute of International Affairs

Brack, D. and Hayman, G. (2001) *Intergovernmental Actions on Illegal Logging, Options for Intergovernmental Action to Help Combat Illegal Logging and Illegal Trade in Timber and Forest Products*, London, Royal Institute of International Affairs

Brunner, J., Talbott, K. and Elkin, C. (1998) *Logging Burma's Frontier Forests: Resources and The Regime*, Washington, DC, World Resources Institute

Callister, D. J. (1999) *Corrupt and Illegal Activities in the Forestry Sector: Current Understandings, and Implications for World Bank Forest Policy*, Washington, DC, World Bank

Cerutti, P. O. and Tacconi, L. (2006) 'Forests, illegality and livelihoods in Cameroon', Working Paper no 35, Bogor, Centre for International Forestry Research

Contreras-Hermosilla, A. (2002) 'Law compliance in the forestry sector: An overview', WBI Working Papers, Washington, DC, World Bank

Environmental Investigation Agency (1996) *Corporate Power, Corruption and the Destruction of the World's Forests: The Case for a New Global Forest Agreement*, London, Environmental Investigation Agency

Environmental Investigation Agency and Telapak Indonesia (2002) *Above the Law: Corruption, Collusion, Nepotism and the Fate of Indonesia's Forests*, London and Bogor, Environmental Investigation Agency and Telapak

Fay, C. and Sirait, M. (2002) 'Reforming the reformists in post-Suharto Indonesia', in Colfer, C. J. P. and Resosudarmo, I. A. P. (eds) *Which Way Forward? People, Forests and Policymaking in Indonesia*, Washington, DC, Resources for the Future

Forests Monitor (2001) *Sold Down the River: The Need to Control Transnational Forestry Corporations: A European Case Study*, Cambridge, Forests Monitor

Glastra, R. (1999) *Cut and Run: Illegal Logging and Timber Trade in the Tropics*, Ottawa, International Development Research Centre

Global Witness (1999) *The Untouchables: Forest Crimes and the Concessionaires – Can Cambodia Afford to Keep Them?*, London, Global Witness

Global Witness (2001) *The Logs of War, the Timber Trade and Armed Conflict*, London, Global Witness

Global Witness (2002) *Logging Off: How the Liberian Timber Industry Fuels Liberia's Humanitarian Disaster and Threatens Sierra Leone*, London, Global Witness

Global Witness (2004) *Taking a Cut: Institutionalised Corruption and Illegal Logging in Cambodia's Aural Wildlife Sanctuary*, London, Global Witness

Global Witness (2005) *A Choice for China: Ending the Destruction of Burma's Frontier Forests*, London, Global Witness

Greenpeace (2000a) *Against the Law: The G8 and the Illegal Timber Trade*, Greenpeace, www.greenpeace.org/raw/content/usa/press/reports/against-the-law-the-g8-and-th.pdf

Greenpeace (2000b) *Plundering Cameroon's Rainforests: A Case-Study on Illegal Logging by the Lebanese Logging Company Hazim*, Spain, Greenpeace

Greenpeace (2002) *Partners in Crime: Malaysian Loggers, Timber Markets and the Politics of Self-Interest in Papua New Guinea*, Amsterdam, Greenpeace

Hempel, L. (1996) *Environmental Governance: The Global Challenge*, Washington, DC, Island Press

Lawson, S. (2001) *Timber Trafficking, Illegal Logging in Indonesia, South East Asia and International Consumption of Illegally Sourced Timber*, London, Environmental Investigation Agency

Newell, J., Lebedev, A. and Gordon, D. (2000) *Plundering Russia's Far Eastern Taiga: Illegal Logging Corruption and Trade*, Vladivostok, Bureau for Regional Oriental Campaigns, Friends of The Earth Japan and Pacific Environment and Resources Center

Obidzinski, K. (2003) *Logging in East Kalimantan, Indonesia: The Historical Expedience of Illegality*, PhD thesis, Amsterdam, University of Amsterdam

Palmer, C. E. (2001) *The Extent and Causes of Illegal Logging: An Analysis of a Major Cause of Tropical Deforestation in Indonesia*, London, Economics Department, University College London, Centre for Social Economic Research on the Global Environment, University College London and University of East Anglia

Scotland, N. (2000) *Indonesia Country Paper on Illegal Logging*, Jakarta, Department for International Development

Seneca Creek Associates and Wood Resources International (2004) *'Illegal' Logging and Global Wood Markets: The Competitive Impacts on the US Wood Product Industry*, Poolsville, MD, American Forest and Paper Association

SGS Trade Assurance Services (2002) *Forest Law Assessment in Selected African Countries*, Draft report, Washington, DC, World Bank/WWF Alliance

Siebert, U. (2001) *Benin's Forest Corruption*, Berlin, Free University of Berlin, Department of Social Anthropology

Tacconi, L., Boscolo, M. and Brack, D. (2003) *National and International Policies to Control Illegal Forest Activities*, Bogor, Centre for International Forestry Research

White, A. and Martin, A. (2002) *Who Owns the World's Forests? Forest Tenure and Public Forests in Transition*, Washington, DC, Forest Trends and Centre for International Environmental Law

Williams, M. (2003) *Deforesting the Earth: From Prehistory to Global Crisis*, Chicago, University of Chicago Press

World Rainforest Movement and Forests Monitor (1998) *High Stakes, the Need to Control Transnational Logging Companies: A Malaysian Case Study*, Cambridge, World Rainforest Movement and Forests Monitor

WWF (World Wide Fund for Nature) (2002) *Forests for Life: Working to Protect, Manage and Restore the World's Forests*, Washington, DC, and Gland, Switzerland, WWF

WWF (2005) *Failing the Forests: Europe's Illegal Timber Trade*, Godalming, WWF–UK

2
The Social Basis of Illegal Logging and Forestry Law Enforcement in North America

Michael R. Pendleton

Introduction

Why is it that illegal logging is flourishing in many countries at a rate that is often equal to half that of legally extracted wood? How is it that the policy response of tough law enforcement can promise so much but apparently deliver so little? The simplicity of these questions obscures the complexities of their answers. In this chapter I offer some insight into the nature of illegal logging and the law enforcement response that may explain why illegal logging persists and law enforcement, as it is currently conceived and practised, will fall short.

This chapter is based on knowledge gained from an ongoing research programme dedicated to the study of crime and enforcement in the natural settings found in forests and parks of North America. This effort began in the mid 1990s at the University of Washington in the US and is an ongoing component of the work conducted by Pendleton Consulting LLC. What is reported here is a compilation of results from various studies conducted as part of this programme that directly inform our understanding of illegal logging and related law enforcement.

The setting for this research was the national forests and parks in the US and Canada, primarily those located in the Pacific North West. Participating agencies included the US National Park Service, the US Forest Service, Parks Canada and the British Columbia Ministry of Forests. Numerous non-governmental organizations and community groups have also participated in this effort.

This is the first research of its kind. While both qualitative and quantitative methods have been used in this programme, the data reported here were produced using the ethnographic method. Literally thousands of hours of field observations and interviews have been conducted to produce this knowledge. Extended stays in remote logging camps, direct observations of late-night enforcement actions, physical arrests of offenders that involved the principle investigator, and jailhouse interviews of arrested loggers highlight just some of the 'on-the-ground' sources of data. Videotaped data of the largest and most protracted anti-logging protest in the history of North America, the theft of cedar trees in progress and the first mass arrest of loggers in the history of the US augment conventional qualitative data found in organizational reports and newspaper accounts.

The strategy that guided collection and analysis of the data was analytical induction, in which a researcher, upon entering the field, builds and revises a conceptual model as empirical evidence is confronted (Strauss and Corbin, 1990). Data stability is based on the triangulation of both methods and sources (Denzin, 1978). Finally, pattern analysis was employed to further refine and understand the data (Strauss and Corbin, 1990).

The chapter first considers the social construction of the concept of illegal logging. It addresses the issue of whether illegal logging is universally regarded as harmful. Tree theft (among the most common violations observed in this study) is then considered in detail, with a focus on the role of its contribution to the forest community. The thresholds that may lead to the criminalization of logging are then addressed. The chapter then turns to analysing how and why tree theft is accommodated by institutions, and the implications for law enforcement. Considerations of the lessons learned from the study and the implications for policy conclude the chapter.

Expanding our thinking on illegal logging: The interactionist perspective and the shared meaning of illegal logging

Much of the literature on illegal logging starts from an assumption that it is universally believed that illegal logging is harmful, wrong and otherwise without merit. This 'harmful act' view of illegal logging is both intuitively compelling and consistent with many of the harms noted throughout the literature. Yet, this view fails to explain why illegal logging persists at such an astounding level. How could something that is so wrong flourish? Data from our research suggest that to fully understand why illegal logging persists, the 'harmful act' view of illegal logging must be amended in at least two important ways: first, illegal logging is not *universally* viewed as wrong, and, second, simply because it is illegal does not necessarily mean that it is without merit.

Illegal logging as functional deviance: A social interactionist perspective

It has long been recognized that deviance is not simply a disruptive social act, but also an important condition for preserving the social system (Durkheim, 1893; Erikson, 1966). One way in which deviance contributes to social stability is through the interaction between deviant persons and the community to include agencies of social control. It is through these interactions that norms are established and maintained that create the social boundaries of the community (Erikson, 1978). In effect, deviance is a way of establishing the social meaning of community.

One of the most powerful 'boundary-maintaining mechanisms' is the social meaning and subsequent labelling of particular deviant acts as crime. As an interactive process, crime is not automatically determined by a deviant act alone, but is also contingent upon the social reaction to the act (Schur, 1971). Through highly discretionary, contingent and selective confrontations with the institutions of law enforcement, particular behaviours and people become eligible for the label of crime. It is through these interactive confrontations that crime becomes a changing, often manipulated, idea based on social reaction. As Becker (1973, p13) notes:

> ... behavior may be an infraction of the rules at one time and not at another; may be an infraction when committed by one person, but not when committed by another; some rules are broken with impunity, others are not. In short, whether a given act is deviant or not depends in part on the nature of the act and in part on what other people do about it.

Often deviance is allowed to continue because it is a visible product of established social structure and, thus, serves certain social functions. To treat these acts as criminal would not only disrupt the social system, it would implicate it as well (Reiman, 1990). To withhold the criminal label, in these cases, is an important method of preserving the shared meaning of community. It is a protective decision.

Is it possible, then, that illegal logging persists because it is actually a stabilizing influence on the social order of the forest community and to criminalize it would disrupt the shared identity of this community?

Preserving the forest community: A shared meaning of illegal logging in the US forest community

Tree theft was among the most prolific violations observed in this study. On virtually every research observation, evidence of tree theft was encountered. While it was beyond the scope of this research to quantify the volume and acreage of trees stolen, one senior administrator estimated that 'literally hundreds of thousands, if not millions, of dollars worth of trees had been stolen' from a particular forest. Quite simply, tree theft is common. Three distinct types of tree theft were identified in this research, each distinguished by a shared meaning *and* social contributions to the forest community:

1. timber trespass as shared community authority;
2. timber theft as shared community identity and social boundary; and
3. tree poaching as shared deviance and selective exclusion.

Type I tree theft: 'Timber trespass' as shared community authority

All timber theft has been officially labelled *timber trespass* and institutionalized as part of the commercial enterprise. As this label suggests, the criminal seriousness implied in the meaning of theft is recast into a lesser act of cutting in the wrong place. While officially applied to all forms of tree theft, the meaning of timber trespass is most consistent with the illegal cutting of trees as part of legitimate commercial timber sales. Typically, this form of theft occurs within the context of an authorized logging operation and involves cutting outside the boundaries of the prescribed cutting area. Although over-harvest is intentional, it is most often considered a 'mistake' based on the view that logging is an imprecise activity compounded by the complexity of boundary marking, thereby making it easy to accidentally cut outside the boundary. The absence of overt criminal intent on the part of the logging contractors is a distinguishing feature of timber trespass.

Many participants view cutting outside the boundaries as normal industrial practice that is formally expected within the forest community. A forest law enforcement officer (LEO) explains the basis of this expectation:

> All the harvest contracts have a 10 per cent over- and undercut provision. Companies are allowed to leave trees or take more because of the 'imprecise nature' of the business. In all my years in the Forest Service, I have never seen an undercut. They always take the extra 10 per cent.

Evident in the contractual over-cut provision is the view that taking 'extra trees' is expected, boundaries are ambiguous and that boundary setting is authority shared between the Forest Service and the logging community. Less obvious in the over-cut expectation is an ambiguity over forest ownership supporting a widely held view that local people, not just the federal government, have entitlement to the forest. To consistently take a 10 per cent over-cut does not simply mean more money for the logging contractor, it also reaffirms a sense of shared authority and entitlement. In effect, it legitimizes the view that the forest belongs to the community.

Type II tree theft: 'Timber theft' as shared community identity and social boundary

The most frequently observed form of tree theft is informally known as timber theft and is distinguished by a clear intent to illegally take trees that are not authorized by formal regulation or protocol. Two categories of timber theft were observed:

1 affiliated timber theft; and
2 unaffiliated timber theft.

Affiliated timber theft occurs within the auspices of an established logging business *and* authorized timber sale. Clear intent to take trees by violating boundaries, bidding rules and/or volume accounting is apparent. However, the clear intent to violate the law is blunted by the legitimacy of commercial affiliation with an authorized sale. Affiliated timber theft is also a collaborative activity requiring the participation of many people. It is the complexity of the commercial logging process that serves both to broaden community participation and to obscure, if not mask, the forms of affiliated theft observed. For a detailed description of the forms of affiliated theft, see Pendleton (1998a).

The most distinguishing feature of affiliated timber theft is the community participation, knowledge, acceptance and support for this form of tree theft. It is a community activity. In the wake of a massive bid rigging and load ticket fraud that occurred four years prior to this research, in which trees worth millions of dollars were stolen, community members recall to a news reporter (Hessburg, 1988a) the community basis of the timber theft:

> *I think there's not a doggone driver, faller or rigging man who doesn't know about this crooked system.* (Retired logger)

> *Sure, what they were doing was wrong, but everyone had been doing it for years and years. He [bid rigger] just got caught. When [he] was growing up, his father was buying timber, and he'd meet in the parking lot with other buyers, and they'd agree to split up the logs. So he grew up seeing his father do that ... he thought that was how they did business.* (Community member)

The open nature of this form of theft was consistently observed during this research. In a case of boundary jumping, a logging company had purchased a state timber sale that was next to the national forest boundary. The fallers were directed by the contractor to continue cutting into the national forest. When the LEO and researcher discovered the theft, all that was left standing in the middle of several acres of down trees was a US national forest boundary sign. The fallers had simply cut around it, leaving it visible from a distance.

The ongoing and blatant nature of these thefts is viewed less as deviance and more as a valuable means of preserving the logging community. A local elected official provides insight on tolerance based on community membership:

> *Say a logger is fourth generation and everybody knows him. If they're aware he's ripping the Forest Service off, that isn't really considered a crime in this area.* (Local mayor)

Affiliated timber theft as a form of deviance serves not as a means of excluding people from the community, but as a vehicle for affirming shared values of family and loyalty. In effect, one role of timber theft is to create community cohesion. Community members provide insight into the stabilizing role of timber theft within the community:

> He was taking care of us. His dad was sick and living with us. His nephew had no income and was living with us. At the time we were six months behind on the rent. (Wife of accused log truck driver)

> Better to [rig bids] and keep some people employed. (Community leader)

> A guy did have a choice – just do it or walk. I should have quit, but I was obligated to my job. I got paid every Friday. There was some loyalty to my company. (Accused logger)

The depth of community loyalty is further evidenced in the way the community responded to the formal charging of a prominent logging company owner in a bid-rigging case. In spite of being convicted of bid-rigging charges, the business owner was allowed to stay on the board of directors of the scaling bureau that measures logs in the area. In addition, the business owner was inaugurated as the president of the 79-year-old Pacific Logging Congress, an international professional logging association. A local community newspaper owner sums up the local status of the offender: 'I have nothing but praise for the man.'

Predictably, the observed and reported finding that community informers are rare in cases of affiliated theft is consistent with the view that the community supports affiliated timber theft. As an involved logger noted:

> I could have squealed, I suppose, but I didn't want to stick my neck out. We were all lucky to hold on to our jobs. The timber industry was caught in a slump; it made crooks out of good guys.

Conversely, timber theft also provides a means for community members to *actively* guard its normative boundaries through overt sanctions. An unwillingness to adhere to the 'don't tell rule' is considered a challenge to the community and is met with resistance. One logger who originally cooperated with an investigation was so intimidated that he later refused to testify. Another logger who did cooperate reported being run off the road in his car and being shot at just four days before he was to testify (Wilson, 1988).

Although community informers are rare within the framework of affiliated theft, they are regularly evidenced in a second classification of timber theft defined as unaffiliated theft. It is in the act of the community telling on violators that the significance of *affiliation* becomes apparent.

Unaffiliated timber theft occurs when a logging company or operation simply goes into the forest without the auspices of a legitimate timber sale and cuts trees. Outside the veil of a formal timber sale, type II tree theft becomes unacceptable. In these cases, community informers are more likely to notify the Forest Service of the violation. An LEO provides insight on why informers are more common in unaffiliated theft:

> Most of my informers come to me about unauthorized timber harvest violations [unaffiliated theft]. They are concerned about unfair advantage. When someone can simply go take trees in large quantities, it creates an uneven playing field for the rest of the industry. They don't like it.

It appears that unaffiliated theft violates the community rules of engagement among the industry, the Forest Service and the community. Unfair advantage signals unpredictability and system destabilization. By avoiding the protocols of acquiring a formal timber sale, non-affiliated theft removes both the Forest Service *and* the logging community from the pattern of accommodation that characterizes affiliated theft. It is the blatant nature of the violation that links the 'honourable' profession with dishonesty while publicly challenging the authority of the Forest Service and the efficacy of the LEOs. The subsequent tarnishing of the logging community image invites increased scrutiny and demands official action.

In one such case a logging contractor had taken over 100 fir trees and was gone before an informant alerted an LEO. The contractor had been able to set up a logging operation, take the trees and leave without detection. In this case, a community member came forward and served as an informer to alert the LEO of both the offence and the offender. It seems that community members were complaining and expected official sanction. In effect, the unaffiliated theft challenged the community customs that ensure stability.

Type III tree theft: 'Tree poaching' as shared deviance and selective exclusion

The third form of tree theft is commonly called *tree poaching*. This form of theft involves the taking of single trees by individuals or small groups. Tree poaching accommodates the sense of community in four important ways. First, it is one mechanism for establishing status and hierarchy within community relationships, along with a sense of local history through social mentoring. Second, it is a means by which any member of the community can gain access to status associated with logging and the local wood-based economy. Third, tree poaching requires the development of a network of trust-based relationships to realize the primary goal of economic reward. Finally, it is the principal means of excluding unwanted community members through the formal labelling of tree theft as a criminal act.

Tree poaching as social learning

Tree poaching is an acquired skill that is taught through family and community relationships. Fathers teach sons, and other community 'folk heroes' take younger tree thieves under their wings to teach them the techniques. One LEO in the study was able to identify three generations of tree poachers who were *not* family members who had been 'raised up' by a local man in his late 70s. These mentoring relationships are ordered in a hierarchy based on age, experience and proven ability (for an in-depth explanation of the types of tree poaching, see Pendleton, 1998b).

It is instructive that the skill of tree poachers is known and admired among LEOs. Numerous unsolved cases of tree poaching were recounted, along with the various techniques used to accomplish the theft. In the case of a man who was carrying cedar blocks downhill on his back, the officers greatly admired his strength and decided not to arrest the man until a more 'controlled time' for fear of the consequences of a fight with such a poacher.

Virtually every district in the research forest has locations where notable thefts have occurred, each with a 'story' involving well known 'locals'. Together, the telling of these stories and the mentoring relationships surrounding tree theft serve to pass on community history as social learning that places in context tree theft techniques, as well as locations in the forest.

Tree poaching as a gateway to the local economy

Tree poaching can be extremely profitable. Depending on the location, an old-growth cedar tree can be removed in one to two days. The tree is sold to local cedar mills for half the market price on a tax-free basis and can bring between US$5000 and $10,000. For the wood to be effectively marketed, however, the mill owner must agree to the illegal relationship because all unprocessed wood transported out of the forest is required to have formal written permits, and records must be kept by the local mill accepting the wood. Poaching relationships are based on a mutual trust that once the illegal transaction occurs, the other will not tell the authorities. In effect, tree poaching requires the development and maintenance of community relationships before profit can transpire. As a deviant subculture, tree poaching provides a gateway for accessing the local economy that otherwise may not exist for some criminal members of the community. Tree poaching provides a means for people to participate in the community without engaging in more traditional types of criminal behaviour and without being publicly identified as criminal. The community tolerates poaching because it can serve as an alternative to more disruptive behaviour.

Community exclusion through selective labelling

Tree poaching, while tolerated by the forest community, enjoys the lowest community status of the three forms of tree theft. Often referred to as 'shake rats' (a general term for both authorized and unauthorized hand harvest of down cedar trees), tree poachers are most closely aligned with the traditional definition of a thief. These offenders are often precluded from legitimate employment because of a criminal history or outstanding warrants for arrest. Stealing trees is a convenient way of earning an income without being tracked by the usual paperwork and process of regular employment. Quite simply, these offenders are more likely to have the characteristics commonly held by criminals. In one of the most notable tree poaching cases in a particular forest, an LEO discovered and subsequently captured one of the most notorious local criminals in the act of stealing an old-growth cedar worth over US$20,000. The criminal, who was known to fight the police, was captured single-handedly by the LEO and subsequently sent to prison for felony theft of federal timber. What is most notable about this case is the degree of subsequent community respect for the LEO for 'bringing in' this 'undesirable' member of the community. The respect was enhanced by the fact that the officer was a woman and that the outcome of the arrest was the expulsion of the offender from the local community.

Accordingly, it was common for community members to inform LEOs of the identity of tree poachers and, in one case observed during this research, the impending theft planned by an offender. It is noteworthy that of the three types of tree theft, tree poaching is the primary focus of LEOs and exclusively accounts for those people who were formally labelled 'criminal' by the Forest Service during this research.

Tree theft as a spectrum of deviance

As Figure 2.1 suggests, the tree theft types form a spectrum distinguished on two dimensions:

1 by the level of publicly exposed deviance, which is closely aligned with
2 intent and the degree of affiliation with a legitimate logging business working an authorized timber sale.

Low	Manifest deviance	High
Timber trespass	Timber theft	Tree poaching
High	Legitimate affiliation	Low

Source: author

Figure 2.1 *Tree theft spectrum*

The normative boundary of the forest community seems to be firmly drawn by the conditions under which the label of 'crime' is applied to tree theft. In all cases, the shared meaning of tree theft as deviance approaches the definition of a crime only when the acts become publicly salient. Two factors seem to determine salience: public consciousness and the extent to which the behaviour moves from affiliation with a legitimate commercial logging operation towards more individual acts of crime committed in relative social isolation. Although all types of tree theft are technically available for the label of crime, in this study only tree poaching was formally treated by the Forest Service as a criminal act. It is in the inclusion and elimination of the tree theft types from the roster of crime that the functional importance of tree theft to the forest community becomes apparent.

The logging community accommodates tree theft only when it contributes to community cohesion and stability. Tree theft enables widespread constructive participation in the community. This participation can be passive, as in the case of knowing but not telling, or it can be active, as in the case of a participant who takes trees in the service of an economic system that demands wood. The litmus test of acceptability centres on contributing to community stability. Community sanction for ostracizing those who tell is justified to preserve the predictability of profitable harvest. Conversely, denigration of tree theft and the expulsion of tree thieves from the community become necessary to prevent heightened police scrutiny while preserving the reputation of honesty that self-defines the community. Subsequently, the status degradation ceremonies of criminalization are reserved for those who are perceived to threaten the symbolic and/or physical safety of the community.

As long as tree theft continues to service the stability of the forest community, it will persist along with a complex system of accommodation. Implicit in the tree theft spectrum, however, is a threshold beyond which the meaning of illegal logging shifts from simply deviance to crime. We now turn our attention to the question: what might cause the shared meaning of illegal logging to shift from simply functional deviance to a crime that requires an enforcement response?

Beyond the threshold: The criminalization of logging in British Columbia, Canada

Evident throughout the literature of occupational deviance is the view that occupational actions that result in harm are not intrinsically criminal. As the findings reported above suggest, illegal logging is such an example of occupational deviance. Kagen (1984) notes that one characteristic that distinguishes regulatory offences such as environmental violations from the clearly 'evil' properties of many crimes is the question: 'How bad are they?' This moral ambiguity tempers the social pressure for enforcement in the face of the economic standing associated with many occupations that at times sponsor harmful behaviour. The fact that logging is a time-honoured profession upon which entire societies depend for wood fibre is a legitimate barrier to applying a criminal label to it. For illegal logging to truly be considered and accepted by all as a criminal act, the moral ambiguity must be removed and the question 'how bad is it?' answered.

In 1994, the province of British Columbia in Canada amended its forest practices law to make illegal logging activities a crime punishable by up to a Cdn$1 million fine and three years in prison. In effect, many standard logging practices, including the taking of unauthorized trees, were criminalized. The Canadian portion of this research programme was launched just prior to the amendment of the law and specifically focused on the social factors that led to the criminalization of logging (Pendleton, 1997a). For this transformation to occur, four fundamental dimensions of environmental harm converged to create a threshold, beyond which the shared meaning of logging as it was practised in British Columbia was redefined as clearly wrong – in effect, a criminal act that demanded 'tough enforcement'.

From forests forever to unsustainable logging: The magnitude threshold

Historically, working loggers have believed it to be impossible to cut trees faster than they could grow back. A former employee of the largest timber company in British Columbia recounted this historic view:

> I don't recall anyone ever speaking of the threats to the environment and the resource that are so widely debated now. Not at work, not at home, not at school. As far as we knew, there would be forests forever. All you had to do was look out your window, or take a drive in the country, and you could see that. (Littleton, 1993, p13)

Numerous participants in this study noted the fallacy of this view. To these participants, the volume and pace of extraction had outstripped the capacity of the forest to regenerate itself. For many participants, the harm of logging centred on the magnitude of the environmental damage defined by the size of clear cuts, the extensive road systems and the extended ecosystem damage. Beyond this magnitude threshold, unsustainable logging practices were transformed into environmental crimes.

In many areas observed in this research, entire mountains and ridges had had their trees eliminated through use of the clear-cut method. Reforestation had been left to the natural reseeding process, and little re-growth was evident. Clear cuts as large as 300ha were frequently observed. Many loggers had declared themselves against these large clear cuts. As one logger noted:

> Some of us recognized a decade ago that too much was being cut. But no one cared. We had no clout, and our warnings fell on deaf ears.

Predictably, the exposed land was further degraded by increased temperatures during the summer and the loss of usable soil during the frequent rain events that occur in the winter.

Because of the amount of logging, a vast road system estimated to be 11,000km in length now crosses the mountains. On virtually every field observation, actual or potential road damage and failure were evident. It became virtually impossible to maintain this road system and road failures are common. During one field observation, a logging company supervisor was observed refusing to comply with a directive given by a Forest Service enforcement officer to repair a failing road. The supervisor remarked:

> It's impossible for us to maintain all these roads. There are just too many... We don't have the equipment, and it's too expensive. If the road comes down, it comes down! We are not going to fix it!

The taking of life: The death threshold

Fundamental to logging is the loss of life to both trees and loggers. This fact was noted often by logging participants in this study. The harvest of old-growth trees was an important threshold for many participants. Loggers admire these giant trees, often referring to them as 'big berries' or 'pumpkins'. For some, their skill, if not identity, had been crafted from their ability to cut down these trees. Fallers are considered the elite in the logging community, commanding large sums of money for their work. They are also the most likely to be killed because their role is the most dangerous. The fact that loggers can cut these giants down does not make them immune from grieving for the death they cause. One participant, a veteran faller, recounted a recent experience when cutting a big berry:

> I sat there for a long time. The tree was huge. It was cedar, and about 12 feet in diameter, I guess about 800 years' old. I just didn't want to cut her. It didn't seem right. She was so ... well, stately. It hurt me; yet I had to do it. It's my job.

Many environmentalists participating in this study, although not opposed to logging *per se*, drew the line at cutting the 'ancient forests'. These participants noted that not only trees die with logging, but also other forms of life. To some, it was simply immoral to take this life.

The importance of biodiversity, the rights of nature and the impending job loss facing loggers notwithstanding, the loss of life was a constant theme in this research. Loggers often referred to logging as 'tree killing', and one Forest Service enforcement officer referred to clear cuts as the 'tombstone effect'. The acres of greying tree stumps reminded him of tombstones in a tree cemetery.

Yet, many participants accepted the loss of life as the inescapable fact of a modern society dependent upon wood. There was a point, however, when all participants would no longer accept tree killing. This threshold between acceptable tree cutting and needless death centred on waste.

A practice known as 'high grading', 'creaming' or 'set jumping' was reported and observed as a common problem. In brief, this practice consists of clear-cutting an area but taking only the tree species that are of the highest value in the current wood market. The other trees are cut, but left to rot. Once the valued trees were taken, the companies moved their equipment (sets) to another area to log again. Large quantities of trees such as hemlock were often observed left to rot after being down for over a year. When loggers were questioned about the practice, they were embarrassed. One logger noted:

> It's really no different than shooting an elk or deer and taking only the horns, leaving the rest to rot... It's waste and should be a crime.

Many participants linked logging damage and the loss of life to other ecosystems. In the Pacific North West region, the viability of 122 stocks of salmon and other fish directly depend upon trees. During the time of this research, the US Pacific Coast salmon fishery for both commercial and sport fishing was closed for the first time in history. The fishery ecosystem had crashed; there were insufficient fish.

It is informative that the impact on the fisheries was of the highest concern to the logging industry and forestry officials encountered in this study. Two factors explain their concern:

1. federal fisheries law provides for large monetary fines and prison; and
2. damage to the fishery invites outside agencies into the forests to monitor logging practices.

In effect, this monitoring opened the door to a secluded profession.

Spatially remote and publicly apparent: The visibility threshold

Many of the areas observed in this research are remote, accessible only by boat, airplane or hours of automobile travel. Predictably, few people actually see, directly, the logging in progress or the effects of the occupation on the environment. In effect, logging is a secluded profession, as one forest official noted:

> *This seclusion was only a matter of time given the volume of the cut. Remember, when it's done, its done. It's here to see and it doesn't go away.*

The visual impact of large-scale clear-cut logging is impressive, particularly to the uninitiated. During this study, the large clear cuts were referred to as 'moonscapes' or 'nuclear test sites'. As one logger noted:

> *There is no such thing as a pretty clear cut. Logging is ugly, and the only thing that saves us is that most people never see it.*

During this research, however, this changed. As trees became limited, the clear cuts made their way closer to areas that were inhabited or frequented by large numbers of people. The visibility threshold was surpassed with the cutting of a particular area in full view of a major highway used to access several townships and a national park. This park alone drew over 700,000 visitors each year. In literally a few weeks the once forested viewscapes surrounding this area were clear-cut.

This highly public display of logging quickly educated the public to the realities of clear-cutting. Plans to extend the clear cuts to the next township met stiff resistance, leading to the first blockade of logging in the history of Canada. Subsequent intentions to continue logging resulted in the largest environmental protest in the history of North America. During a five-month period, over 850 protestors were arrested and criminally charged with blockading logging roads.

The logging industry countered these protests with a one-day protest that placed over 15,000 people in front of the parliament building protesting for the right to log.

Central to these events was unrelenting media exposure of the clear-cut method of logging and its impacts. So profound was this exposure that during the height of what became known as the 'Clayquot summer' protest, the editor of the leading newspaper announced that the issue had generated more letters to the editor than any other issue in the 135-year history of the paper.

The high profile of the protest exposed the realities of logging and accelerated the momentum for the criminalization of the forestry law. When *National Geographic* devoted an issue to logging in British Columbia, characterizing Canada as the 'Brazil of the North', one Forest Service enforcement officer observed:

> *It was my district ... all the atrocities that have been aired internationally that led to the 'Brazil of the North' label, it occurred in my area. The Black hole, Uce Hill and the Escalante were all mine. It made me sick; but there was nothing I could do to stop it. We have been wanting this [tougher enforcement].*

The decision by the government to prosecute the environmental protestors in criminal rather than civil court (a discretionary choice) publicly posed the question: 'Who are the real criminals?' The answer was quick to come. One top Canadian diplomat warned: 'Canada is getting a reputation as an environmental outlaw', while an industry analyst noted that the logging companies were now considered criminals in need of some rough justice (Mitchell, 1993; O'Neil, 1993).

From industrial citizens to amoral calculators: The legitimacy threshold

Kagan and Scholtz (1984) have argued that corporations become eligible for tough enforcement when the shared view of the corporation is transformed from legitimate citizen to an amoral calculator. In the former view, the organization violates the rules because the rules are unreasonable. Because the company is basically honest, non-compliance is more a matter of a mistake than an intention. In the latter view, the corporation is viewed as a culpable criminal.

According to numerous participants, the logging industry had, over time, adopted an 'industry attitude' that favoured profit at any cost. Eventually, this view led to a growing disregard for both employees and the Forest Service. The result was resentment, as described by a local woodworker union leader:

> *We hate each other! The company has always been cruel in its treatment of our workers. They think nothing of laying us off [without] warning or concern for our families. The big corporations are a disease, with a huge appetite for trees. We [loggers] hate the huge clear cuts.*

If employees refuse to comply with harmful or even illegal directives from the logging companies, they are confronted with termination. A company forester explains this problem:

> This is the company attitude: 'You work for us ... look who signs your paycheque. If you won't approve this harvest plan, we will find someone who will.' It was well known that if you were fired from the company, you would be blackballed, never to work again.

Disrespect of Forest Service enforcement officers was observed directly and reported by participants, as noted by this officer:

> It is was not uncommon for the privates to tell us to leave ... to go ... off when we tried to come in and scale. They owned the areas now and we all knew it. We simply left with our tails between our legs.

Indeed, scaling (measuring of cut trees to manage the number of trees to be taxed and harvested) fraud was constantly reported and observed directly during this research. On one observation of a surprise scaling check, more than half of the trees were unmarked. Among those that were marked, many were marked with a switched brand from other areas with a lower stumpage tax rate.

These data suggest that the logging companies crossed a threshold beyond which they were viewed as criminals. This transformation was not, however, solely based on their track record of degrading the environment, but also on the corporations' willingness to demean the local people. The response was the criminalization of logging through changes in the Forest Practices code. As one government official noted: 'The code should restore public confidence and international faith in BC products' (Lavole, 1994).

However, the distance between the law-making and the law-enforcing components of the system often obscures the distinction between the passage of hard law and the enforcement of such law. This distinction is often lost in the literature on illegal logging. In fact, most countries today have criminalized illegal logging, only to have these potential sanctions neutralized by corrupt or ineffective enforcement. It is important that we punctuate this distinction by now turning to the nature of law enforcement as it is applied (or not) to illegal logging.

Looking the other way: The institutional accommodation of tree theft

One of the most enduring themes in the literature on illegal logging is the lack of effective law enforcement. One persistent explanation for non-enforcement has been chronic corruption among officials responsible for enforcing logging laws. Quite simply, tough laws and penalties are neutralized by the financial gain that comes to officers who simply look the other way and do not arrest offenders and/or confiscate illegally acquired logs. Corruption in this view is an intentional act by individuals for personal gain. Implicit in this view of non-enforcement is the assumption, if not hope, that if the corrupting influences associated with financial gain were eliminated, or if more honest people were retained as enforcement officers, effective enforcement would occur. But would this really be the case?

Once again, the interactionist view of deviance suggests that this 'rotten apple' theory does not *fully explain* the lack of meaningful enforcement. Social reactions to deviant acts such as avoidance and/or toleration through inaction often occur in order to allow the social community to live with deviance (Rubington and Weinberg, 1973). Acts of accommodation, such as police non-enforcement, are employed to avoid overt acknowledgement and the disruption to the social order associated with the formal labelling (arrest) of deviance (Black, 1971). In effect, non-enforcement will occur as a means of preserving community stability, even without the presence of the financial gain that defines non-enforcement as a corrupt act. In many situations, it becomes a legitimate response. Such instances include use in less serious cases, use to mitigate the overreach of the law, to secure a commitment to long-term compliance or as a resource to secure cooperation in obtaining other enforcement goals (Pendleton, 1997b, 1998b).

Is it possible that non-enforcement of forestry law can occur even in institutions considered as 'model agencies' where individuals are selected and known for their professionalism and integrity? We turn now to just such an agency (Kaufman, 1960) – the US Forest Service – and the question: what are the patterns of accommodation associated with tree theft and the organizational beliefs that support these behaviours?

Chronic tree theft and a non-enforcement response

As stated earlier, tree theft was among the most prolific violations observed in this research programme. In face of these findings, the observed prevalence of non-enforcement at the field level or at the investigative level was striking and inconsistent with existing data on both regulatory and traditional enforcement patterns (Black, 1971). It simply did not occur as a formal Forest Service action. Participants uniformly reported that the Forest Service, as an organization, did not support formal law enforcement, but preferred that LEOs interact with the public using the education and information methodologies identified by Kaufman in his landmark study of rangers (Kaufman, 1960, pp195–197).

Some LEOs voluntarily conformed to the organizational emphasis on non-enforcement. Compliance was based on organizational rationales used to neutralize the seriousness of the offences. For these participants, non-enforcement was normalized by rejecting enforcement as a means of obtaining organizational approval. Other participants resisted pressures to voluntarily comply with the non-enforcement preference of the organization. These LEOs generated cases in the field and forwarded them for investigation and formal prosecution. In effect, these participants forced the agency to demonstrate an unwillingness to prosecute these cases that was consistently evidenced in acts of non-enforcement at the administrative level. These cases of non-enforcement were the source of great frustration for these officers, which often resulted in acts of defiance.

Pattern analysis of the non-enforcement data at the field and administrative level reveals three distinct patterns of accommodation. Each of these types of accommodation is linked to organizational beliefs within the Forest Service and to operational behaviours, which together form a structure of accommodation.

Assimilatory accommodation

The most visible patterns of accommodation were observed in the day-to-day practices that are assimilated within the officers' routine behaviours. These practices are based on what officers reported as the agency's 'good host doctrine'. As a widely shared organizational philosophy, the good host doctrine is designed to provide the forest visitor with a pleasant work or recreational experience. The key to a pleasant experience is minimal contact with other people, particularly Forest Service officers, whose authority might be viewed as constraining the freedom of the outdoor experience. If contact is necessary, the preferred focus is one of tolerant understanding emphasizing education over formal action. Failing at this approach, officers should know when to 'back off and let it go'. As one senior supervisor noted:

> We don't want them [LEOs] out there playing cowboy. We prefer the social relations approach which emphasizes public relations. We have evolved from the days where we cuffed and stuffed them [offenders].

Shaped by the good host doctrine, two types of assimilatory accommodation were observed:

1 pre-enforcement cues; and
2 predictable patrol.

Pre-enforcement cues

One of the most feared situations in the forest is an unanticipated and uncontrolled discovery of a crime, such as tree theft, in progress. In these cases, officers are routinely outnumbered, often by armed offenders, while faced with the awkward reality of a crime that should result in formal enforcement. Officers consistently commented on the possibility of injury. To accommodate the possibility of unanticipated crime, pre-enforcement cues were used to signal the presence of the officer.

In all of the districts observed, officers routinely parked their patrol vehicle in view of passing citizens, often in front of the ranger station. When the truck was visible, offenders would know that officers were 'in the district'. In one case, the officer lived near the ranger station and would leave the truck parked in front of his house, signalling his presence. On the officer's day off, the truck would be visible, while the officer's boat and private truck would be missing, signalling the day off. A state forest enforcement officer reported that offenders whom he had arrested indicated that they would routinely drive by the Forest Service officer's home to check for vehicles. If the boat and private pick-up were missing, they would proceed to steal trees.

In addition to visual cues, officers would routinely tell the district front desk and phone receptionist of their location and schedule for the day. It was routine for the receptionist to advise callers of the whereabouts of the officer, including the location and schedule of long vacations or trips out of the area.

One situational habit observed among the officers most known for their non-enforcement style was the practice of briefly turning on their siren prior to entering an area where contacts were likely. The result was a brief but sharp noise revealing the approaching officer. When questioned about the practice, officers would consider the event to be an inadvertent error. Yet, the practice was observed often among various officers. During one occasion, an officer noted: 'It's not necessarily bad that they know we are coming.'

Predictable patrol

Contacts with tree thieves were minimized through the use of highly visible patrol routines in the forest. Using clearly marked patrol trucks, LEOs patrolled in a routine manner. Some officers would patrol their districts in a cycle that began during the patrol week at one end of the district, driving all of the roads until the whole district had been patrolled. This commitment to a comprehensive *horizontal patrol* was based on the continuous systematic display of the visible police symbols on the patrol truck. This approach creates a *spatial regularity* that becomes predictable. These officers explained this method as a means of expanding the image of their presence over the large areas to be patrolled. The purpose was not to catch offenders, but to deter them. Yet, on virtually all of these patrols, evidence of tree theft was encountered and not once were any offenders observed or apprehended.

The routine of patrol was further defined by the *temporal regularity* of the patrol. It is common for officers to establish the same days off during the week. These days off were easily known in the forest community. In one case, an officer was consistently finding evidence of tree theft, yet never encountered the offender. During a routine patrol, the officer contacted a tender of a small dam within the forest to learn more about the thefts. When questioned, the dam tender recalled a man cutting trees in the theft area. When pressed, the tender noted that the man came every Tuesday and Wednesday. These days were the weekly days off for the LEO. In spite of this information, no attempt was made to change the days off or times of patrol in order to apprehend the offender.

Another patrol routine that enables theft can be classified as *vertical patrol*. This patrol strategy enables enforcement predictability by patrolling the same areas in the forest during the same time each year. This method has the effect of routinely saturating some areas during the year, while others receive little or no attention. When combined with both the visual cues of a visible patrol and the pre-enforcement cues discussed above, the method of patrol becomes easily predictable. Officers who would employ this method would build up a bank of patrol hours during periods of saturation and then take extended periods of time off. In one case, the officer would work many hours during the summer and then take four to six weeks off during the end of the year, which coincided with the Christmas tree-cutting season. In this forest, the six weeks after Thanksgiving comprise the period when most visitors come to the forest. The purpose of these visits is to cut trees. Yet, one officer was observed to always take this time off.

Patterns of agency oversight were consistently cited to explain the persistent nature of tree theft. As indicated by this Forest Service official:

> It's a joke. People in the logging industry laugh about how little the Forest Service is on the ground. Our people never used to work weekends – only a core of about 10 am to 3 pm, Monday through Friday. (Hessburg, 1988b, pA4)

Together, these practices have become assimilated within the daily behaviour of the officers. These patterned behaviours increase the likelihood that tree thieves can learn these routines to avoid apprehension.

Anticipatory accommodation

A second type of accommodation is shaped by the agency's anticipation of the potential or future consequences of timber theft based upon past situations and their understanding of how others will respond to these thefts. A tentative tone surrounds this type of accommodation in anticipation of how timber companies, the US Attorney's Office *and* others within the Forest Service are likely to respond to enforcement. A central feature of anticipatory accommodation is the perceived level of support from supervisors and colleagues. A misreading of this support creates the risk that officers may lose professional standing by ignoring the implications of formal enforcement action. Consequently, agency policy *and* patterns of non-enforcement reflect a minimization of tree theft, which constrains the eligibility of these acts from the formal label of crime.

Policy of pre-emptive avoidance

This form of anticipatory accommodation is based on what one officer termed the 'honest industry theory'. The honest industry theory of enforcement rests on the assumption that logging companies and wood industry workers are honest people who would not intentionally steal trees. A Forest Service officer sums up this prevailing view:

> These [forest workers and loggers] are not bad people … they are not criminals. These are good, hard working people simply trying to make a living. You can't blame them for trying to make ends meet.

When unauthorized trees are taken it is assumed that the imprecise nature of logging and the complexity of timber-cutting boundaries are responsible. During this research, LEOs operated under an informal agency policy that discouraged patrol of active logging sites. Boundary transgressions and other harvest violations were not treated as criminal offences but were handled within the provisions of the harvest contract. During the research, LEOs were observed to visit active logging sites only twice, both times after working hours. In both cases, officers found unsecured log load ticket books that, historically, have been used to steal loads of logs. Formal enforcement action was not observed.

One reason cited by officers to explain non-enforcement of tree theft laws was the organizational belief that the amount of harm associated with the thefts was 'no big deal'. Even though the theft of one tree could net up to US$2000 a day, the amount was considered to be insignificant. A senior administrator sums up this view:

> *These thefts [single trees] are nothing ... all the real theft occurred ten years ago. Literally hundreds of thousands, if not millions, of dollars have been stolen. What happens now is insignificant. There are only single trees left now. Forest product offences are no big deal... I mean, the theft of five trees on the boundary is no big deal.*

The petty nature of tree theft is reinforced by the federal prosecutor, who refused to take many of the cases eligible for prosecution. This refusal was based, in part, on the limited budget of the prosecutor's office, which was 'already over-taxed by the tidal wave of drug cases. Forest Service cases are simply not a high enough priority.'

In anticipation of the refusal of the US Attorney to prosecute timber theft cases, LEOs operated under an informal agency threshold of US$50,000 as the criterion for formal investigation and submittal for prosecution. Cases with less than US$50,000 worth of stolen trees were not pursued. Consequently, numerous thefts below this threshold occurred without a formal enforcement response. In one district, the same offender had established a well-known pattern of stealing trees valued below the threshold.

Reactive tolerance
In spite of the policy emphasis on non-enforcement, some LEOs refused to comply. The LEOs would write reports and/or citations, which are referred to the supervisor's office (headquarters) for investigation and formal referral to the US Attorney for prosecution. One of the most persistent findings in this study is absence of investigation and referral for prosecution of cases that were *beyond* the US$50,000 threshold. It was common to let cases sit in the files or to conduct modest follow-up without formal referral.

When questioned about the lack of investigation, investigators uniformly reported that there was no support from federal prosecutors or Forest Service administration for formal action. Officers supported their views by noting that the formal reports of crime in this national forest chronically underreported criminal activity. A subsequent comparison between the annual report compiled by the Forest Service staff and data collected directly from a single forest district revealed a 95 per cent underreporting of documented violations, even though the LEO had forwarded the data to headquarters. One officer explained the underreporting in this way:

> They [administration] don't want an image of crime in the forest. If that happens, there will be pressure to put money into the programme [law enforcement].

Atrophic accommodation

The third category of accommodation is defined by those institutional decisions that emaciate the law enforcement programme. Atrophic accommodation refers to those patterns of programmatic decline within the law enforcement programme that reduce the threat of formal enforcement.

Patterns of atrophic accommodation are based on two fundamental features of the Forest Service organization. First, the agency is not predominately an enforcement agency. Rather, the Forest Service is a science-based agency, in which law enforcement is viewed as a secondary activity, at best. The result is a professional separation between the law enforcement programme and the remainder of the agency. A high-ranking Forest Service administrator notes this distance:

> I don't want to be that close to it [enforcement]. I don't understand it ... and I have no experience with it. Consequently, I am very uncomfortable with it. There is no output – how are we supposed to measure it?

Second, this professional separation is magnified by the budget system in place during this research. Unlike the timber harvest programme, which generates the budget for the agency, law enforcement does not produce revenue. Rather, funds for the programme were taken from other agency programmes depending upon the amount of enforcement effort in the respective programme areas. To the extent that forest crime includes timber theft, enforcement not only interrupts harvest (revenues) but takes funds from the harvest programme.

Structural disarticulation

A principle concern of LEOs in the study is the declining structure of the law enforcement programme. The result is a systematic weakening of the programme. These reports are supported by the following observations:

- During this study there was no applicable written forest-wide law enforcement plan. LEOs reported that they were 'told not to do a district plan' by headquarters staff, a directive that is inconsistent with the Forest Service manual and standard agency practice.
- Even though equipment for a forest-wide radio communication system had been purchased and installed in the headquarters, it was not operational.

- During the research the number of law enforcement staff was reduced from 6.5 to 4 full-time equivalents, representing a 38 per cent reduction in staff. Among the positions cut were two investigator positions that operated out of the headquarters.

Together, these data suggest that the law enforcement programme is declining within the organizational environment. Subsequently, tree theft may proceed unimpeded because of the lack of integration of the law enforcement programme within the overall agency.

The 'fatal flaw' hypothesis

When participants were questioned about the systematic non-investigation of timber theft violations, a theme common to all cases emerged. Virtually all of the timber theft cases initially investigated and/or referred to headquarters by LEOs were, after an initial review, found to have administrative errors and/or operational mistakes by employees. These errors were related to the process of forest management and/or investigation that creates a legal flaw precluding prosecution. For example, one large timber theft case was excluded from prosecution when it was learned that the survey to establish the National Forest boundary was not only inaccurate, but fraudulent. The case could not be prosecuted because federal jurisdiction was not established. Faulty boundary surveys were common in several cases.

In another case, the LEO attempted to make a case on a man who had built an illegal road into the forest and was stealing trees. Upon investigation, the officer learned that a Forest Service manager had given the man verbal permission to build the road and take the trees. In yet another case, it was discovered that an administrative staff officer was systematically rewriting LEOs' initial case reports to make them more 'consistent with the charges', thus legally compromising the cases.

The prosecutor assigned Forest Service cases confirmed the 'screwed up' nature of the cases and the fact that it is rarely possible to pursue formal charges.

Looking the other way and the preservation of a forest culture

It is tempting to view the patterns of accommodation observed in this study as evidence of occupational ineptness, if not deviance through complicity. Such a view, while satisfying the urge to vilify, obscures a more fundamental dynamic. Avoidance and/or toleration of deviance are increasingly a preferred response and are facilitated by 'cultural closeness' (Black, 1993) to maintain the prevailing social order (Reiman, 1990).

The cultural closeness of this resource system is illustrated in the following quote from the former director of the Forest Service's national timber theft taskforce:

> [Tree theft] took time to recognize... We came west together. We followed the railroads west. The industry and the Forest Service settled rurally together in a very rural and wild environment where resources were plenty. Our kids went to school with industry kids, churches together. We grew up and lived together in the Wild West. And so, culturally, we're geared that way... Plus you're dealing on a regular basis with the people we grew up with culturally. So to accept possibly that those people are dishonestly dealing with us in some cases – please add in some cases because I believe the vast majority of operators are honest – that's a cultural transition. (Taylor, 1994, p39).

Those who comprise the 'forest community', which includes loggers, their families and Forest Service professionals, as well as tree thieves, view tree theft as a minor cost in service of the larger objective of a profitable industry. Acts of accommodation reinforce the stabilizing role of tree theft by creating a *socially recognized certainty* that the threat of sanction is not real. To formally treat tree theft as a crime not only undermines social certainty, it also challenges the view that the forest is a desirable place and that those who come to the forest are honourable people. Quite simply, tree theft persists, in part, because the cultural integrity of the forest community is more valuable than the trees lost to theft. In effect, looking the other way preserves the forest community.

Prognosis and policy implications: Lessons learned

It may occur to some who read this chapter that knowledge gained from North America has little relevance to illegal logging in other parts of the world. Clearly, there are cultural differences that might make such a view valid. Yet, recent reports from Brazil suggest that such differences may be minimal, at best. An in-depth report by *The New York Times* revealed findings stunningly similar to the research reported above (see also Chapter 10 on Brazil). Tree theft is epidemic and law enforcement agencies are 'chronically short of staff and money, its employees often threatened'. Logging plans and permits are often non-existent. As one worker noted: 'No, we don't have any management plans. Nobody here does.' Most interesting are the views reflected in the comments of a former Brazilian mayor:

> If you're going to bust me, you're going to have to bust everybody, because nobody here has authorizations. We're just trying to survive. Who is going to give me money to pay my employees and educate my children?... Who cares about the law: what am I supposed to do, go hungry? (Rother, 2005)

So, how might the knowledge gained from the research reported here inform the future of illegal logging and corresponding policy options?

Lesson one: The forest community matters. It seems abundantly clear that illegal logging persists not simply because of financial incentives, but because it also performs critical social functions essential to the preservation of the forest community. Policies designed to stem illegal logging must reflect an intimate understanding of the social nature of the forest community and the corresponding needs that are addressed by illegal logging. Community identity, community stability and community values are but a few of the powerful social factors that will determine the future of illegal logging.

Lesson two: The criminalization of illegal logging is more than the passing of law; it requires direct linkage of the harms it produces to those involved to qualify for meaningful social sanction. It is difficult for local loggers to care about global warming when they are simply trying to provide their next meal or send their children to school. The harmful effects of illegal logging must have personal meaning in order to be viewed as a wrong that requires the immediate enforcement of coercive sanctions. In effect, the application of sanctions must have the moral authority from the forest community that only comes from the personalization of illegal logging as an activity to be stopped. As long as it is viewed as acceptable deviance rather than a criminal act that produces real harm, it will continue.

Lesson three: The rhetoric of law enforcement must be reflected in the reality of an independent community-based law enforcement system. The social and organizational pressures to accommodate illegal logging are powerful and currently neutralize any real 'on-the-ground' impact. Additional data from this research programme suggest that law enforcement systems that are embedded in existing forest management agencies are clearly impotent. In effect, pronouncements of tough enforcement are little more than rhetoric designed to preserve existing extraction levels by neutralizing the effects of real enforcement (Pendleton, in progress).

Meaningful law enforcement can come only from independent community-based systems. Such an independent system can capitalize on the 'cultural closeness' of the forest community if it truly reflects the moral authority of the community. If the objectives of law enforcement are to stop the harm as it is experienced within the local community, illegal logging will cease. The community-based policing and problem-oriented policing models in the US provide one example of how an independent enforcement system can reflect community values and needs.

Our future: The coming anarchy

The vilification of loggers as criminals promises to fall short of the structural changes necessary for effective management of the socio-natural paradox: to live, we must take life; yet, without sustaining life, we cannot live. Failure to manage this paradox effectively promises a future of social disruption and deepening reliance on coercive sanctions. In Port Renfrew, British Columbia, a community observed during this research, a well-known criminal economy has emerged in the wake of the once vibrant logging community. Quite simply, the loss of all the surrounding forests left few legitimate opportunities for local residents. So brazen is this criminal economy that local offenders have stolen the Parks Canada ocean rescue boat, robbed the local fishing fleet doctor of his drugs and forcibly looted a marooned sailboat of its gear while the owners were still aboard. In the absence of structural reform to ensure sustainable resource extraction, hard enforcement (prison and fines) may be expanded from the defence of depleted forests to manage the conflicts that come from displaced members of the forest community.

Finally, it is important to remind the reader that illegal logging, by definition, will result in the elimination of forests worldwide. Legal extraction levels are based on a sustainable yield policy that synchronizes logging levels with regrowth rates. To flaunt our disregard of sustainable yield will result in the collapse of ecological, societal and social systems as we know them. Yet, lessons learned from the collapse of former societies are available, should we somehow find the leadership and social resolve to accept these lessons (Diamond, 2005). The real question is: are we willing to stop illegal logging or have we accepted the inevitable?

References

Becker, H. (1973) 'Outsiders', in Rubington, E. and Weinberg, M. (eds) *Deviance*, New York, Macmillan

Black, D. (1971) 'The social organization of arrest', *Stanford Law Review*, vol 23, pp1087–1111

Black, D. (1993) *The Social Structure of Right and Wrong*, San Diego, Academic Press

Denzin, N. (1978) *The Research Act: A Theoretical Introduction to Sociological Methods*, 2nd edition, New York, McGraw Hill

Diamond, J. (2005) *Collapse: How Societies Choose to Fail or Succeed*, New York, Viking

Durkheim, E. ([1893] 1965) *The Division of Labor*, translated by George Simpson, New York, The Free Press

Erikson, K. (1966) *Wayward Puritans*, New York, John Wiley

Erikson, K. (1978) 'Notes on the sociology of deviance', in Rubington, E. and Weinberg, M. (eds) *Deviance*, New York, Macmillan

Hessburg, J. (1988a) 'New ethics in the forest', *Seattle Post-Intelligencer*, 13 June, pA4

Hessburg, J. (1988b) 'Forest Service learns from its mistakes', *Seattle Post-Intelligencer*, 15 June, pA4

Kagan, R. (1984) 'On regulatory inspectorates and police', in Hawkins, K. and Thomas, J. (eds) *Enforcing Regulations*, Boston, Kluwer-Nijhoff Press

Kaufman, H. (1960) *The Forest Ranger: A Study in Administrative Behavior*, Baltimore, Johns Hopkins Press

Lavole, J. (1994) 'Forest code to restore faith', *BC, Times-Colonist*, 31 May, pA-1

Littleton, E. (1993) 'Goodby, Mac Blo', *Monday Magazine*, 30 November, p17

Mitchell, D. (1993) 'Proposed new forest code offers more stick than carrot to our major resource sector', *Business in Vancouver*, 30 November, p8

O'Neil, P. (1993) 'Canada getting environmental outlaw label, diplomat warns', *Vancouver Sun*, 21 August, pA-1

Pendleton, M (1997a) 'Beyond the threshold: The criminalization of logging', *Society and Natural Resources*, vol 10, pp181–193

Pendleton, M. (1997b) 'Looking the other way: The institutional accommodation of tree theft', *Qualitative Sociology*, vol 20 no 3, pp325–340

Pendleton, M. (1998a) 'Taking the forest: The shared meaning of tree theft', *Society and Natural Resources*, vol 11, pp39–50

Pendleton, M. (1998b) 'Policing the park: Understanding soft enforcement', *Journal of Leisure Research*, vol 30, pp552–571

Pendleton, M. (in progress) *Regulatory Dramas: The Neutralization of Risk and Uncertainty of Coercive Environmental Enforcement*

Reiman, J. (1990) *The Rich Get Richer and the Poor Get Prison*, 3rd edition, New York, Macmillan

Rother, L. (2005) 'No letup in rainforest logging: Laws flouted', *The Seattle Times*, 16 October, pA26

Rubington, E. and Weinberg, M. (1973) *Deviance: The Interactionist Perspective*, New York, Macmillan

Schur, E. (1971) *Labeling Deviant Behavior*, New York, Harper

Strauss, A. and Corbin, J. (1990) *Basics of Qualitative Research: Grounded Theory Procedures and Techniques*, Newbury Park, CA, Sage

Taylor, S. (1994) *Sleeping with the Industry: The US Forest Service and Timber Interests*, Washington, DC, Center for Public Integrity

Wilson, D. (1988) 'Bid rigger has respect in Forks', *Seattle Post-Intelligencer*, 14 June, pA-4

3

From New Order to Regional Autonomy: Shifting Dynamics of Illegal Logging in Kalimantan, Indonesia

Anne Casson and Krystof Obidzinski

Introduction

In the year 2000, the then Ministry of Forestry acknowledged the rise of illegal logging in Indonesia by releasing a statement saying that its data indicated that illegal logging had damaged 1.6 million hectares of forest between January and July 2000. This activity had apparently caused the Indonesian government to lose some US$360 million in annual tax (*Jakarta Post*, 2000). By 2001, illegal logging was thought to be one of the most critical threats to Indonesia's forest capital (ITTO, 2001).

This chapter highlights some of the shifting dynamics of the illegal timber sector at the turn of the 20th century in the districts (*kabupaten*) of Berau, East Kalimantan and Kotawaringin Timur, Central Kalimantan – two of Indonesia's main suppliers of illegal timber. By drawing on these two case studies, we argue that illegal logging is not necessarily a phenomenon driven by macro-economic considerations (such as processing overcapacity, inefficiency, flawed pricing and rent seeking) and general socio-political ills, such as patronage and corruption, alone. While these are important causal factors, they do not provide a complete answer to the question of why illegal logging has been such a vibrant, adaptable and virtually unstoppable force in Indonesia during recent years. A key factor adding to the resilience and dynamism of illegal logging in East and Central Kalimantan is the fact that, since the fall of Suharto in May 1998, illegal logging has been operating in a greater variety of forms and guises. While well-entrenched networks of patronage and corruption remained, the decentralization process initiated in 2000 blurred the distinction between legal and illegal logging by giving rise to locally sanctioned timber extraction. In other words, local governments officially legitimated timber extraction by issuing timber permits.

The mass media and a number of environmental non-governmental organizations (ENGOs) have predominantly portrayed illegal felling as criminal acts by unscrupulous business groups and individuals who openly sponsor encroachment into protected areas, remote border regions and other inaccessible parts of the country (EIA and Telapak, 1999, 2000; *Kaltim Post*, 2000; *Kompas*, 2000a, 2000c). Over-cutting or cutting out of blocks by forest concessionaires and 'hit-and-run' operations by phoney plantation companies have also been recognized as acts of illegal logging (Kartodiharjo, 2000; *Kompas*, 2000b). The following discussion will show, however, that while these forms of illegal logging still exist, illegal logging has also undergone several operational transformations since the onset of the 1997 economic crisis and that most of these logging activities were 'legalized' by district governments. This came as a result of the fact that decentralization regulations, particularly regarding small concession schemes in the forestry sector and the ability to generate district taxes, were successfully misused to bestow a degree of formalization, or even outright legalization, upon hitherto illegal logging.

Moreover, these new forms of illegal logging activities have important implications – socially, ecologically and economically – for forest policy in Indonesia, primarily because they usher in a new set of daunting challenges as the process of formalization obscures and effectively dissolves the distinctions between what is lawful and what is not. The increasing complexity of the illegal logging problem renders illegal loggers progressively less responsive to oppressive law enforcement measures.

Before discussing the shifting dynamics of the illegal logging sector in Kalimantan in further detail, we provide a general introduction to the illegal logging problem in Indonesia prior to the fall of Suharto and attempt to explain the intensification of illegal logging activities since 1997. We then go on to discuss the shifting dynamics of the sector in the two districts of Berau, East Kalimantan, and Kotawaringin Timur, Central Kalimantan.[1] A general discussion of the social, economic and environmental consequences of illegal logging activities in these two districts is provided before we conclude the chapter.

Illegal logging during the Suharto era

Illegal logging, particularly by local communities, is not a new phenomenon in Indonesia (Callister, 1992; Scotland et al, 1999; McCarthy, 2000a). Peluso (1992) and McCarthy (2000a) have both pointed out that tensions between state and local interests over the control of forest resources in 'outer island' Indonesia have a history that extends back to the colonial period. The policies of the Suharto government merely sharpened and intensified these tensions.

Under the Suharto government, all Indonesian forests were declared state forests and the outer island forests were opened to large-scale timber extraction in order to generate much needed revenue. This centralized control of Indonesia's forestry sector followed the 1967–1970 period of a relatively relaxed policy that allowed district authorities and village communities to engage in small-scale logging activities. This period was significant in the sense that, similar to what happened after the fall of Suharto, it allowed small-scale concessions to be issued locally (Ruzicka, 1979; Peluso, 1983).[2] The Suharto government in its early days took this measure to appease regional politicians – many of whom had military backgrounds. Once the centralization and consolidation policy began to take effect, however, provincial and district timber enterprises were marginalized in favour of multinational corporations linked to central government elite and key military figures. This effectively pushed local timber operations underground, giving rise to the formation of an informal timber sector.

At the same time, conglomerates with close connections to the Suharto family and the army were able to obtain 20-year logging licences in order to exploit these forests. According to Brown (1999), there were 585 timber concessions covering a total of 62 million hectares of forestland by the end of 1995. These concessions were primarily divided among 64 timber groups. The five largest private groups holding concessions were Barito Pacific (6.1 million hectares), Djajanti (3.6 million hectares), Alas Kusuma (3.4 million hectares), Kayu Lapis Indonesia (3.0 million hectares) and the Bob Hasan group (2.4 million hectares). Together, these five timber companies controlled 18.5 million hectares, or 30 per cent of Indonesia's total timber concession holdings of 62 million hectares. In addition to this, Indonesian state forest corporations (*Perseroan Terbatas Industri Hutan Indonesia,* or *PT Inhutani*) controlled 3.9 million hectares of forestland, or 6 per cent of the total forest area allocated for production. Despite controlling extensive areas of forest, these large-scale timber companies have long been harvesting timber over the Ministry of Forestry's approved level (20 million cubic metres per annum) and obtaining timber from illegal sources in order to meet growing demand. During the late 1980s, for instance, Schwarz (1990) estimated that around 2 million cubic metres of timber were being illegally removed from protected forest, conversion forest and reserve areas each year.

During the Suharto era, local people living in or around logging concessions received very little from logging activities and were forced to enter into covert agreements with local entrepreneurs and concessionaires in order to obtain some benefits. In doing so, local leaders allowed loggers to cut into community forests and local people often assisted concessionaires in logging outside their boundaries, or within protected areas, in exchange for salary or rent. Some local people also became engaged in illegal logging by stealing logs from concession areas (Potter, 1990; McCarthy, 2000b). The marginalization of provincial and district timber entrepreneurs throughout the New Order period never really managed to completely eliminate well-established military–bureaucratic–entrepreneurial networks at lower administrative levels that predated the current forest concession period. Throughout the 1970s, these networks expanded and solidified (as a result of the progressive militarization of the New Order bureaucracy) in order to take advantage of growing international demand for Indonesian timber after timber supplies became exhausted in countries such as the Philippines (Ross, 2001).

The intensification of illegal logging after the fall of Suharto

After the fall of Suharto in May 1998, political developments and changes to legislation created conditions that contributed to a boom in illegal logging (Khan, 2001; Wadley, 2001). This was particularly the case in the districts of Berau, East Kalimantan, and Kotawaringin Timur, Central Kalimantan. The boom in illegal logging can be attributed to a number of factors, including changes arising from the economic crisis, a decline in law and order, regulatory changes arising from *reformasi* – a movement calling for democracy, reform and change – and the new decentralization laws. These four factors are discussed in further detail below.

The economic crisis

After the economic crisis hit Indonesia in mid 1997, local communities and people who lost their jobs in the manufacturing and industry sectors began to increasingly rely upon forest resources to meet their daily needs. In the era of *reformasi*, local governments have been sympathetic to local community needs and are turning a blind eye when it comes to illegal logging activities. Moreover, the economic crisis has severely affected the operations of some of the large logging companies, which were plagued with substantial debts. Large logging companies, such as the Kayu Mas Group in Central Kalimantan, were forced to leave their concessions idle after the fall of Suharto, and local people moved into these concessions to conduct so-called illegal logging. These local people had long been denied the right to benefit from their own natural resources and now felt that it was their time to profit from timber extraction.

While large-scale timber companies suffered during the period of economic crisis, the devaluation of the rupiah allowed small and medium-sized sawmills to take their place. These sawmills were able to take advantage of small investment requirements, low operational costs and an abundance of cheap raw materials sourced illegally. For instance, in the Berau district, only two large mills using band saws were opened for production between 1998 and 2000, bringing the total number of large sawmills to 19 in the district. Dozens of circular blade sawmills, however, sprang up throughout the area over the same time period. Despite an overall drop in international prices for timber in the key markets of Japan, Korea and China, the price for roughly sawn timber in the main transshipment states of Sarawak and Sabah remained attractive for Indonesian sellers (US$250 per cubic metre) in the year 2000. Shipping sawn timber internally, primarily to Java, although comparatively less profitable at US$120, was also considered to be worthwhile.

Decline in law and order

While Suharto was in power, a number of conglomerates and individuals with close connections to the Suharto family or the army obtained large forest concessions. The Indonesian army was paid to protect these concessions and to ensure that no one else logged them. It also played some role in preventing excessive illegal logging in Indonesia's national parks by only allowing those with close connections to Suharto and the army into protected areas. When Suharto resigned, the role of the Indonesian army was significantly drawn back during the period of the Habibie government. Local communities, co-operatives, entrepreneurs and outsiders soon realized that they no longer had to fear going into forbidden forest zones.

Our field observations included several instances of communities flouting police controls. For instance, when fieldwork was carried out in Central Kalimantan in June 2000, loggers had burned a police car because the police had attempted to confiscate some of the illegal timber they were transporting. In Tanjung Puting National Park in Central Kalimantan, two NGO activists were assaulted after they attempted to investigate illegal logging in the park (EIA and Telapak, 2000). The unstable security situation made it difficult for NGOs to return to the area and investigate what was happening on the ground. Illegal logging continued within the park while activists continued to wage a campaign against it in Jakarta and abroad.

Regulatory changes arising from *reformasi*

After the fall of Suharto in mid 1998, the Indonesian government was forced to show an intent to reform the forest sector. In an attempt to facilitate more just and equitable management of forest resources, the Ministry of Forestry issued instructions allowing communities residing in or near forest areas to be actively involved in forest exploitation through co-operatives, work groups and associations.[3] Environmental NGOs that had long been arguing for local communities to have a greater role in forest management and to be allowed to carry out low-impact extraction activities, primarily of non-timber forest products, initially welcomed these instructions. This idealistic formula did not work out as expected, however. It soon became apparent that some rural communities were not going to be content with increased access to non-timber forest products. They also wanted ownership rights to traditional forest areas, as well as equal standing vis-à-vis forest concession holders regarding the extraction of timber. Increased awareness about the value of forests among local communities arose from a marked increase in timber agents scouring rural areas to gain access to customary forest areas. This development coincided with the release of the 1999 Forestry Act, which acknowledged customary rights to land and forest areas. While the acknowledgement of customary rights in the new Forestry Law was a significant step forwards, it unfortunately resulted in a flurry of forestland claims in rural areas for the establishment of community-based logging concessions. Simultaneously, a lot of emphasis had been placed on border delineation of claimed locations.

Decentralization: Automoney versus autonomy[4]

After the fall of Suharto, President Habibie's interim government passed new legislation on regional governance and on fiscal balance between the central government and the regional governments.[5] These laws were meant to give greater financial and decision-making powers to local government, particularly at the district and sub-district level. Shortly after these new laws were released, the central government initiated the decentralization process in natural resource management by releasing legislation that devolved elements of authority to manage forests from the central government to provincial and district authorities.[6] This legislation gave governors and regents the authority to issue permits for small forest concessions, known as *Hak Pemungutan Hasil Hutan* (forest products harvesting right, or HPHH) or *Izin Pemungutan dan Pemanfataan Kayu* (permit to harvest and use wood, or IPPK).

Within weeks of putting these policies into effect, most regent offices in the province of East Kalimantan were flooded with applications for small-scale forest concession permits. In the Berau district, for instance, small-scale forest concession permits increased from virtually none in early 2000 to more than 30 by the middle of that year for a total of over 11,000ha. Towards the end of 2000, more than 100 applications had been submitted to the district government and an average of five new permits of 100ha each were being issued every month.[7] In Malinau district, there were 150 small-scale forest concessions in operation by the end of 2000, and there were 220 IPPKs and nearly 50 HPHHs in the newly formed district of Kutai Barat.[8] Kutai Induk[9] had also issued over 200 IPPKs and around 50 HPHHs. Cumulatively, these concessions covered hundreds of thousands of hectares of forestland in East Kalimantan. Whether working as a *Hak Pengusahaan Hutan* (forest concession contractor, or HPH) or operating jointly with Malaysian timber interests, former illegal loggers have been joining the HPHH/IPPK system in scores.[10] Community-based small concessions had thus provided a venue for the formalization of illegal logging. What was once illegal consequently became legal.

Similarly, in Central Kalimantan, the new decentralization laws were modified to allow the Kotawaringin Timur district assembly to issue a regulation that effectively legalized illegal logging in order to generate revenue through local government taxes. This new district regulation was legitimized through Article 80 of Law No 22, which states that 'the sources of the region's revenues shall consist of regional tax income'. This was strengthened by the issuance of Law No 34/2000 on regional taxes and regulations. This law enables local governments to create their own taxes through district regulations, provided that they have the approval of the district assembly and explain the idea to the local community.

Growing recognition of illegal logging in East and Central Kalimantan after the fall of Suharto

As mentioned above, illegal logging is not a new phenomenon. After the fall of Suharto in mid 1998, however, district, provincial and central governments all started to officially acknowledge its existence and to even document it in official statistics. In East Kalimantan for instance, local economic and political interests provided the means for such recognition in the form of exceedingly lax and permissive legislative frameworks aimed at facilitating an easy inclusion of informal logging activities within the formal sector. Lax regulations usually came in the form of district regulations that vaguely outlined the rights and obligations of small-scale concession operators. Invariably, these regulations concentrated on business and operational details, paying virtually no attention to the problems of monitoring or verifying logging activities. These district regulations provided no mechanism to prosecute those who violated the regulation by logging outside the 100ha boundaries, and only a few, such as the one ratified by the Pasir district assembly, mentioned that whoever violated the regulation would be fined. Since the fine was very small, it was unlikely to deter small-scale concession holders from logging outside the area granted.

As the trade in illegal logging became increasingly recognized as a legitimate practice, however, data on illegal logging became easier to gather and document, which made it simpler for researchers and NGOs to monitor the problem. For instance, field research in the districts of Berau, Malinau and Pasir of East Kalimantan during 2000 showed 72 sawmills operating in these three areas by May 2000. Earlier government reports had stated that there were only 30 active sawmills operating in the area in 1995 (Bappeda and BPS, 2000). Together, these 72 sawmills produced approximately 133,000 cubic metres of sawn timber in 2000. While all of these sawmills were officially recognized, all could also be said to be illegal because they lacked some of the required permits, relied on illegal timber for stock, or both. Supplies of illegal timber to sawmills were also secured through a network of logging camps that were operating independently or sponsored by sawmills or timber traders. In the majority of cases, these logging groups employed manual logging techniques. In 2000, 331 illegal logging camps were found in the districts of Berau (186), Malinau (31) and Pasir (114). Together, these logging camps were thought to have produced around 271,000 cubic metres of timber in 2000, or close to one quarter of the official log production of 1.3 million cubic metres reported for that year (see Table 3.1). Individually, illegal sawmills operating in these areas were also thought to have produced 89,000 cubic metres of timber in Berau, 15,000 cubic metres in Malinau and 28,000 cubic metres in Pasir during the 1999/2000 period.

Table 3.1 *Estimated legal and illegal log production in East Kalimantan, 2000*

Industry	Official legal log production		Illegal log production		Estimated real production	
	Total units (forest concession contractors, or HPHs)	Official production (m³/year)	Total units (camps)	Illegal production (m³/year)	Total units	Production (m³/year)
Berau*	8	798,000	186	160,000	194	958,000
Malinau	10	422,540	31	17,000	41	439,540
Pasir	3	74,578	114	94,000	117	168,578
Total	21	1,295,118	331	271,000	352	1,566,118

Note: *The number of HPHs in Berau does not include the five companies that took over the PT Alas Helau concession.

The illegal production figures are based on direct investigation of sampled illegal logging camps in three districts. These figures do not include production from small-scale concessions. Estimated real production was calculated by combining statistics on official log production with available statistics on illegal log production.

Source: Bappeda and BPS (2000)

Similarly, because the provincial and district governments in Central Kalimantan were losing revenue from illegal logging, government officials started to document and acknowledge its existence. Until recently, provincial and district forestry offices kept production and export statistics only on legal logging activities. Official statistics released by the provincial forest department in Palangkaraya stated that Central Kalimantan produced 1.5 million cubic metres of timber products in 1998/1999. In 2000, it became clear, however, that real production was much higher than the previously reported production after the provincial government began to document illegal logging activities at the request of the Ministry of Forestry.

In March 2000, the provincial forest department reported that there were six legally recognized mills producing plywood, 315 sawmills producing sawn timber and 22 mills producing moulding in Central Kalimantan. Together these mills consumed a total of around 1.5 million cubic metres of timber between January 1999 and January 2000 (Departemen Kehutanan dan Perkebunan, 1999). Like the sawmills in East Kalimantan, most of these sawmills sourced their timber from illegal logging operations. Moreover, the provincial government reported that there were, at the very least, 190 illegal sawmills operating in Central Kalimantan, excluding Barito Utara and Barito Selatan districts. Most of these sawmills could be found in Kotawaringin Timur, Kotawaringin Barat and Kapuas districts. These mills were thought to have consumed at least 155,750 cubic metres of timber between January

Table 3.2 *Estimated legal and illegal production in Central Kalimantan, 2000*

Industry	Production capacity (m³/year)	Official legal production		Illegal production		Estimated real production	
		Total units	Official production (m³/year)	Total units	Illegal production (m³/year)	Total units	Production (m³/year)
Sawmill	1,660,706	315	757,569	190	155,750	505	913,319
Plywood	495,000	6	628,325	n.a.	n.a.	6	628,325
Moulding	276,070	22	92,851	n.a.	n.a.	22	92,851
Total	2,431,776	343	1,478,745	n.a.	155,750	533	1,634,495

Notes: n.a. = not available.

Estimated real production was calculated by combining statistics on official log production with available statistics on illegal log production.

Source: Departemen Kehutanan dan Perkebunan (1999) and personal communications with Kotawaringan Timur district assembly staff, June 2000

1999 and January 2000 (personal communications with Kotawaringan Timur district assembly staff, June 2000). This amount is approximately 11 per cent of the total timber volume consumed by official sawmills operating in Central Kalimantan between January 1999 and January 2000 (see Table 3.2).

These statistics are unlikely to accurately portray the real productivity of legal and illegal sawmills in Central Kalimantan. The real production output of timber from Central Kalimantan between January 1998 and January 1999 was probably higher, given that the production capacity was reported to be 2.43 million cubic metres per year. These figures do, however, illustrate the recognition of the importance of illegal logging in overall production output and regional income.

Having argued that there was a growing recognition of illegal logging after the fall of Suharto, we shall now discuss ways in which the illegal timber sector underwent a transformation after the fall of Suharto. In doing so, we argue that illegal logging increasingly became a legitimate practice supported and encouraged by district and provincial governments. This transformation is illustrated by discussing the shifting dynamics of the sector in the districts of Berau, East Kalimantan, and Kotawaringin Timur, Central Kalimantan.

Shifting dynamics of the illegal logging sector in Berau, East Kalimantan

Berau is located in the north-western part of East Kalimantan and now borders the districts of Bulungan, Kutai Timur and Malinau (see Figure 3.1). Berau is

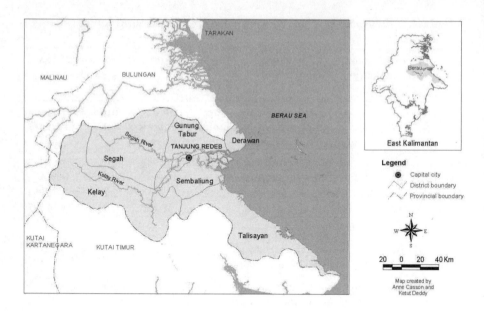

Figure 3.1 *The district of Berau, East Kalimantan*

Source: author

one of the largest, least populated and resource-rich districts in East Kalimantan. With a land area of about 24,000 square kilometres, it has a population of around 100,000 people.

While per capita income in 2000 in the district was over twice the national average, the district's physical infrastructure and industrial facilities were extremely limited. Most of the roads in the district are unpaved and large segments of the urban population, particularly in the more remote sub-districts of Sambaliung and Gunung Tabur, do not have running water or electricity. Education levels are also poor. While there are a number of primary, secondary and high schools in the area, teachers have always been in short supply.

Berau's economy is centred on mineral and natural resources, of which logging and coal mining are the most important industries. Together, these two industries generated close to half of the district's gross domestic product in 1997. Two large-scale investment projects also dominated the economy. These were PT Berau Coal, which operated coalfields in the Sambaliung and Gunung Tabur sub-districts, and PT Kiani Kertas, which constructed a multibillion dollar pulp and paper plant in the eastern part of the Sambaliung sub-district near the estuary of the Berau River. The Berau regency has also tried to diversify into the agro-industry sector by putting emphasis on the development of large-scale oil palm plantations. By the year 2000, plantation permits had been issued for around 480,000ha. The realization of the licensed oil palm projects has been extremely slow, however. In 2000, no more than 4000ha of plantations had been established.

Illegal logging in the regional autonomy era

While the Berau government had no official statistics on illegal logging activities in the district, information could easily be collected in the field since the trade had become increasingly legitimized. As of May 2000, there were approximately 186 illegal logging operations in Berau. Most of these activities were found in the middle and lower sections of the Segah River and its tributaries. Illegal logging activities could also be found in the upper watershed area of the Kalay River and around the coastal towns of Lati and Kasai on the Berau River. Overland roads connecting Tanjung Redeb to Tepian Buah and Berau to Bulungan and Samarinda were also found to be rife with illegal logging activities. Together, all of these camps were estimated to be capable of producing around 150,000 cubic metres of timber per annum.[11] This amount was approximately 21 per cent of the official log production reported for 1997 (698,000 cubic metres).

Between 1998 and 2000, these camps increased in number and size, and they tapped into legitimate logging sources. This coincided with a transformation in the sector that arose after the Ministry of Forestry released its regulation allowing district governments to issue 100ha concessions to local communities.[12] These regulations had effectively created a new source of inexpensive raw timber for sawmills in the form of community timber. The process was facilitated by sawmills or independent timber entrepreneurs who surveyed areas around villages in order to assess their timber stocks. If survey results and enquiries with the locals were promising, they would contact the village head and make a proposal in which they offered to provide a certain amount of money, seedlings and/or village facilities in exchange for low-priced timber. Once the agreement was sealed, the village formed associations of village cooperatives (*Koperasi Unit Desa*) or farmer groups (*Kelompok Tani*), usually with the official purpose of organizing smallholder ventures such as cocoa, coffee and coconut plantations. These associations allowed entrepreneurs to help villagers open forest areas that were purportedly needed for plantations. The plantations rarely, if ever, materialized.

The small-scale concession scheme became very popular and gained the support of district officials, entrepreneurs and local communities. District officials in Berau supported the scheme (veiling it in the politically correct rhetoric of regional autonomy) because they were able to generate considerable amounts of revenue from it. This revenue generation was made possible by Berau district regulation no 48/2000, which stated that all parties operating small-scale forest concessions (IPPK) were to pay one third-party contribution (*sumbangan pihak hetiga*) of US$1 per cubic metre of timber to the district government. This tax allowed the district government to generate approximately US$444,000 in 2000. In comparison to the taxes imposed by Kotawaringin Timur, the revenue that this tax generated in 1999 was relatively insignificant because the tax rate was much lower than that imposed by the authorities in Central Kalimantan. Still, this modest amount constituted nearly one half of the entire gross domestic product generated in the district in 1999.[13] The revenue generated through this scheme was also complemented by a number of informal payments originating from the initial allocation of IPPK payments and sawmill permits. For instance, an IPPK permit for 100ha required an unofficial contribution to the regent's office in the amount of US$1500. Each extension or addition of a new 100ha area required further payments. As a result, 50 permits totalling 8300ha generated at least US$125,000 in informal income in 1999. Revenue from sawmills[14] was also obtained through issuance of a letter to validate forest harvesting permits, which are bought at the price of US$22 per cubic metre. The estimated annual production of timber by sawmills in Berau was 89,000 cubic metres of sawn timber.[15] Since sawmills made arrangements with forestry officials to have legal transportation documents for only about 20 per cent of their real production, each month the district forestry office in Berau collected around US$33,000 in unofficial payments from the sawmill industry.

Entrepreneurs were supportive of the scheme because it offered considerable economic benefits and required much less capital than forest concessions. The holder of a large forest concession, for instance, had to pay approximately US$43 per cubic metre of timber in formal and informal taxes, while a small-scale concession holder had to pay around US$25 per cubic metre of timber. The holder of a large concession also invariably had to go through a great number of bureaucratic hurdles before it could obtain a permit, while small-scale forest concession holders easily obtained their permits from district governments. The overall investment in a large-scale forest concession was also considerably higher than that required for a smaller forest concession. This is because large sawmills were much more capital intensive than small or medium-sized ones. For instance, it cost up to US$15,000 to make one large band saw operational (exclusive of costs related to auxiliary structures and facilities), and the initial start-up cost was estimated to be in the order of US$70,000 to $80,000. The total start-up costs for a medium-sized sawmill, however, was estimated to be in the vicinity of US$40,000 – approximately half the amount required for large sawmills.

The only obstacle that a former illegal or small company faced when considering a small-scale forest concession was the start-up capital and operational funds, which, although not nearly as high as in the large forest concession, were considerably larger than what was required for illegal logging. One way of overcoming this difficulty was to join forces with companies already possessing small-scale forest concession permits in the capacity of a supplier or contractor. Large-scale forest concessionaires were also using this arrangement to escape criticism and mounting resentment against their operations. For instance, a state logging concession holder – PT Inhutani II, a government entity – recently adopted this approach in the district of Pasir since the company effectively ceased to work on its own, handing over logging operations to small-scale contractors. Another way of securing start-up capital for logging in small concessions was through partnerships with Malaysian entrepreneurs eager to obtain cheap Indonesian timber. For example, in Malinau district, Malaysian timber financiers from the cross-border town of Tawau were the single largest external business group (*Suara Kaltim,* 2001a). Thus, the net revenue obtained from a small-scale logging concession was estimated to be around US$59 per cubic metre of timber, while it was only US$37 for a large forest concession.[16] What is interesting about this new arrangement is that some parties, particularly army and police officers, were left out from the unofficial payroll in the small-scale forest concession system. The small-scale forest concession system had been legalized and therefore did not require as much protection. The main beneficiaries at the bureaucratic level were the regent's office and the district forestry office since it cost approximately US$1500 to secure a small-scale forest concession permit for 100ha.[17]

Finally, local communities were willing to participate in the scheme because it offered them the opportunity to legitimately benefit from their surrounding forest resources and to increase their well-being. Communities residing in or near forest areas participated in the system of small-scale timber concessions by providing forest areas for logging and obtaining fees, either per hectare of forest cleared or per cubic metre of timber extracted. Fees per hectare were very rare and were associated with clearing highly degraded forest (with little valuable timber) for what is claimed to be plantations, primarily oil palm. The most common type of fee was the one based on the amount of timber harvested. In Berau, these fees averaged around US$3 to $4 per cubic metre. Similar fees were also found in the neighbouring districts of Bulungan and Malinau (*Suara Kaltim,* 2001b). In the Mahakam area, where these small logging concessions were first introduced, community fees were higher, reaching approximately US$16 per cubic metre (Casson, 2001). While the income generated from these concessions was relatively small, it was nevertheless much more than local people were able to obtain from other forms of forest exploitation during the Suharto era. As communities were able to benefit from this new form of income, they became more vigilant about delineating village forest borders and enforcing these borders. Thus, as they entered into joint venture agreements for logging, the area available for freelance felling of timber decreased. This shortage of quality areas for conventional logging, combined with the prospects of working through small-scale forest concessions, was a driving force behind the formalization of illegal logging in Berau.

Illegal logging in Kotawaringin Timur, Central Kalimantan

Kotawaringin Timur is the largest district in Central Kalimantan (see Figure 3.2). Located in the centre of Central Kalimantan, it had a land area of 50,700 square kilometres and a population of around 500,000 people in 1999. The district was extremely rich in terms of its forest resources, which made it particularly attractive to illegal loggers. According to Central Kalimantan's

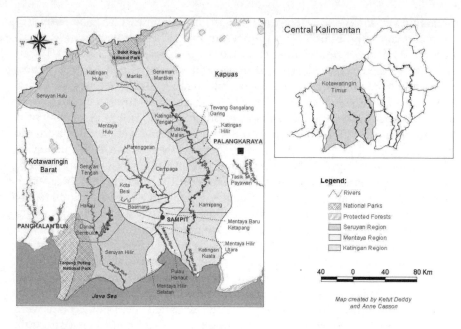

Figure 3.2 *The district of Kotawaringin Timur, Central Kalimantan*

Source: author

reconciled land-use plan, 2.7 million hectares of Kotawaringin Timur were classified as forestland. This forestland amounted to more than half of the district's total land area.

While Kotawaringin Timur is an extremely wealthy region in terms of its forest resource potential, the district's physical infrastructure and industrial facilities are limited and most of its inhabitants have subsistence livelihoods. There is only one asphalt road in the region, which runs through Sampit, connecting Pangkalanbun to Palangkarya. The road is in poor condition due to heavy traffic from logging trucks. While most of the villages in Kotawaringin Timur are extremely poor, Sampit, the capital city of Kotawaringin Timur, has some infrastructure. For instance, the capital city hosts a small airport, where a daily flight from Palangkaraya to Pangkalanbun stops over. All of the government offices were moved to new quarters on the road out of Sampit to Pangkalanbun soon after the fall of Suharto. However, while newly built, they were very basic and only a few had computers or resources at the time of the fieldwork. Many of the employees who staffed these offices also had limited skills and poor education.

Kotawaringin Timur's local economy revolves largely around the timber and mining industries, as evidenced by the amount of revenue generated from various sectors during the period 1994 to 1999. Over this period, US$1.9 million was generated through the forest sector and US$500,000 through the mining sector. The only district to generate more income than Kotawaringin Timur through the mining sector during the same period was Barito Utara, which generated US$2.3 million. While the plantation sector generated only US$110,000 in Kotawaringin Timur, the only district to generate more income from this sector was Kotawaringin Barat, which generated US$130,000.

Illegal logging in the era of regional autonomy

In March 2000, a new regent was elected in the district of Kotawaringin Timur. Shortly after being elected, he formed a special task force called the Integrated Service Team. Led by the vice regent of Kotawaringin Timur, the team was requested to investigate illegal logging in the region and devise ways of collecting revenue from the trade. Specifically, the team was ordered to find out how much timber was leaving the district illegally.

On 6 May 2000, the team reported that 178 ships carrying illegal timber had been found on the Mentaya River alone. These ships were carrying approximately 77,100 cubic metres of sawn timber. Instead of confiscating the timber on these ships and prosecuting those responsible, the Kotawaringin Timur assembly decided that ships carrying illegal timber would be permitted to leave ports in Kotawaringin Timur if they carried a paper stating that a 'contribution for forest product retribution' had been paid to the district income office. The letter had to state whether or not they had deposited money into the regional capital bank in the name of the Kotawaringin Timur district assembly. Shiploads would subsequently be checked to verify that they carried the load stated on their letter of receipt (*Surat Keterangan Lunas*) when they passed through Samuda – the final port on the Mentaya River.

During the year 2000, the regent was able to generate a great deal of revenue through this new initiative. For instance, in the months of April, May and June 2000, the regent was able to generate approximately US$2.5 million by taxing illegal carriers of timber coming out of Kotawaringin Timur.[18] During that period, 170,641 cubic metres of Meranti logs were shipped out of the district illegally, but with the knowledge of the district government. When compared with the official production figures for 1998, this figure is significant. According to the Central Kalimantan provincial department of the Ministry of Forestry (*Dinas Kehutanan*), 1,259,580 cubic metres of logs were officially produced in 1998 in Kotawaringin Timur (see Table 3.3). In other words, the recorded volume of illegally harvested logs for the three-month period of April to June 2000 amounted to 14 per cent of the district's total legal production for that year. If we assume that the production of illegally sourced logs continued at this pace throughout the year 2000, 511,823 cubic metres of illegally harvested logs would have been shipped out of the district. This amount constitutes close to half the district's total legal production for the same year.

The ability of the regent to generate income through this new tax increased his popularity and consolidated his political position in his own province. His decision to tax carriers of illegal timber, however, did not escape criticism from provincial officials, legal concession holders and the central government.

Table 3.3 *Total log production in Kotawaringin Timur*

	m³
Production of logs from illegal timber sources in Kotawaringin Timur, April–June 2000	170,641
Production of logs from illegal timber sources in Kotawaringin Timur, July–December 2000*	341,282
Estimated total production of logs from illegal timber sources in Kotawaringin Timur, April–December 2000	511,823
Official production of logs from Kotawaringin Timur, 1998	1,259,580
Expected real log production from Kotawaringin Timur, 2000	1,771,503

Note: * The figure is an estimate. It was calculated by dividing by three the total amount of timber to come out of the region in April to June 2000 and then multiplying this amount by six – the remaining six months of the year.

Source: personal communications with Kotawaringin Timur district assembly staff

Provincial officials primarily voiced concerns about the environmental consequences of the new regulation, but were more worried by the fact that they had lost control of the way in which natural resources were being managed in the district. They also voiced concerns about the new regulation contradicting provincial and national laws and uncertainty about how much of the revenue generated would be distributed to the provincial and national levels. The Kotawaringin Timur government felt that it should retain a minimum of 80 per cent of the revenue in accordance with the new decentralization laws. When fieldwork was conducted for this study, however, few discussions had been held with the provincial or central governments about this issue. In fact, the central government was largely being kept in the dark about the tax and revenue obtained from small-scale logging because the practice contradicted national law.

Legal concession holders raised the most vocal opposition to the new regulation. They complained that the regulation encouraged illegal logging in their concession areas and was unjust because they had invested large sums of money to secure these rights. They also raised concerns about the tariffs that illegal carriers were charged compared with the tariff that legal carriers were charged. When the tax was first introduced, illegal carriers were charged around US$14 per cubic metre, while legal carriers paid around US$21 per cubic metre. They therefore requested that the tariff be the same for both legal and illegal carriers to ensure that they were not disadvantaged by the new tax system. The Kotawaringin Timur government responded by raising the tariff from US$14 to $18 to appease the concerns of legal concessionaires (personal communications with Kotawaringin Timur district assembly staff, June 2000).

Illegal timber companies and ships also raised concerns about the legitimacy of their activities and requested that the government of Kotawaringin Timur issue a district regulation to legitimize the tax that illegal carriers were required to pay. The Kotawaringin Timur government responded positively to this request and decided to issue District Government Regulation No 14 on retribution for logs and processed timber to legitimize the issuance of letters to illegal carriers, stating that they have paid a tax to the regional government.

The decision to charge illegal timber carriers was ratified by the district assembly and the governor of Central Kalimantan in late April 2000. The governor agreed to the regulation in May 2000, provided that some of the revenue generated was distributed to the provincial and central governments. The amount to be distributed was left to the regent's discretion. When fieldwork was carried out for this study, no decision had been made about how much of the revenue collected through this initiative would go to the central or provincial government. In fact, the district government did not seem too keen to give any of the revenue to the central or provincial government. This was undoubtedly causing both upper levels of government some concern and adding fuel to their protests against the initiative.

The regulation has also met with disapproval from the central government. In a workshop on illegal logging in Indonesia held in Jakarta in late 2000, the secretary general of the Ministry of Forestry said that 'some regents are legitimizing the trade in illegal timber by issuing district government regulations. This contradicts national legislation and will not be tolerated.' The ministry looked into ways to control the situation and indicated that it was considering legal action against regents that had issued regulations that conflicted with national forestry laws. Despite these threats, however, the central government did not have authority under Law No 22 to revoke district government regulations, and Article 80 stated that the sources of regions' revenues would consist of regional tax income. This principle was also strengthened by the issuance of Law No 34/2000 on regional taxes and regulations, which enabled local governments to create their own new taxes through district government regulations, provided that they had the approval of the district assembly and explained the idea to the local community.[19] In this regulation, the central government was only given the authority to cancel a district government regulation within 30 days of receiving notice of it. District government regulation no 15 was, however, released before this regulation was issued, so the central government was unable to revoke it.

In Kotawaringin Timur, local officials were adamant that they would continue to collect revenue from the trade in illegal timber and defy the wishes of a central government that had exploited the district's forest resources for over 30 years. Following the central government's hard-line stance against the regent's policies, the district government became more cautious about revealing how much money it collected from this trade. When fieldwork was conducted in June 2000, the regional government was very open and also extremely proud of how much revenue it had been able to collect. By the following September, district officials were much more cautious and indicated that special permission from the regent himself would be required before data concerning revenues collected from the illegal timber trade could be released.

The economic, social and environmental consequences of illegal logging

Having discussed the nature of illegal logging in the two districts of Berau and Kotawaringin Timur, we now discuss some of the economic, social and environmental implications of these new forms of logging activities. In this section we argue that while an increase in community-based illegal logging has a number of short-term economic and social benefits, these benefits may be marginal in the long term. This is so because illegal logging, together with a number of other unsustainable land-use practices, is contributing to a degraded environment that will be unable to sustain the livelihoods of forest-dependent people in the near future.

Economic consequences

In the district of Berau and Kotawaringin Timur, new initiatives that arose after the fall of Suharto and an attempt to decentralize power to the district governments allowed district governments to benefit from the surge in the informal timber sector. This was most obvious in the district of Kotawaringin Timur, where the local government generated a considerable amount of revenue by taxing illegal carriers of timber. In the three months of April, May and June 2000, the regent was able to generate approximately US$2.5 million by taxing illegal carriers of timber coming out of Kotawaringin Timur. In Berau, the district government was able to generate a comparatively small amount of US$444,000 from small-scale forest concession holders because the imposed tax rate was far lower than that in Kotawaringin Timur. But while less income was generated, the US$444,000 accounted for 50 per cent of Berau's gross domestic product of 2000. The Berau government was also able to generate a great deal of informal income through the initial IPPK application process (US$125,000) and from issuing transportation documents to sawmills (US$55,000).

While both district governments undoubtedly benefit from the legalization of illegal logging, the rate of exploitation is far from sustainable. Consequently, the local governments will only be able to financially benefit from timber exploitation, both illegal and legal, for the next five to ten years. They will then have to find other ways of generating district income. In both of the districts discussed in this chapter, the local governments hope to convert forestland to oil palm to ensure that there is an established revenue-generating industry once production forests have been depleted. They are therefore making efforts to attract investors to the area in the hope that oil palm will provide a key revenue source after the region's national capital has been depleted. But while the political and economic situation in Indonesia remains unstable, investors have been reluctant to establish oil palm plantations in remote areas such as Kotawaringin Timur and Berau.

Moreover, at the national level, various forms of illegal logging are thought to be costing the national government US$2 billion per annum, without considering losses in terms of ecological costs (ITTO, 2001). Some estimates place the total annual loss to the country from illegal logging at about US$3.5 billion (EIA, 2001). This estimation is a stark increase over the US$1.2 billion that the World Bank said the Indonesian government lost due to illegal logging over the period of 1980 to 1985 (Callister, 1992). While some of the revenue generated from illegal logging is going to local people and district governments, the great majority is continuing to fall into the hands of a privileged and well-connected elite (see Chapter 5).

Social consequences

Similar to the economic consequences of illegal logging, there are also positive and negative social consequences arising from a marked increase in illegal logging. On the positive side, the illegal logging industry is generating a great deal of employment and income for local communities and newcomers. It also constitutes the focal point for a variety of small businesses, and because the work is extremely strenuous, it has a high turnover rate, which means that jobs are always available and comparatively well paid. For instance, in Kotawaringin Timur, at the time of fieldwork there were thought to be more than 300 illegal sawmills employing people along the Mentaya, Cempaga, Katingan and Seruyan rivers (personal communication with the Kotawaringin Timur head of Forestry, June 2000). Approximately 1500 people were also thought to be employed by illegal sawmills operating in Berau. While the latter figure seems fairly insignificant at first glance, such interpretation is misleading. It must be remembered that nearly 100 per cent of the labour force are temporary workers and the turnover rate is high – on average, workers do not stay in these jobs longer than one or two months. This means that the total labour turnover per annum could have been in the vicinity of 12,000 people, out of an approximate population of 100,000 people. This pool of jobs constitutes an important source of part-time employment for the rural poor.

In the informal logging sector, wages can vary from a monthly wage of US$25 for permanent employees to US$100 for output-measured work. These wages compare favourably with wages from other legal forms of physical labour. They therefore attract migrants, jobless poor and subsistence farmers. With a payroll of over US$100,000 per month, the sector also rivals wages given in most of the district civil service departments. Moreover, in addition to direct employment in sawmills, a number of indirect work opportunities are created. Families of sawmill workers frequently reside near or on sawmill grounds, where they perform auxiliary tasks such as cooking and laundry for small remuneration. Itinerant traders of foodstuffs and chemicals make brisk business as well. Local suppliers of fuels (gasoline, diesel) and the lubricants used in sawmills are also benefiting from the new boom in logging. Finally, since many small-scale forest concession holders are also sawmill operators or deliver timber to sawmills to be processed locally, their activities result in additional forms of informal payments that constitute an important source of income for other government officials, particularly the police, army and forestry officials. In the district of Berau, for instance, small sawmills set aside about US$44,000 for security officers each month.[20] For many of them, this pool of money constitutes the main source of income, far outweighing their official government allowances. As a result, small-scale illegal logging operations can be viewed as a lifeline supporting numerous segments of society in rural Kalimantan.

While the illegal timber sector is undoubtedly generating new work opportunities, local people are easily exploited for their labour. Prior to 1998, local people were denied rights to exploit forest resources and they are now willingly taking the opportunity to do so while they can. The wages that local people do receive for their labour compete well with other income-generating activities, which means that local people are often keen to engage in illegal logging for a fee. They are often still reliant on agents, however, who provide chainsaws and transport. This situation causes them to become dependent upon agents and vulnerable to exploitation.

In Kotawaringin Timur, local people tended to cut down the surrounding forest and sell it to brokers, primarily Chinese Indonesians, who arrange for it to be transported downriver to one of many illegal sawmills. The timber is then sawn, primarily by Malay people, who have moved to Central Kalimantan from South Kalimantan to work in the mills. These sawmill operators are attracted to the relatively high wages that this job has to offer. The work is extremely strenuous and workers are generally paid by the month according to the mill's output. While their salary can be anywhere between US$4 and $6 a day, it is often reduced by equipment breakdowns, accidents and adverse weather conditions (personal communications with illegal sawmill operators along the Mentaya River, Kotawaringin Timur, June 2000). Most can also only endure working in the mills for six months to one year because the work is hard labour and extremely dangerous. Many a worker manning the band saw has lost a hand as no safety precautions are taken. It is precisely for this reason that loggers often work on a temporary or trial basis when other sources of income (such as fishing, shifting cultivation and small-scale plantations) do not suffice.

Thus, the people who primarily benefit from the illegal logging of forests in East and Central Kalimantan are local Chinese-Indonesian brokers, Jakartan and Malaysian (again mainly Chinese) buyers and the end buyers in Japan, China, Taiwan, Hong Kong, Singapore and Europe. Unless local people can be given rights to forestland, they will not be sufficiently empowered to reverse this situation and they will continue to be dependent upon brokers for logging equipment, transport and market access.

Environmental consequences

Increasing illegal logging activities have a significant environmental impact, particularly if illegal logging occurs in already logged-over areas that have not had a chance to recover, in watershed areas or areas of high biodiversity. The regulation passed by the regent of Kotawaringin Timur effectively permitted anyone to log wherever and whatever they like, including active and inactive logging concessions, watershed areas, protected forest areas and national parks. These illegal logging activities are unregulated and consequently pose a significant threat to biodiversity and forest regeneration. Illegal logging in already logged-over areas may also result in many logged-over forest areas being unable to recover a commercially viable volume of timber for a second cutting cycle (Scotland et al, 1999).

The loss of Indonesia's forests is of concern because Indonesia is one of the most biologically and culturally diverse countries in the world. Although the Indonesian archipelago represents only 1.3 per cent of the Earth's land surface, it contains an estimated 25 per cent of the world's mammals, 11 per cent of the world's known flowering plant species, 15 per cent of all amphibians and reptiles, 17 per cent of all birds, 37 per cent of the world's fish species, and an unknown number of species of invertebrates, fungi and micro-organisms (Adisoemarto, 1992). If these species are to be protected, there is a great need to regulate illegal logging and to promote more sustainable and equitable forestry practices in Indonesia.

Conclusions

From the foregoing discussion it is clear that the illegal extraction and processing of timber in Kalimantan is an extensive and deeply entrenched system, with economic as well as socio-political dimensions. It provides both direct and indirect income opportunities, constitutes a focal point for a multitude of spin-off businesses and provides employment opportunities that are both flexible and well paid in comparison to conventional labour. Perhaps most importantly, the illegal timber sector has progressively become institutionalized as a result of two concurrent processes. First, the legalization of hitherto illegal forms of logging contributed substantially to district budgets; and, second, whether illegal or formalized, the informal timber sector has continued to be an important source of income for both civilian and military bureaucracies in the districts. As such, illegal logging can be viewed as a structurally important element of life in rural Kalimantan; in some cases, it is no longer considered illegal.

Given the above, the central government is facing the daunting task of addressing a problem that has a long history of entrenchment and possesses an operational structure that cuts across all levels of society. Stamping out corruption and instilling occupational professionalism, accountability and transparency as elements of democratic governance are all long-term prospects for which the forest can ill afford to wait. They are also unlikely to occur given the deeply entrenched nature of the system and the fact that responsibility and accountability are now so diffuse as to make it difficult to target offenders.

One of the more immediate options to deal with the problem could be a fundamental restructuring of HPH operations. This step would mean a fundamental redesign of the HPH paradigm in the direction of co-operative/partnership arrangements with local communities. It would have to include both economic incentives for local communities (at least as attractive as IPPKs) and the opportunity for them to be actively involved in logging operations. Formal recognition of local people's rights to forest resources would also have to be facilitated in order for local people to have a stake in protecting and conserving natural capital.

In the short term, it is also essential that district and national governments communicate with each other and agree on some of the legislative discrepancies arising from the decentralization process. Once this has been done, both levels of government must work together to implement good forest governance and effective law enforcement. In doing so, it may be necessary for both levels of government to provide a range of incentives to those who protect and sustainably manage forest resources. The question, as always, is implementation and political will.

Unfortunately, despite these options, the process of regional autonomy in Indonesia created conditions that are conducive to the continuation and/or further differentiation of illegal logging activities. This situation has arisen because the autonomy legislation remains weak and there is a lack of clarity and consistency in virtually all aspects of forest policy and management. This deficiency has largely resulted from the initial period of the regional autonomy era which focused on political bargaining at various administrative levels, securing economic bases and continual readjustment to changing circumstances. The resultant fluidity of the situation prevents consistent planning and policy-making from taking shape. The illegal logging problem is therefore likely to continue until these conditions stabilize and district governments are able to adopt a longer-term perspective in policy and management.

Acknowledgements

Both authors benefited a great deal from the support and assistance of the following people while carrying out fieldwork for this study: Abdi, Imam and Kiki (YTT Kalteng); Enra Rositah (BIOMA); Hendrik Segah (Centre for International Cooperation in Sustainable Management of Tropical Peatland, CINTROP); Lone and Odom (Nyaru Menteng Orang-utan Rehabilitation Centre); Neil Scotland (DFID); Rona Dennis (CIFOR); Pak Suwido (CINTROP); Mary Stockdale, Jon Corbett and Iman (PIONER); Ian and Tri in Tanjung Redeb; and the Berau Forest Management Project and all of the local inhabitants of Kotawaringan Timur and Berau that gave up their time for us to interview them. During the time of writing of this chapter, the following people also offered a great deal of assistance, support and encouragement: Chris Ballard (Australian National University, ANU); Chris Barr (CIFOR); Colin Filer (ANU); Daju Pradnja Resosudarmo (CIFOR); Enrica de Mello and Erik Meijaard (ANU); Grahame Applegate (CIFOR); Hidayat Al-Hamid (ANU); John McCarthy (Murdoch University); Joyotee Smith (CIFOR); Liz Chidley (Down to Earth); Patrice Levang (CIFOR); Patricia Shanley (CIFOR); Peter Kanowski (ANU); and Rod Taylor (WWF International). Ketut Deddy

(WWF International) deserves special thanks for his assistance with the maps. Finally, both authors would like to thank the Centre for International Forestry Research (CIFOR) for its assistance and support for this chapter. Despite all the assistance we have received from the above people, any errors are our own. The ideas expressed are also the sole responsibility of the authors and do not represent an official view on behalf of CIFOR or any other organization. This chapter is reprinted, with slight editing, from Casson, A. and Obidzinski, K. (2002) 'From new order to regional autonomy: Shifting dynamics of illegal logging in Kalimantan, Indonesia', *World Development*, vol 30, no 12, pp2133–2151, with permission from Elsevier.

Notes

1 Krystof Obidzinski collected information on the informal timber sector in Berau in 2000, while Anne Casson collected information on Kotawaringin Timur during the same year. Various newspaper articles and other secondary sources supplemented and supported the field information. In the course of fieldwork, both authors also consulted numerous people, including government officials, local community members, timber buyers and suppliers, company representatives, academics and NGO staff at the provincial level and in Jakarta. Both authors undertook fieldwork for this study while based at the Centre for International Forestry Research.
2 There are numerous technological similarities between *banjir kap* logging and current illegal timber operations (e.g. *kuda-kuda*). The crucial difference is that small-scale timber concessions of today employ mechanized means of exploitation.
3 PP No 62/1998 and SK MenHutBun No677/1998.
4 This phrase was used by the director of *Badan Planologi* at the Ministry of Forestry and Estate Crops when Casson discussed the issue with him in August 2000.
5 Laws No 22/1999 and No 25/1999.
6 PP No 6 1999, SK MenHutBun No 310/Kpts-II/1999 and SK MenHutBun No 317/KPTS II/1999.
7 In early 2001, the four northern districts of Nunukan, Malinau, Bulungan and Berau in East Kalimantan had issued approximately 500 HPHH/IPPK licences (*Suara Kaltim*, 2001a).
8 By mid 2000, the Kutai Barat government had issued more than 600 small-scale logging licences.
9 Now known as Kutai Kartanegara.
10 Personal communications with timber brokers in Berau, Bulungan and Malinau in East Kalimantan, September 2000.
11 This figure was derived through the following calculation: daily timber production per illegal logging camp is estimated at roughly 3 cubic metres per day (1.5 cubic metres per chainsaw, each camp having two chainsaws). Considering the fact that there are about 25 effective workdays per month when logging is actually carried out, the monthly timber production from illegal logging groups in Berau should be in the vicinity of 13,950 cubic metres, or 153,450 cubic metres per annum.
12 A series of regulations, not a single decision, was actually responsible for the emergence of the community-based timber extraction activities. PP No 62/1998 and SK MenHutBun No 677/1998 laid the groundwork for this new trend

by recognizing the concept of community forests (*hutan kemasyarakatan*). Subsequently, SK MenHutBun No 310 and 317 of 1999, as well as PP No 6 of 1999, served as the foundation for the HPHH and IPPK licence system (*Hak Pemungutan Hasil Hutan* and *Izin Pemungutan dan Pemanfaatan Kayu*).

13 In the year 2000, Berau's total budget was 47 billion Indonesian rupiahs (US$4.7 million). Only 9 billion rupiahs (US$900,000) were generated within the district, however. The remaining 38 million rupiahs (US$3.8 million) came from the central government in Jakarta.

14 Nearly all of the sawmills in Berau are illegal in the sense that they do not possess complete, or valid, permits. Nevertheless, all are allowed to operate for the sake of the people (*rakyat*). It is argued that the industry generates employment, income and cheap building material for the poor and therefore cannot be eliminated.

15 This estimate is based on data obtained directly from a survey of sample sawmills in Berau.

16 These log prices are based on industry standards for Red Meranti. It must be remembered that the profit margin for HPHH/IPPK and illegal logging is likely to be higher considering that they concentrate on high-value timbers for export, such as ironwood. These species can fetch prices much higher than Red Meranti.

17 This unofficial charge varies depending upon the quality of timber stock in the area in question. In the Kutai Barat district, IPPK/HPHH licences have been known to command prices of up to US$5000 per 100ha (*Suara Kaltim*, 2001c).

18 The regent's income generation performance was often compared to the poor performance of the former regent, who was only able to generate US$550,000 in one year.

19 The *Perda* must also address certain criteria. These 'good tax' criteria assert that:

- tax objects must be located within the administrative area of a particular local government and possess relatively low mobility across local government boundaries;
- the tax does not contradict the public interest;
- the tax does not constitute a national or provincial tax;
- the tax has sufficient revenue potential;
- implementation of the tax will not negatively affect the local economy;
- development of the tax takes into consideration issues of fairness to, and capacity of, local residents; and
- the tax guards environmental conservation.

While *Perda* no 14 clearly does not guard environmental conservation, the district government is more likely to focus on the fact that the central government has legitimized its ability to tax illegal carriers and to continue to reap the benefits.

20 This estimate is based on data collected from 40 operating sawmills in Berau – 13 large sawmills (with a monthly unofficial budget of US$2780), 6 medium sawmills (with a monthly unofficial budget of US$556) and 21 small sawmills (with a monthly unofficial budget of US$222).

References

Adisoemarto, S. (1992) *Indonesian Country Study on Biological Diversity*, Prepared for the United Nations Environment Programme (UNEP), Jakarta

Bappeda and BPS (2000) *Kalimantan Timur dalam angka 1998,* Kerjasama Bappeda Provinsi Kalimantan Timur dengan Badan Pusat Statistik Provinsi Kalimantan Timur, Samarinda, Bappeda and BPS

Brown, D. (1999) *Addicted to Rent: Corporate and Spatial Distribution of Forest Resources in Indonesia: Implications for Forest Sustainability and Government Policy,* DFID/Jakarta, ITFMP

Callister, D. (1992) *Illegal Tropical Timber Trade: Asia-Pacific: A Traffic Network Report,* Sydney, Traffic and WWF

Casson, A. (2001) *Decentralization of Policies Affecting Forests in Kutai Barat, East Kalimantan,* Case Study 4, Bogor, Centre for International Forestry Research

Departemen Kehutanan dan Perkebunan (1999) *Statistik kehutanan dan perkebunan propinsi Kalimantan Tengah, tahun 1998/99,* Palangkaraya, Kantor Wilayah Propinsi Kalimantan Tengah

EIA (Environmental Investigation Agency) (2001) *Illegal Timber Trade in the ASEAN Region: A Briefing Document for the Forestry Law Enforcement Conference Preparatory Meeting,* Jakarta, 2–3 April

EIA and Telapak (1999) *The Final Cut: Illegal Logging in Indonesia's Orang-utan Parks,* Jakarta, Environmental Investigation Agency and Telapak

EIA and Telapak (2000) *Illegal Logging in Tanjung Puting National Park: An Update on the Final Cut Report,* Jakarta, Environmental Investigation Agency and Telapak

ITTO (International Tropical Timber Organization) (2001) *Strengthening Sustainable Forest Management in Indonesia,* Report submitted to the International Tropical Timber Council by the mission established pursuant to Decision 12 (XXIX), 31st session, 29 October–3 November, Yokohama, Japan

Jakarta Post (2000) 'Illegal logging involves central government officials', *Jakarta Post,* 19 August

Kaltim Post (2000) 'Pasokan kayu di Kaltim cukup, karena pencurian', *Kaltim Post,* 4 August

Kartodiharjo, H. (2000) 'Forest management by concessions', Paper for a CGI seminar on Indonesian forestry, Jakarta

Khan, A. (2001) 'Preliminary review of illegal logging in Kalimantan', Paper presented at the Resource Management in Asia-Pacific Conference on Resource Tenure, Forest Management and Conflict Resolution: Perspectives from Borneo and New Guinea, 9–11 April, Canberra, Australian National University

Kompas (2000a) 'Penjarahan kayu makin merajela', *Kompas,* 9 August

Kompas (2000b) 'Itu namanya maling masuk ke ruah kita. Pangdam tanggapi soal pencurian kayu di perbatasan', *Kompas,* 20 August

Kompas (2000c) 'Peredaran kayu illegal. Dilema ekonomi rakyat', *Kompas,* 5 September

McCarthy, J. (2000a) 'Draft paper on the implications of regional autonomy for forest management', Paper presented to the second Insela Conference on Decentralization and Environmental Management in Indonesia, 31 May, Jakarta

McCarthy, J. (2000b) *'Wild logging': The Rise and Fall of Logging Networks and Biodiversity Conservation Projects on Sumatra's Rainforest Frontier,* Occasional paper no 31, Bogor, Centre for International Forestry Research

Peluso, N. (1983) *Traders and Merchants: The Forest Products Trade of East Kalimantan in Historical Perspective,* MSc thesis, Ithaca, New York, Cornell University

Peluso, N. (1992) *Rich Forests, Poor People: Resource Control and Resistance in Java,* Berkeley, University of California Press

Potter, L. (1990) 'Forest classification, policy and land-use planning in Kalimantan', *Borneo Review*, vol 1, no 1, pp1–128

Ross, M. (2001) *Timber Booms and Institutional Breakdown in Southeast Asia*, Cambridge, Cambridge University Press

Ruzicka, I. (1979) *Economic Aspects of Indonesian Timber Concessions, 1967–1976*, PhD thesis, University of London

Schwarz, A. (1990) 'A saw point for ecology', *Far Eastern Economic Review*, vol 148, no 16, p60

Scotland, N., Fraser, A. and Jewell, N. (1999) *Roundwood Supply and Demand in the Forest Sector in Indonesia*, Report number PFM/EC.99/08, Jakarta, Indonesia–UK Tropical Forest Management Programme

Suara Kaltim (2001a) 'Hutan di utara Kaltim terancam. IPK dikeluarkan, pengusaha Tawau banyak masuk', *Suara Kaltim*, 19 February

Suara Kaltim (2001b) 'Polda tangkap WNA di Malinau. Diduga support dana bagi pengusaha local', *Suara Kaltim*, 28 February

Suara Kaltim (2001c) 'Masyarakat dapat duit, jangan foya-foya', *Suara Kaltim*, 7 March

Wadley, R. (2001) 'Community co-operatives, illegal logging, and regional autonomy: Empowerment and impoverishment in the borderlands of West Kalimantan, Indonesia', Paper presented at the Resource Management in Asia-Pacific Conference on Resource Tenure, Forest Management and Conflict Resolution: Perspectives from Borneo and New Guinea, 9–11 April, Canberra, Australian National University

4
Turning in Circles: District Governance, Illegal Logging and Environmental Decline in Sumatra, Indonesia

John F. McCarthy

Introduction

In the wake of the economic and political crises that struck the country from 1998, Indonesia experienced an epidemic of illegal logging. As the Indonesian government implemented decentralization and associated fiscal reforms, many critics worried that these reforms would only exacerbate the rapid exploitation of the nation's dwindling timber resources (Hafild, 1999; Lay, 1999). At the same time, despite the large amount of policy discussion about unregulated logging, the complex set of incentives and interests driving logging at the local level remained poorly understood.

In a number of studies in several developing nations, several authors have described the logic-determining patterns of power and interest at the district level that are pertinent to understanding Indonesia's illegal logging phenomena. These studies have suggested that, in a range of contexts, political, economic and social exchanges and accommodations among regional politicians, officials, businessmen and key local figures generate the *de facto* institutional arrangements governing the allocation of resources at the local level (Moore, 1973; Oi, 1989; Migdal, 1994; Kahn, 1999).[1] Following these analyses, this chapter seeks to understand how the informal logging system operated in one district during the 1990s. It analyses how, for a range of political and economic motives, actors at the district, sub-district and village levels became involved in mining the most valuable resource found in the district – tropical timber. As these actors entered into exchanges and accommodations around logging, they created the *de facto* institutional arrangements governing timber operations at the district level. Based on an exploration of the impact of outside interventions and economic and political changes on this phenomenon, the chapter concludes that extra-legal logging constitutes a complex, multidimensional and over-determined phenomenon that allows for no straightforward remedy. Moreover, the chapter concludes that, since many of the dynamics described here continued to predominate as Indonesia implemented new decentralization laws, the system of exchange and accommodation surrounding logging at the district level was set to irrevocably drive the felling of Indonesia's once great forests.

The study was carried out in the mountainous and heavily forested district of South Aceh, located on the west coast of the island of Sumatra. It involved applying qualitative research methodologies over 12 months of fieldwork spaced out over three years (1996–1999). The research derives its conclusions from extended contacts and repeated interviews with journalists, officials, community leaders and village heads, NGO workers, loggers, and farmers before and during the crisis that marked the end of the Suharto era and just prior to the onset of separatist conflict in Aceh.[2] The research area consisted of a narrow but heavily cultivated coastal plain that faces the Indian Ocean to the west, backed to the east by the steep Bukit Barisan Mountains. Much of this area falls within Gunung Leuser National Park (GLNP), one of the great national parks in South-East Asia. GLNP itself is nested in a wider area of state-claimed forestland that forms the 'largest contiguous expanse of undisturbed rainforest of the western Indo-Malay type in the world' (Rijksen and Griffiths, 1995). Known as the Leuser Ecosystem, from the mid 1990s this area was subject to a European Union-funded Integrated Conservation and Development Project (ICDP), the Leuser Development Programme.[3] The so-called Leuser Ecosystem was designed to contain the ranges of the major elements of the biological diversity of northern Sumatra – including the Sumatran tiger, the elephant, the orang-utan and the Sumatran rhinoceros. Extending over approximately 2 million hectares, the Leuser Ecosystem was said to constitute the largest rain forest reserve in the world (Rijksen and Griffiths, 1995, p30).

The following section considers the logic leading to extra-legal timber networks across the district. The third section examines how logging networks accommodate village elites and ordinary villagers. The fourth section discusses how outside factors – including NGO activities, as well as economic and political changes – affect logging. Finally, the chapter considers the implications of this study, including the impact of decentralization on illegal logging networks.

The district level

Tensions between the state and local interests over the control of forest resources in 'outer island' Indonesia have a history that extends back to the colonial period (McCarthy, 2000). The policies of Suharto's New Order regime (1966–1998) sharpened and intensified these tensions. Under the New

Order, forestry policy shifted towards bringing in urgently needed revenue from the large economic rents that could be generated from Indonesian forests (Tjondronegoro, 1991). Towards this end, the state embarked on a territorial strategy that classified most of 'outer island' Indonesia as state 'forest zone' (Peluso, 1995). In this fashion, 88 per cent of the mountainous and heavily forested district of South Aceh was categorized as 'forest zone', leaving some 39 per cent of a total area of 8910 square kilometres classified as 'production forest' (Bappeda, 1992). Under this system, conglomerates and interests with close connections in Jakarta obtained 20-year logging concessions to selectively log these forests. As local people watched surrounding forests being logged by outside interests irrespective of longstanding local notions of territoriality, resentment emerged among communities, often leading to various forms of overt and covert conflict between logging concessionaires and local people.

At the same time, local entrepreneurs (known as *cukong*) were also interested in finding ways of exploiting the lucrative opportunities offered by logging operations. Yet, without the capital or connections, local entrepreneurs could not obtain logging concessions from the Department of Forestry in Jakarta. At the district level, however, there were other avenues to gain access to the forest. For instance, those opening timber plantations or agricultural plantations could obtain smaller leases in the state forest zone. If the land to be cleared still contained productive stands of timber, they could acquire a timber use permit, or *Izin Pemungutan dan Pemanfataan Kayu* (IPPK), to remove and process valuable logs from district and provincial authorities. This enabled actors wishing to open timber operations to obtain permits locally.

According to a businessman who attempted to open a logging operation in South Aceh, the process for acquiring the authority to manage wood legally was long and involved. Timber operators needed to obtain two other types of permits in addition to the IPPK, each entailing several approvals from various government agencies, and each approval facilitated by payments (McCarthy, 2006). The long, convoluted process for obtaining these permits led to widespread abuses. The informant estimated that entrepreneurs would need to spend more than 40 million Indonesian rupiahs and could not expect to complete the business within one year: 'So nobody is going to do it this way', he said; 'I tried myself and gave up.' He concluded that it was better to find a shortcut: paying local officials to turn a blind eye.[4] Even for entrepreneurs who managed to complete the process, the cost would be so high that they would need to recoup the investment – by disregarding regulations and working in a way that suited them. According to a 1994 report in *Serambi Indonesia* (1994a), 'as well as cutting in their own area, they also cut in the area around them' – including inside the neighbouring GLNP.

While many operations did not obtain all of the legal permits, entrepreneurs preferred to operate with a semblance of legality. The nature of the sawmill licences granted by the local department of industry office also facilitated this approach. According to the regional industry office, a sawmill permit would be issued after the raw materials were available. It should be based on the IPPK issued by the forestry office. While the IPPK was issued for only a single year, the local industry office issued sawmill licences for the life of the sawmill.⁵ Thus, the sawmill could continue to operate when the IPPK had expired, thereby generating a demand for illegally obtained logs. Moreover, entrepreneurs could obtain a licence from a sawmill whose logging area was exhausted, or move a sawmill with a valid permit into a new location. In 1994, some observers estimated that 50 of the 85 sawmills operating in South Aceh were operating illegally (*Serambi Indonesia*, 1994b). However, once the wood was processed, to carry it to the market in the city of Medan, a permit for transporting wood was also required. At that time, legal sawmills could issue their own transport permits, and other sawmills bought transport permits illegally from legal sawmills or directly from the provincial Forestry Service.

As a result, this system opened up opportunities for extra-legal exchanges between entrepreneurs and forestry staff. In short, in order to operate, entrepreneurs needed to seek the patronage of a rogue official prepared to act corruptly (known in Indonesian as *oknum*),⁶ who would offer protection to the entrepreneur's operations. At the same time, the operations of the logging network were connected to the need for rogue officials to engage in exchanges with clients, peers and patrons, and as we will see this logging also served local government priorities.⁷ Accordingly, the pattern of collusion that emerged between rogue officials and logging operations generated a set of accommodations that operated according to a predictable logic.

Logging and revenue

Devas (1989) has observed that regional governments have always desperately tried to augment local revenue. Setting ambitious targets for each year, each district typically has levied a large number of local taxes, the vast majority exacted under regulations (*perda*) set by the district administration itself (Devas, 1989). For a district head (*bupati*) and his administration, the raising of district-generated income has long been a central preoccupation. Under the New Order (1966–1998), the final decision regarding the appointment of a district head rested with the central government. Yet, even before obtaining the office of district head, an aspiring local politician needed to be successfully nominated by the local district assembly. A large entourage of supporters sustained via extensive patronage would help a figure to secure office. At the same time, it would also assist a serving district head to obtain this office for a second time. For these purposes, the district head and other district politicians could enter into exchanges with the bureaucratic peers and colleagues, and with local powerful figures and entrepreneurs. They could also use their discretion over budgetary allocations, utilizing contacts at the centre and other assets at their disposal. In these circumstances, besides requiring funds to service the debts of the district, a district head needed to find revenue to support projects and distribute favours – either within or outside the formal district accounts.

If a district government was unable to raise sufficient self-generated income, it would have to primarily depend upon funding allocated by the central government. Most of these funds were already earmarked for paying the incomes of local officials, or allotted to specific budget lines and development priorities. Therefore, district heads have embarked on ambitious revenue-raising drives to build up district revenues by levying district taxes (*retribusi*). It is hardly surprising that, at times, the local assembly has evaluated the performance of the local administration in terms of its ability to raise revenue and initiate high-profile projects. In this context, for a district head's administration, the ability to generate its own funds has been one measure of the effectiveness of the district head's period in office. Consequently, as Tables 4.1 and 4.2 show, while self-generated district income constitutes a small component of the total district government budget, it has played a significant role in the considerations of local government.

In South Aceh, each district head has faced the problem that, unlike many areas of Aceh that produce oil and natural gas, this district has few industries upon which to levy taxes to generate revenue. While South Aceh has had a timber industry, most of the taxes levied on logging concessions accrued to the central government, and only a small proportion has been returned to the provincial government. As a former district head of South Aceh pointed out in 1998, government revenue generated from timber included the receipts from timber royalties; the Reforestation Fund and Tax on Land and Building

Table 4.1 *District government incomes, South Aceh*

Fiscal year	Self-generated district income (rupiahs)
1987/88	85 million
1989/90	560 million
1990/91	469 million
1997/98	2881 million

Note: This table contains figures for those years for which the author has data.

Source: *Kompas* (1991); Badan Pusat Statistik Kabupaten Aceh Selatan (1992); *Analisa* (1996); Badan Pusat Statistik Kabupaten Aceh Selatan (1997)

Table 4.2 *Total district government budget, South Aceh*

Fiscal year	District budget (rupiahs)
1987/88	1.5 billion
1990/91	12.2 billion
1995/96	24.4 billion

Note: This table contains figures for those years for which the author has data.

Source: *Kompas* (1991); Badan Pusat Statistik Kabupaten Aceh Selatan (1992); *Analisa* (1996); Badan Pusat Statistik Kabupaten Aceh Selatan (1997)

amounted to an 'immense sum', as large as 2–2.5 trillion rupiahs. Yet, he observed, according to district budgets, the district only obtained 15 billion to 20 billion rupiahs each year (*Waspada*, 1998a). The district head over the period of 1988–1993 proved a successful lobbyist and fundraiser. As Table 4.1 illustrates, he oversaw a sixfold increase in district income, and during his period of office the district budget also grew significantly. This district head also waged a populist struggle against logging companies from outside the area, whose logging of watershed forests, he believed, was endangering the long-term future of the district. At one stage, he even threatened to resign if the Department of Forestry agreed to new concessions in the district (*Kompas*, 1991).

Under the man who held the position of district head over the successive period (1993–1998), the district policy changed. While this district head did not have the same ability to lobby for funds in Jakarta, he proved to be adroit at raising district income. Over his period of office, self-generated district income increased sixfold, while total district budgets approximately doubled (see Tables 4.1 and 4.2). Under this district head, the strategy of levying district taxes expanded to include taxes on logging operations often operating with only a semblance of legality. As a World Wide Fund for Nature document politely put it, the district head 'was more interested in increasing regional government income than efforts to protect the environment' (Kelompok Kerja WWF ID 0106, undated, p3). Over this period, sawmill operators enjoyed close connections with local government. According to informants, the district head gave permission for officials to gain revenue from their operations. As one source put it, 'in the name of the region, Pak [district head] made his own regulations to use the forests for funds to develop the region'.[8] According to a local journalist interviewed in the course of this study, in South Aceh, during this period, the district gained an estimated 60 per cent of its self-generated income from fees, both formal and informal, levied on the timber industry, mostly illegal logging.[9] Local government fees levied on timber included:

- the Industry and Natural Resource Tax, a charge imposed on timber;
- the Third Party Tax, a tax levied on products exported from the district (on the provincial border with north Sumatra, timber trucks would pay 500,000 rupiahs to leave the district);
- the Wood Collection Location Charge, levied on all companies engaged in logging operations (*Analisa*, 1996);
- 'scores of illegal fee collection posts along the whole length of the highway' set up by the police or army (*Serambi Indonesia*, 1995).

To some degree this revenue-raising strategy could claim legitimacy. As noted earlier, outside interests and central government gained the lion's share of economic rents derived from legal logging concessions. Meanwhile, local communities had to bear the negative externalities of the logging. This gave populist local leaders scope to criticize the way in which the official state forestry regime facilitated access for outside corporate interests and to condemn the lack of revenue accruing to the district budget from these logging operations.

As a journalist revealed, however, the impact of the local policy that permitted this logging on the local environment did worry some observers:

> *It was an open secret that the district government was obtaining the bulk of its finance – increasing district generated income – from wood... At the time, we wrote articles warning against the dangers of this... Many saw the need not to become dependent on this sector alone. Lots of people suggested that the district government seek out other sources of revenue that didn't destroy the environment. But it is not easy to move away from this dependence on wood revenue.*[10]

Moreover, this system lacked transparency. The annual district accounts published by the district statistics office each year failed to show how much revenue was raised in this fashion. In these circumstances, as several informants interviewed during 1996–1999 observed, a large amount of the funds collected remained in the personal pockets of tax collectors and their superiors.[11] This observation was confirmed by a 1996 report in the Banda Aceh daily *Analisa*. The report carried a statement from a member of the regional assembly complaining about the operations of the district government office responsible for collecting the district-generated income in South Aceh. It was estimated that if there was no 'leakage' or 'corruption', district receipts from timber alone would have amounted to a huge sum. However, a journalist estimated, less than one sixth of the money collected entered the official accounts.[12] One of the problems the *Analisa* article noted was that before collected funds entered regional government accounts, the district revenue office placed the money into the personal account of a certain local figure (*Analisa*, 1996).

Logging in Menggamat

The influence of district logging networks extended into villages where significant timber reserves could be found. One such area was Menggamat.[13] A range of factors was involved in the emergence of logging operations in Menggamat. During the 1980s, a paved road to Menggamat was constructed, connecting the area to sawmills. This, together with the arrival of chainsaws, made the felling of trees and the transport of timber profitable activities. Another causal factor in the emergence of logging was a shift in power relations. From historical times, Menggamat has consisted of a league of villages (known as a *Kemukiman*), a customary (*adat*) authority structure that governed the cultivated lands, and forest resources within community territory (Nababan, 1996).[14] As historical sources reveal, *adat* authorities controlled access to, and use of, the forests by outsiders. A trader buying forest products from the area or a collector gathering products directly would need to ask permission from the *adat* head of the community in whose territory the forest was found. The collector or trader would then be subject to a 'tax' paid to this *adat* head.[15]

Following the implementation of a new Village Government Law (Act No 5/1979) some years earlier, the principal *adat* leader (*imam mukim*) had lost most of his autonomous authority. The new law failed to recognize the Acehnese system of village leagues, and villages were arranged into administrative sub-districts. This arrangement meant that village heads were directly responsible to the sub-district head rather than the principal *adat* leader. Despite the key role that the principal *adat* leader played in both the traditional structure of community government and the implementation of *adat*, this *adat* head no longer had a significant role in the government structure (Mattugengkeng, 1987). In Menggamat, the position of the principal *adat* leader was particularly weak. He was a newcomer who had married into the area and had been elevated to this position following the intervention of district and sub-district officials. Thus, he was easily co-opted by outside logging networks. When approached by superiors involved in logging, this leader began to operate as 'an agent for sub-district officials':

> *The imam mukim, as well as operating four chainsaws, also has a role as the surrogate of officials in the sub-district. Several community leaders say that for his role the imam mukim receives a 'share' from the sub-district head for securing the interests of civil and military officials who are his superiors.* (Nababan, 1996, pp3–4)

After negotiating at this level, brokers also needed to approach the village heads who were responsible for enforcing *adat* rules in the village territory. However, as the broker had the support of local army officers, sub-district head and police, the village heads were unable to enforce *adat* rules that previously regulated where and when the cutting of trees could occur. Initially, the village heads just let the logging happen. As an NGO fieldworker later explained, 'the village heads were afraid of coming into conflict with the sub-district head and wouldn't really act'.[16] Over time, however, village heads, in collaboration with other village decision-makers, attempted to reassert some control over access to the local forest territory. In several cases, particular village heads, together with their village councils, came to an agreement to impose fees on those logging community territory. They charged loggers a 'development fee' for access to the forest under their authority. As the name suggests, the revenue raised in this way would then be used for developing village facilities, such as mosques and schools. As one village head later recalled:

> *At first there was no 'development fee' because we were still stupid. We could still be provoked by outsiders who said that the state owned the forest, not the village... We didn't know the regulations at the time, and still could be tricked ... and the apparatus had given permission.*[17]

By April 1995, according to a WWF report from this time, village heads were imposing fees on loggers, requiring that they report first before carrying out logging (WWF-LP, 1995).

Village heads played a much more ambiguous role than this suggests, primarily because the position of village head involved a conflict of interest. This circumstance was at least partly related to the nature of village finances. Village heads receive only a small honorarium from the government, and at least one village head complained that this was in no way commensurate with the work involved in carrying out their duties. Therefore, to support themselves, it has been common practice for village heads to levy fees for certain services and permits (Devas, 1989, p37).[18] These include administrative charges for letters of recommendation that are required for different purposes. Usually the scale of these fees was not set down in any formal way, but depended upon either the generosity of the person seeking the services of the village head or negotiation between the parties. As well as acting to maintain some control over village territory and to levy a 'development fee' on behalf of the community, by charging fees on their own behalf, village heads themselves directly benefited from logging.

Since rogue officials within the military, the police and regional government were involved, even this community control of the logging was weak. With more than 200 chainsaws operating around Menggamat by 1995, the community could not maintain its *adat* regulations before the tide of loggers.[19] While the boss of a logging team may, at first, have reported to the village head, he would then invite his friends, who would proceed to the forest without requesting permission. Thus, many logging teams were operating without reporting to the village head. In one case, there was an army official who carried six or seven chainsaws, then logged in areas protected by *adat* customary rules (watersheds and hills). Members of the community could not forbid this. This logger entered without permission and without paying a 'development fee' to the village head (WWF-LP, 1995).

In this situation, if village leaders were not to reap the rewards of logging for themselves, like other villagers who did not join in the logging of community forests, they would end up sitting by while outside parties enriched themselves. As Peluso (1992, p217) found in a similar case from West Kalimantan, in this context, village leaders – along with villagers involved in logging – were practical about the implications of not participating: 'a total loss of benefits as opposed to the enjoyment of benefits in the short term'. Therefore, village heads adopted opportunistic strategies, allowing logging and obtaining benefits from their position.

Gradually, village heads also began to extract rents from the forests for themselves. In 1995, WWF reported that timber brokers were cooperating with village heads. Soon, in addition to levying a 'development fee' for the village, many village heads operated as brokers themselves, buying chainsaws and providing capital to those carrying out the logging. One village head even operated 16 chainsaws on behalf of another broker. As an NGO project worker recalled: 'Of the village heads there, only two were not involved, and I am only certain that one of them was not involved.' The village heads built houses, and bought satellite television sets and new motorbikes from the profits of logging.[20] As village heads followed the logic of the situation, many were absorbed into the webs of power and interest that involved rogue officials, brokers and entrepreneurs.

Consequently, logging networks operating at the district level had extended downwards to embrace *adat* leaders and village heads, and this development was connected with a shift in local power relations. The rearrangement of village government in accordance with village government law, together with the injection of capital by outside entrepreneurs, weakened the ability of the *adat* community facing outside intervention. The principal *adat* head became an instrument of outside interests upon whose patronage he depended. After the restructuring of village administration, village heads were no longer under the tutelage of the *adat* elders; they were now directly responsible to a sub-district head (*camat*) who (informally) benefited from logging. *Adat* sanctions could not be brought to bear on loggers violating *adat* principles or local leaders who used their office corruptly. To some degree, the community had lost its ability to control logging that damaged forests and endangered surrounding farming lands. Nababan (1996, p5) concluded:

> *Observing the complexity of the problems connected with the uncontrolled logging shows that the destruction of forest resources is a consequence of the weak bargaining position of the community towards various outside interventions into the area, both as a result of formal government policy implementation as well as the injection of capital.*

Nonetheless, *adat* leaders did not sit back idly and watch. On behalf of the community, they reasserted community property rights by imposing a tax on timber extracted from what was considered to be the *adat* territory of the community.

The emergence of logging was also connected to the situation faced by villagers. Outsiders had first begun to introduce chainsaws into Menggamat in 1992, and the first uncontrolled logging began at that time. Before 1995, however, villagers primarily worked their own gardens.[21] It was only in 1995, when the price of the major cash crop at that time – patchouli (an oil produced from *nilam* leaves) – fell sharply, that things changed on a large scale.[22] At this time, people moved to other occupations, primarily the lucrative business of working in logging teams. Brokers offered villagers a strategy of survival at a critical moment. They employed local people to carry out the logging, organizing teams of 6 to 15 loggers, including a sawyer skilled in operating a chainsaw, usually from outside the area, and teams of bearers, often labourers from surrounding villages. By joining a logging team, villagers became the clients of brokers who paid expenses and provided loans for their families. This money formed a debt that would be paid off when loggers produced the timber (WWF-LP, 1995).

As one Menggamat resident noted, when logging became widespread, a large web of people benefited from logging operations. First, logging operations directly employed villagers and migrant workers from surrounding areas. These included the chainsaw operator, the logging team, timber carriers, those floating the wood down the river, the workers loading and unloading trucks down to the sawmill, truck drivers to Medan, sawmill operators and administrative staff. In addition, these operations indirectly employed a wide range of others, from the boat operators taking logging teams upriver to kiosk attendants, mechanics and motorcycle vendors. Counting the extensive number of people involved, a village school teacher estimated that one chainsaw led to the employment of up to 200 people.[23] And this number is in addition to those benefiting beyond Menggamat itself: the entrepreneurs and brokers, and a wide network of district officials raising taxes and rogue officials receiving payments from these operations.

Consequently, to the extent that the outside entrepreneurs and brokers recruited local villagers to work in logging teams and paid 'taxes' to local leaders, the local community benefited from the unsustainable mining of what was considered local property. In contrast to large logging concessions, the logging networks recognized local property rights over forest territory: by allowing members of the community to gain a share of the stream of benefits derived from logging, they were able to co-opt the *adat* community. Local *adat* authorities now allowed rapid exploitation, sacrificing the long-term values of the forest for short-term gain. Since villagers took up logging at a time when the price of the major cash crop (patchouli) had fallen so drastically, timber extraction in Menggamat became legitimate, and most local residents were sanguine about the results.

External interventions, economic crisis and reform

Over this period a number of outside factors – including NGO activities and economic and political changes – had various effects on local patterns of resource use. In 1994, WWF chose Menggamat as a site for community-based conservation activities. The project aimed to create a new community conservation regime that would guard the limiting conditions under which local agriculture operated while generating income for local villagers from the sustainable extraction of non-timber forest products.[24] Building upon aspects of the *adat* regulations that had related to the forests, WWF helped the community to craft rules in a form that could nest within the wider state legal framework.

However, this intervention faced difficulties that were ultimately insurmountable.[25] First, key officials and many community leaders and village heads supported the logging, while the district head valued its contribution to regional budgets. These decision-makers and rogue officials collectively had a tacit, but effective, veto over the working of the community forest initiative. Without their active support, the community leaders that backed the WWF plan could not implement the new regime effectively. At best, *adat* heads could defend *adat* property rights over surrounding forests by charging a fee to the loggers. Second, many villagers had become economically dependent upon logging, and the rapidly shifting economic forces continued to drive local villagers to mine forest resources. In the short term, at least, the proposed conservation regime could not guarantee village livelihoods and therefore struggled to get off the ground.

However, during 1997–1998, a range of changes affected the scale of logging activities. These changes transformed the situation in Menggamat, causing the tide of uncontrolled logging to begin to recede. While logging ultimately continued across South Aceh, for several reasons logging in Menggamat was now on a much smaller scale. From 1997, an economic crisis led to wide fluctuations in the dominant socio-economic realities shaping local life. As elsewhere in Indonesia, the crisis increased commodity prices (in rupiah terms) and made their production more attractive, leading to a temporary increase in forest clearing (Sunderlin et al, 2000). This began in late 1997 as the value of the rupiah sank and, coincidentally, the US dollar value of the product extracted from the area's key cash crop, patchouli oil, skyrocketed.[26] These twin influences led to a drastic increase in the local price of patchouli oil from around 35,000 rupiahs per kilogramme two years earlier, first to 150,000 rupiahs per kilogramme in early 1997 and then to around 1.08 million rupiahs per kilogramme at the beginning of 1998. At the height of the boom, a single hectare would produce a profit of around 20 million rupiahs, enough to build a house or buy land.[27] Those villagers who had abandoned logging and switched back to *nilam* cultivation some months earlier began to reap windfall profits. As the crisis reduced the purchasing power of villagers, it had a devastating impact on local people who were not producing agricultural commodities fetching high prices (Sunderlin et al, 2000). Spiralling prices and the collapse of jobs in the towns led many to open *nilam* plots: '*nilam* fever' spread across South Aceh. As long as the boom lasted, villagers who had been involved in logging switched over to *nilam* farming. Many loggers sold their chainsaws to obtain capital for opening *nilam* plots. As wood supplies dried up, entrepreneurs now used their capital to buy *nilam* oil and to trade it with wholesalers from Medan.[28] However, the *nilam* boom was based on a commodity price fluctuation. By mid 1998, *nilam* prices began to fall, just as many of the crops planted during the previous wet season began to be harvested.[29] *Nilam* production began to increase sharply just as demand slackened, with the consequence that *nilam* crops were hardly worth harvesting. As early as October 1998, villagers began to turn back to logging. Yet, in the meantime, other factors had changed, and at least in the Menggamat area logging now occurred on a much smaller scale.

The economic crisis also affected the sawmill operations on a broad scale, initially leading to a collapse in demand for wood, as the international price for plywood fell by 40 per cent during 1997–1998 (Sunderlin et al, 2000).[30] With the fall in the value of the rupiah, sawmills faced rising costs of components for imported sawmill machinery.[31] The economic crisis hurt the entrepreneurs, and they no longer had the capital to support large-scale logging operations.[32] Logging in Menggamat was becoming increasingly less viable because of the difficulty of obtaining high-quality timber. A former WWF fieldworker noted that previously 'there was a lot of timber close to the river, and it didn't entail operational costs in getting it. Before loggers only needed to put the wood into the river... Now they must pull it long distances as well. One log needs two to three people.'³³ He concluded that, as logs became less available, sometimes there was not enough wood even for one sawmill to function, and operational costs were too high compared with the profits that could be gained. Consequently, the number of sawmills in the sub-district dropped from the seven that had operated there at the height of the wood boom to only two, and even these worked sporadically – when there was wood.

Political changes also affected the situation. Following the resignation of Suharto in May 1998, in parallel with a nationwide movement against corrupt collusion and nepotism, the political environment became less tolerant of clientelist logging networks. News reports more openly discussed the networks of power and interest involved in illegal logging. NGOs, community groups and others opposed logging concessionaries and rogue officials who colluded with illegal logging networks. On several occasions, NGOs and community groups vigorously protested against logging operations, threatening logging operations and, in one case, even burning down a logging camp (*Serambi Indonesia*, 1998; *Waspada*, 1998b). For some time, these changes altered the behaviour of officials throughout the two districts, suggesting that a conducive political environment and a vital civil society will increase the chances of positive environmental outcomes.

The selection of a new district head in early 1998 also had an effect, demonstrating that office holders able to muster the political will can help to curb the behaviour of local officials tempted to form accommodations with logging networks. According to local sources, the new district head was a relative of a senior figure involved in the Leuser Development Programme and the Indonesian organization formally supporting this high-profile project, the Leuser International Foundation (YLI).³⁴ A district forestry official asserted that YLI had used its considerable influence in Jakarta to ensure the selection of the district head. Accordingly, the new district head supported YLI initiatives. He discontinued the previous policy of tolerating illegal logging in the name of raising local revenue and also attempted to have the leases of timber concessions terminated.

As further developments demonstrated, however, the favourable political situation alone was not enough to curtail unsustainable logging. The central role that economic demand plays in stimulating illegal logging became evident again during 1999, proving that the changes that had worked against the logging networks during 1997 to 1998 were temporary. The crisis had shaken loose the previous livelihood strategies of regional communities and elites. By 1999, the demand for wood had picked up again and, with few economic opportunities, timber networks again began to extend across the district. By the middle of 1999, the uncontrolled logging of lowland rain forest was occurring in areas adjacent to South Aceh, and reports of widespread logging activities in the Leuser Ecosystem became common. By March 1999, the logging teams had even moved into the Suaq Balimbing Research Area, an internationally renowned orang-utan research site located within the boundaries of the GLNP (Newman et al, 1999).

The problem was exacerbated by the crisis of legitimacy facing central government agencies and their policies following the excesses of the Suharto era. Under the New Order, the exploitation of natural resources in the region had generated large rents. Yet, the fruits had been inequitably shared across the nation, leaving underdeveloped remote districts from where the resources had been extracted. Under the form of decision-making that operated during the New Order, powerful conglomerates and politico-business families in Jakarta had gained privileged access to regional resources. Since this occurred in the name of national development, it discredited the state regime. Centralized top-down management of natural resources was associated with poor environmental outcomes: the previous three decades had produced rapid deforestation and forest fires. In short, from a regional perspective, cronies at the centre had enjoyed the benefits, while regional populations stood by, bearing the negative externalities of resource extraction (McCarthy, 2000).

While this resentment had existed during the New Order period, in the early heady days of the reform era following the fall of Suharto it contributed to a further weakening in the capacity of state agencies to enforce laws. Resentment against the state law extended to the state forestry regime, presenting a context for logging networks to flout the law with seeming impunity. In Aceh, hostility against the central government was particularly endemic and has been exacerbated as the Acehnese separatist conflict spread to South Aceh. As the security situation began to deteriorate in the first six months of 1999, this further undermined the legitimacy of, and compliance to, state laws relating to forest use (Newman et al, 1999).

Conclusion

This chapter has examined the networks of exchange and accommodation surrounding illegal logging in one district in Sumatra, suggesting that the roots of illegal logging can be found in the various interests of a diverse array of actors. Officials exploiting their positions could generate significant rents, both for their own personal use and for political purposes. Key local politicians could increase their popular support by expanding district budgets to provide for projects and programmes that offered opportunities to clients and followers. At the same time, entrepreneurs and their agents could maintain timber operations by entering into exchanges that involved extra-legal gifts and favours with certain local politicians and state functionaries. As these external networks entered the village arena, the pattern of exchange also needed to accommodate customary (*adat*) assumptions and authority structures. In the face of external interventions backed by powerful district figures, rather than merely watching the depletion of local forests from the sidelines, community leaders chose to compromise: loggers could cut in community forests if they agreed to pay unofficial taxes to village leaders. In this way, villager leaders attempted to reassert community property rights over community forest territory by collecting funds for community purposes, at the same time extracting rent for their own benefit. In the face of severe fluctuations in the price of agricultural commodities, villagers were also attracted by the short-term livelihoods that could be gained from the unsustainable mining of local forests. In this way, village actors enjoyed a portion of the flow of benefits gained from logging community territory.

This case suggests that the logging endemic has complex multidimensional causes and consequently allows for no easy remedy. A wide range of conditions would need to change before the tide of logging might recede. These conditions include a slackening in the demand for timber; the existence of an economy that provided stable economic livelihoods and offered viable alternatives to logging; steady revenue streams for local district government budgets outside of district timber taxes; greater local awareness of the long-term impact of unsustainable logging; an accountable political system with an active civil society able to control the behaviour of local officials and politicians; the renegotiation of customary rules governing resource access and use, taking ecological values into account; and the reconciliation of those customary rules with legitimate and enforceable state rules. A change in one factor without simultaneous changes in several others may not sufficiently alter the dynamics driving illegal logging.

This chapter has also considered the relationship between extra-legal logging and the wider state institutional framework. In this respect, as Ostrom (1990) has suggested, state agencies working with nationwide organizational rules are often unable to accommodate the interests of diverse groups and the variety of variables involved in local settings. Consequently, it is unlikely that externally imposed rules will reflect the circumstances of resource users or that they will enhance the development of effective governance regimes (Ostrom, 1990). Indeed, as the foregoing discussion revealed, the formal administrative system created difficulties for local actors wishing to gain access to the forest through official channels. This helped to legitimize the emergence of an informal system of resource extraction that circumvented the 'contradictions inadvertently created by the formal political and economic systems' (Oi, 1989, p229). This extra-legal system provided 'the flexibility needed to survive the myriad rules of a bureaucratic state that has failed to satisfy the needs of its citizens' (Oi, 1989, p228). Clearly, the legitimacy of laws pertaining to forest uses might be improved by reforming resource governance to ensure that the regulation of forest access and use reflects local needs and interests more closely, thereby avoiding the contradictions that legitimated the emergence of the informal system of resource extraction.

The decentralization reforms undertaken after 1999 for a short time seemed to mark a step in this direction. Yet, in Bolivia, devolution of power to local governments strengthened local elite groups in some municipalities who had vested interests in unsustainable harvesting of forest resources. Moreover, cash-starved municipal governments needed to increase revenues. Since there were usually few incentives for sustainable use of resources over the short term at this level, in some municipalities decentralization was associated with increased exploitation (Kaimowitz et al, 1998). By 2000, there were indications that similar dynamics were operating in many areas of Indonesia during the transition to a more decentralized system (*Jakarta Post*, 2001; McCarthy, 2001a, 2001b; Obidzinski, 2001). Indonesia's decentralization laws granted district governments greater powers to arrange their own regulatory regime and gave them more responsibility for generating their own revenues to run decentralized services. Now, as in South Aceh during the period of this research, powerful entrepreneurs involved in logging – and even sitting in the district legislature – continued to be well placed to affect the formal decision-making process of district governments, ensuring that their interests were protected while the district set out to accrue significant revenue for their cash-starved budget. For example, in South Barito district, Central Kalimantan, during 2000 the district government implemented a regulation that levied a new tax on forest products.[35] Previously, in accordance with national laws, the police in the district had confiscated and auctioned illegally obtained timber. Now, under the new district regulation, timber would merely be 'processed', subjected to district and other taxes, and allowed to be shipped on. In other words, the district administration used its new powers to pass a regulation that 'legalized' an informal system similar to that which had operated in South Aceh during 1996–1999, paying attention to whether the timber had been subject to district taxes rather than to whether it had been harvested in accordance with national laws. In nine days in May 2000, district officials began to implement the new local regulation, processing some 30,839 cubic metres of illegal timber and collecting almost 2 billion rupiahs for the district government treasury (*Kalteng Pos*, 2000). By creating a district legal framework for illegal timber, in effect, the new revenue-raising initiative worked to give a more formal status to the system of extra-legal logging and associated taxes described in this chapter.[36]

In light of the earlier discussion, however, it is important to avoid overstating the impact of decentralization on the illegal logging phenomenon by placing such developments in context. First, if we recall the environmental problems associated with the New Order period, it is clear that the centralized system did not function well in terms of social equity or environmental outcomes. This failure, together with the economic crisis, helped to create the complex set of interests and incentives that led to the subsequent logging epidemic. It also impelled the ensuing move to a more decentralized system. Second, as the foregoing discussion indicated, even during the New Order period, the central government had limited capacity to control resource use: there was a considerable gap between the formal authority of the state and its ability to impose *de facto* control. At the district level, informal decisions were often made outside the legal framework and were largely beyond the scrutiny of higher-level government agencies (Gunarso and Davie, 1999; Niessen, 1999).[37] Before decentralization, in districts remote from central control, a highly localized (and largely under-researched) system of resource extraction worked contrary to state regulations, allowing rent seeking by district elites operating with a large degree of independence from central government supervision.[38] After 1999, during a period of crisis and reduced state capacity, it was unrealistic to expect a set of largely administrative reforms (decentralization), which primarily focused on redistributing power and financial resources from the central to regional governments, to change such entrenched patterns in a positive way. At the same time, as national attention shifted onto the performance of regional governments, such patterns became more apparent. Yet, in assessing this, we need to bear in mind the enormous volumes of timber extracted by timber companies over the New Order period: during this time, timber companies extracted timber from the forest well in excess of their entitlements under concession licences and usually in disregard of state forestry regulations (Barr, 1998, 2000). Consequently, in the complex array of interests and incentives driving illegal logging discussed in this chapter, it remained rather disingenuous to blame decentralization for the ongoing liquidation of Indonesia's timber reserves.

In conclusion, many of the considerations described in this chapter continued to drive the degradation of Indonesia's forests during the post-Suharto years. The major district actors involved in logging continued to enter into exchanges and accommodations in order to ensure that their interests were served. Local elites, including timber interests, made sure that district politicians dependent upon local support did not threaten their business operations. In many cases, local business elites influential in local legislatives worked to ensure that decentralized district administrations create new formal decisions regarding the management of local resources that favour their interests and those of their business partners from outside the region. In this sense, implementation of regional autonomy helped to formalize and even extend a system that had already operated widely. This development was hardly surprising because legal sociologists have long observed the way in which pre-existing social arrangements affect the outcome of new reforms. All too often, when governments attempt to implement new laws, the pre-existing social obligations are stronger than a new law: such arrangements tend to dominate how the laws are invoked and the consequences of that invocation (Griffiths, 1995). While, with regional autonomy, many of the dynamics that affected the district scene in South Aceh persisted, with more open political conditions and with more decisions being made locally, in some places there was still a glimmer of hope that these dynamics might be somewhat ameliorated over time. If this is to occur, however, there is a need for more than the redistribution of power and finance between levels of government that was so extensively discussed. It entails a more fundamental transformation in the relationship between local government and local citizenry, a transformation that ensures better political representation and participation and, consequently, a more transparent and accountable system of governance.

Acknowledgements

Thanks to the Asia Research Centre, Murdoch University and CIFOR for funding the research and the Indonesian Institute of Science for sponsoring the research application. Thanks also to David Edmunds; Lini Wollenberg; Chris Barr; Daju Pradnja Resosudarmo; Carol Warren; Sue Moore; Tuti Hendrawati; informants and NGO fieldworkers in Aceh Selatan; WWF Indonesia Programme; the Leuser Management Unit; and many people in Aceh Selatan for their patience, hospitality and invaluable assistance. This is an edited version of a paper published in McCarthy, J. (2002) 'Turning in circles: District governance, illegal logging and environmental decline in Sumatra, Indonesia', *Society and Natural Resources* vol 15, no 10, pp867–886. An earlier version of this chapter was published as a Centre for International Forestry Research (CIFOR) working paper.

Notes

1 For further discussion, see McCarthy (2002).
2 Between August 1996 and January 1999, when this research was undertaken, South Aceh was peaceful. During 1998, following the fall of Suharto, a sense of outrage and injustice mixed with economic grievances against the Jakarta government generated a passion for Acehnese independence. By the end of January 1999, demonstrations in northern Aceh were already escalating into conflicts, and by the middle of 1999, the conflict began to creep south towards South Aceh, and by August that year South Aceh was drawn into a shadowy but violent conflict in which the protagonists often remained unclear.
3 The *Leuser Development Programme Masterplan* (Rijksen and Griffiths, 1995) envisaged that the management problems associated with the Leuser Ecosystem area would be addressed by delegating management responsibility for the Leuser Ecosystem area to a new organizational structure, the Leuser Management Unit (LMU). Under a later presidential decree (Keppres No 33/1998), the Leuser International Foundation (Yayasan Leuser International, or YLF) delegated responsibility for programme management and activities to the LMU, which was entrusted with the implementation of the Leuser Development Programme (LDP).
4 Interview, entrepreneur, Tapaktuan, South Aceh, 5 January 1999.
5 Interview, Dinas Kehutanan, Tapaktuan, South Aceh, 15 January 1999.
6 An *oknum* is literally 'a person acting in a certain capacity', a euphemism for individuals abusing their position or otherwise acting contrary to their official responsibilities.
7 As the system developed, some *oknum* could act as entrepreneurs themselves, backing a broker who would then carry out logging operations at arm's length from his patron.
8 Interview, village informant, Menggamat, South Aceh, 9 January 1999.
9 Journalist, interview, Tapaktuan, South Aceh, 5 January 1999. It is difficult to confirm this account from official figures because local government figures appearing in reports are not transparent. In provincial budgets, district taxes – including those levied on forestry operations – are not disaggregated but tend to be grouped together under the category of 'other' (Devas, 1989, p81).
10 Interview with journalist, Tapaktuan, South Aceh, 5 January 1999.
11 Interviews with district officials and journalists, Tapaktuan, South Aceh, February 1999.

12 Interview with journalist, Tapaktuan, South Aceh, 5 January 1999.
13 Menggamat is sometimes written Manggamat. Kemukiman Menggamat is one of four community leagues where people of the Kluet ethnic group predominate. The Kluet are indigenous to the upper end of the Kluet River valley, a small pocket between the mountains stretching inland from the Indian Ocean.
14 The term *adat* encompasses a range of English meanings, including 'custom', 'customary law' and 'customary management system'.
15 At one time, the fee had been incorporated within colonial legislation governing taxes on forest use (Adatrechtbundels, 1938). When loggers wanted to begin logging operations in Menggamat, they needed to approach the head of the *adat* community in whom (according to village norms) authority over what was still considered community property was still vested.
16 Interview, Tapaktuan, South Aceh. The sub-district head and the district council (*Muspika*) vetoed several initiatives of the community NGO (YPPAMAM) set up to facilitate a community conservation forest and Kemukiman decisions supporting this.
17 Interview, village informant, Menggamat, South Aceh, 23 January 1999.
18 As Devas (1989) notes, this is common practice in the outer islands of Indonesia. Indeed, the *adat* fee on access to forests (*pantjang alas*) had been this sort of charge.
19 In theory, it should not be difficult to control access to the Menggamat forests. There is only one road out of the area and one river to float logs down. Over time, villages set up posts on the road to enforce the collection of village taxes (*uang pembanguan*).
20 Interview, Tapaktuan, South Aceh, 23 January 1999.
21 Interview with *nilam* farmers, Desa Mersak, South Aceh, 6 February 1998.
22 *Nilam* (*Pogostemon cablin*) is a cabbage-sized leafy plant that grows to a height of 30–70cm. By distilling the dried *nilam* leaves, farmers produce patchouli oil, a product used in cosmetics and perfumes and which is now popular with aromatherapists (LPTI Bogor Sub Bagian Publikasi/Dokumentasi, 1970).
23 Interview with village informant, Menggamat, South Aceh, 10 January 1999.
24 WWF set about creating a community conservation forest (CCF). This entailed creating a community management structure and a regulatory regime based on 'revitalized' *adat* regulations relating to the forest. Under this regime, while leaving the ecological functions of the CCF intact, farmers would derive a livelihood from the CCF by collecting forest products (principally *damar* resins collected from *Shorea* spp.) on a sustainable basis.
25 A full discussion of the problems faced by this project is beyond the scope of this chapter, but has been discussed elsewhere (McCarthy, 2006).
26 The US dollar value of the rupiah declined from 2400 rupiahs per US$1 in July 1997, reaching lows of 16,000 to 17,000 rupiahs during 1998. During most of the 1997–1999 period, however, the rupiah traded in the 8000 to 9000 rupiah range (Sunderlin et al, 2000).
27 Interview with *nilam* farmers, Menggamat, South Aceh, 6 February 1998.
28 Like other commodities, patchouli oil prices fluctuate widely on the world market. According to a trader who deals in patchouli, the El Niño event of 1997 caused a drought that affected patchouli production in Sumatra. The actions of speculators and futures traders exacerbated this fluctuation. In 1997, patchouli oil had shot up to US$100 per kilogramme, but by November 1998 it had fallen to only US$25 per kilogramme (telephone interview, John Fergeus, Australian Botanical Products, Hobart, Tasmania, 30 November 1998).

29 On sloping land, *nilam* needs rain, and farmers must plant before the wet season.
30 Towards the middle of 1998, demand for timber began to pick up again (CIFOR, 1998).
31 *Waspada* (1998c).
32 Interview, village informant, Menggamat, South Aceh, 5 January 1999.
33 Interview, NGO fieldworker, Tapaktuan, South Aceh, January 1999.
34 Yayasan Leuser International (YLI).
35 Bupati Barito Selatan, *Peraturan Daerah kabupaten Barito Selatan nomor 2 Tahun 2000 Tentang Retribusi Hasil Hutan, hasil Hutan Bukan Kayu dan Hasil Perkebunan*. Similar district legislation has been implemented in other districts in Kalimantan (*Jakarta Post*, 2001).
36 On 2 July, the South Barito district government invited all actors with timber operations – legal or illegal – to a meeting in the district capital, Buntok. The purpose was to 'socialize' the new regulation (interview, Central Kalimantan, 21 July 2000). The occurrence of this meeting indicated that the district administration had knowledge of all those active in extra-legal logging. This demonstrated how, when it was politically feasible and seen to be in the interests of the district, the district government could craft organizational rules that reflected the needs of key local actors, including district entrepreneurs active in the timber sector, local politicians and officials concerned with the budgetary needs of the district government. However, this was to the long-term benefit of neither the forests nor the poorer Dayak communities. For further discussion of decentralization in Central Kalimantan, see McCarthy (2001a).
37 Gunarso and Davie (1999) call this phenomenon 'pseudo-autonomy'.
38 For an earlier analysis of a similar problem in the context of state forestry in Java, see Peluso (1992).

References

Adatrechtbundels (1938) 'De pantjang-alas in Atjeh (1931)', in *Adatrechtbundels; bezorgd door de commissie voor het adatrecht en uitgegeven door het koninklijk instituut voor de taal-, land en volken-kunde van Nederlandsch-Indie*, Gravenhage, Martinus Nijhoff, pp136–138

Analisa (1996) 'Pengawasan Ketat Diperlukan agar tidak Terjadi Kebocoran di Dispenda', *Analisa*, Medan, 13 March

Badan Pusat Statistik Kabupaten Aceh Selatan (1992) *Aceh Selatan Dalam Angka*, Aceh Selatan, Badan Pusat Statistik Kabupaten

Badan Pusat Statistik Kabupaten Aceh Selatan (1997) *Aceh Selatan Dalam Angka*, Aceh Selatan, Badan Pusat Statistik Kabupaten

Bappeda (1992) *Profil Kabupaten Daerah Tingkat II Aceh Selatan*, Aceh Selatan, Badan Perencanaan Pembangunan Daerah, Kabupaten Daerah Tingkat II

Barr, C. (1998) 'Bob Hasan: The rise of Apkindo, and the shifting dynamics of control in Indonesia's timber sector', *Indonesia*, vol 65, pp1–36

Barr, C. (2000) *Will HPH Reform Lead to Sustainable Forest Management?: Questioning the Assumptions of the 'Sustainable Logging' Paradigm in Indonesia. Banking on Sustainability: A Critical Assessment of Structural Adjustment in Indonesia's Forest and Estate Crop Industries*, Bogor, Washington, DC, CIFOR and WWF-International's Macroeconomics Programme Office

CIFOR (Centre for International Forestry Research) (1998) *The Economic Crisis and Indonesia's Forest Sector*, Bogor, Indonesia, CIFOR, www.cifor.cgiar.org, accessed November 1998

Devas, N. (ed) (1989) *Financing Local Government in Indonesia*, Athens, Ohio University Center for International Studies

Griffiths, J. (1995) 'Legal pluralism and the theory of legislation – with special reference to the regulation of euthanasia', in Peterson, H. and Zahle, H. (eds) *Legal Polycentricity: Consequences of Pluralism in Law*, Aldershot, Dartmouth

Gunarso, P. and Davie, J. (1999) *How Decentralization Can Improve Accountability of Forest Resources Management in Indonesia*, Queensland, Australia, School of Natural and Rural Systems Management, University of Queensland

Hafild, E. (1999) 'A new paradigm for rights based advocacy strengthening NGO alliances in the post-Suharto era', Paper presented at the Australian Council for Overseas Aid Annual Council Meeting, Canberra, September

Jakarta Post (2001) 'Whir of chain saws goes on in East Kalimantan rain forests', *Jakarta Post*, Jakarta, 20 March

Kahn, J. S. (1999) 'Culturalising the Indonesian uplands', in Li, T. M. (ed) *Transforming the Indonesian Uplands: Marginality, Power and Production*, Australia, Harwood Academic

Kaimowitz, D., Vallenos, C., Pacheco, P. B. and Lopez, R. (1998) 'Municipal governments and forest management in lowland Bolivia', *Journal of Environment and Development*, vol 7, no 1, pp45–60

Kalteng Pos (2000) 'Rp 1,7 M "Tercecer" di DAS Barito', *Kalteng Pos*, Palangkaraya, 12 June

Kelompok Kerja WWF ID 0106 (undated) *Konflik Kepentingan Dalam Pengelolaan Sumber Daya Alam Pedesaan, Studi Kasus: Kemumiman Manggamat, Kec. Kluet Utara - Kab. Aceh Selatan*, Report prepared for Kelompok Kerja WWF, Jakarta

Kompas (1991) 'Drs Sayed Mudhahar Akhmad Bupati Pencinta Lingkungan Hidup', *Kompas*, 26 February

Lay, C. (1999) 'The management of natural resources for the strengthening of the local base: The social and political perspective', Paper presented to the Seminar Workshop Towards the Management of Natural Resource for Strengthening of the Local Base, Jakarta

LPTI Bogor Sub Bagian Publikasi/Dokumentasi (1970) *Pedoman Bertjotjok Tanam Nilam (Patchouly)*, Bogor, Departemen Pertanian, Direktorat Djenderal Perkebunan, Lembaga Penelitian Tanaman Industri

Mattugengkeng (1987) *Agama dan adat dalam konlfik birokrasi pemerintahan tingkat pedesaan di Aceh*, Badan Penelitian dan Pengembangan Agama, Banda Aceh, Departemen Agama RI

McCarthy, J. F. (2000) 'The changing regime: Forest property and *reformasi* in Indonesia', *Development and Change*, vol 31, no 1, pp91–129

McCarthy, J. F. (2001a) *Decentralization, Local Communities and Forest Management in Barito Selatan*, Centre for International Forestry Research, www.cifor.cgiar.org/highlights//Decentralization, accessed April 2004

McCarthy, J. F. (2001b) *Decentralization and Forest Management in Kapuas District*, Centre for International Forestry Research, www.cifor.cgiar.org/highlights//Decentralization, accessed April 2004

McCarthy, J. F. (2002) 'Power and interest on Sumatra's rainforest frontier: Clientelist coalitions, illegal logging and conservation in the Alas valley', *Journal of Southeast Asian Studies*, vol 33, no 1, pp77–105

McCarthy, J. F. (2006) *The Fourth Circle: A Political Ecology of Sumatra's Rainforest Frontier*, Stanford, CA, Stanford University Press

Migdal, J. S. (1994) 'The state in society: An approach to struggles for domination', in Migdal, J. S., Kohli, A. and Shue, V. (eds) *State Power and Social Forces: Domination and Transformation in the Third World*, Cambridge, Cambridge University Press

Moore, S. F. (1973) 'Law and social change: The semi-autonomous social field as an appropriate subject of study', *Law and Society Review*, vol 7, no 4, pp719–746

Nababan, A. (1996) 'Pemerintahan Desa & Pengelolaan Sumberdaya Alam: Kasus Hutan Adat Kluet-Menggamat di Aceh Selatan', Paper presented to the Analyisis Dampak Implementasi Undang-Undang No 5 Tahun 1979 tentang Pemerintahan Desa terdadap Masyarakat Adat: Upaya Penyysybab Kebijakan Pemerintahan Desa Berbasis Masyarakat Adat, Wisma Lembah Nyiur – Cisarua

Newman, H., Ruwindrijarto, A., Currey, D. and Hapsoro (1999) *The Final Cut: Illegal Logging in Indonesia's Orangutan Parks*, Indonesia, Environmental Investigation Agency and Telapak

Niessen, N. (1999) *Municipal Government in Indonesia: Policy, Law and Practice of Decentralization and Urban Spatial Planning*, Leiden, Leiden University, CNWS Publications

Obidzinski, K. (2001) 'Operational nature of illegal logging in Indonesia: Its transformations and prospects in times of regional autonomy – perspectives from East Kalimantan', Unpublished paper, The Netherlands, University of Amsterdam

Oi, J. C. (1989) *State and Peasant in Contemporary China: The Political Economy of Village Government*, Berkeley and Los Angeles, CA, University of California Press

Ostrom, E. (1990) *Governing the Commons: The Evolution of Institutions for Collective Action*, Cambridge, Cambridge University Press

Peluso, N. L. (1992) 'The ironwood problem: (Mis)Management and development of an extractive rainforest product', *Conservation Biology*, vol 6, no 2, pp210–219

Peluso, N. L. (1995) 'Whose woods are these? Counter-mapping forest territories in Kalimantan, Indonesia', *Antipode*, vol 27, no 4, pp383–406

Rijksen, D. H. D. and Griffiths, M. (1995) *Leuser Development Programme Masterplan*, Supported by the European Union, Report prepared by the Integrated Conservation and Development Project for Lowland Rainforest in Aceh

Serambi Indonesia (1994a) 'Puluhan Kilang Kayu Liar Masih Beroperasi di Aceh Selatan', *Serambi Indonesia*, Banda Aceh, 27 October

Serambi Indonesia (1994b) 'Menyingkap Aksi Pencurian Kayu (1) Cukong = Curi Kayu Kong Kalikong', *Serambi Indonesia*, Banda Aceh, 28 November

Serambi Indonesia (1995) 'Oknum Aparat Diduga Terlibat Rusak Hutan, Pemerintah Lumpuh', *Serambi Indonesia*, Banda Aceh, 7 August

Serambi Indonesia (1998) 'Bila Izin HPH tak Dicabut, Rimueng Lamkaluet akan Mengamuk', *Serambi Indonesia*, Banda Aceh, 17 September

Sunderlin, W. D., Resosudarmo, I. A. P., Rianto, E. and Angelsen, A. (2000) *The Effect of Indonesia's Economic Crisis on Small Farmers and Natural Forest Cover in the Outer Islands*, Occasional Paper 28, Bogor, Indonesia, Centre for International Forestry Research

Tjondronegoro, S. (1991) 'The utilization and management of land resources in Indonesia, 1970–1990', in Hardjono, J. (ed) *Indonesia: Resources, Ecology, and Environment*, Singapore, Oxford University Press

Waspada (1998a) '20 Pengusaha HPH Dipanggil', *Waspada*, Medan, 24 September

Waspada (1998b) 'Seluruh HPH di Aceh Selatan Diancam Akan Dibumihanguskan', *Waspada*, Medan, 26 September

Waspada (1998c) '50 Persen Kilang Kayu di Aceh Barat dan Selatan Tutup', *Waspada*, Medan, 6 October

WWF-LP (1995) *Laporan dari hasil survei lapangan, Maret 1995*, Unpublished report, Jakarta, WWF-LP

5

Illegal Logging, Collusive Corruption and Fragmented Governments in Kalimantan, Indonesia

Joyotee Smith, Krystof Obidzinski, Sumirta Subarudi and Iman Suramenggala

Introduction

The symbiotic relationship between illegal logging and corruption has been widely discussed in the literature (Callister, 1999; Scotland et al, 2000; Contreras-Hermosilla, 2001; Palmer, 2001). This affinity is particularly relevant in the context of Indonesia, where illegal logging is rampant and corruption is entrenched. The Ministry of Forestry in Indonesia estimates, for example, that Indonesia is suffering a financial loss of US$3.7 billion annually due to illegal logging and exports (*NRM Headline News*, 2003). Transparency International's Corruption Perceptions Index ranks Indonesia as the seventh most corrupt country, out of 102 countries, with a score of 1.9 out of 10 for a highly clean country (Transparency International, 2003).

Corruption or the abuse of public office for private gain exacerbates illegal logging by allowing it to occur in the first place and letting it go unchecked and unpunished. Corruption also poses a corrosive challenge to improved governance. Influential government officials benefiting from corruption strive to prevent or undermine policy and institutional changes that could combat illegal logging. Therefore, illegal logging is unlikely to be controlled unless tools for fighting corruption are simultaneously developed and implemented.

A number of government, non-government and donor agencies are implementing strategies to control corruption in Indonesia. Among these strategies are an analysis of Transparency International's corruption fighting tools to determine their relevance in combating forest sector corruption (FIN,

2003) and a nationwide survey of corruption in Indonesia (Partnership for Governance Reform in Indonesia, 2001). This chapter complements these efforts by drawing on the literature on the political economy of corruption to highlight two factors that, arguably, have received inadequate emphasis in the design of policies: the importance of, first, distinguishing between types of corruption and, second, analysing their interface with the political/institutional environment, its history and its dynamics. In the case of Indonesia, we argue that although corruption in relation to the timber industry was pervasive during the Suharto regime, a more insidious type of corruption, that is harder to root out, has exploded after his downfall and that strategies for combating it will require wider reform.

The chapter first considers types of corruption and their dynamics during periods of political transition. It then reviews illegal logging and corruption in Indonesia during the Suharto regime before reporting the results of the case study on illegal logging and corruption in the post-Suharto period. This is followed by a consideration of possible strategies for controlling corruption and illegal logging. The concluding section notes the challenges faced in controlling the type of corruption that has taken hold in the post-Suharto period.

Types of corruption and their dynamics during periods of political transition

Following Shleifer and Vishny (1993) and Bardhan (1997), we distinguish between collusive and non-collusive corruption and analyse the impact of transitions from strong to weak governments on each type of corruption.

Collusive and non-collusive corruption

In non-collusive corruption, the government demands a bribe for a legal activity, such as obtaining a logging permit. Non-collusive corruption thus drives up costs for the private sector, which now has to pay a bribe in addition to the official cost. In this way, non-collusive corruption pits the briber against the individual or organization being bribed.

Collusive corruption is a more decentralized type of corruption, in which individual government officials and the private sector collude to rob the government of revenues. Government officials, for example, may let exports through without permits, or overlook tax evasion, logging outside authorized areas or the violation of harvesting regulations, in return for a bribe. Bribes are insurance policies taken out to avoid paying penalties for illegal activities, the amount of the bribe being equal to the penalty multiplied by the probability of being caught and punished (Cohen, 1999). Where surveillance is poor and the likelihood of paying a penalty, if caught, is minimal, levels of bribes for collusive corruption would therefore be only a fraction of the cost of carrying out the activity legally. Unlike non-collusive corruption, therefore, collusive corruption reduces costs for the private sector. In collusive corruption, neither the briber nor the individual or organization being bribed has an incentive to report or protest. Thus, collusive corruption is insidious and difficult to detect and therefore more persistent than non-collusive corruption (Shleifer and Vishny, 1993; Bardhan, 1997).

Transitions from strong to weak governments and the impact on corruption

By strong governments we mean regimes characterized by political stability and governments that are powerful enough to maintain law and order and enforce contracts throughout the country. Weak, fragmented governments, by contrast, have a precarious hold on power and are characterized by political instability, anarchy and local fiefdoms (Frye and Shleifer, 1997).

Impact on non-collusive corruption

Strong governments attempt to maximize total bribe revenue from a number of complementary legal transactions. Thus, the level at which the bribe for one transaction (say, obtaining a logging permit) is set takes into consideration the impact on complementary transactions (such as obtaining a timber export permit). This is possible because the government is sufficiently powerful to coordinate bribes from complementary transactions in order to prevent the total demand for permits from falling (Shleifer and Vishny, 1993; Bardhan, 1997). As a result, total bribe revenue from non-collusive corruption is often staggeringly high, particularly for states with valuable natural resources, such as oil or timber (Ascher, 1999).

Under weak governments, corruption becomes decentralized (Shleifer and Vishny, 1993; Bardhan, 1997). Independent fiefdoms set bribes for (say) logging permits, without considering the impact of the bribe level on complementary permits, such as timber export permits, which are granted by other independent fiefdoms. In addition, free entry into this game leads additional agencies to create needs for new permits. The result is an anarchic system of bribery, with multiple bribes being paid to different independent agencies for carrying out legal activities.

Impact on collusive corruption

Strong governments tend to favour non-collusive corruption over collusive corruption, particularly for activities where the potential loss in government revenue from collusive corruption is high, such as evasion of timber taxes. In cases where collusive corruption breaks regulations that were motivated primarily by 'cosmetic' environmental or social objectives, however, collusive corruption is also likely to be pervasive under strong governments. An example would be bribes taken to overlook violations of good logging practices.

Under weak, fragmented governments, the private sector has a better opportunity to lower costs through collusive corruption. Although multiple anarchic bribes also have to be paid for collusive corruption, bribes are likely to be well below official fees because surveillance is likely to be poor under weak governments. While the government loses revenues from fees, it is too weak to control independent fiefdoms. Collusive corruption also suits government officials who, given the political instability that characterizes weak governments, are anxious to maximize short-term personal benefits rather than build up government revenues. In the case of logging, the implication is that under weak governments the level of collusive corruption for activities such as tax evasion or illegal exports would be higher than under strong governments.

Countries undergoing political transitions

A number of countries in recent years have experienced a political transition from strong to weak governments, accompanied by burgeoning decentralized corruption. Examples are Indonesia after the fall of Suharto, the Philippines after the fall of Marcos and post-Communist Russia. Although in all the above examples the transition from strong to weak governments coincided with the overthrow of authoritarian rule, and in some cases also a change from centralization of authority to decentralization, the impact on corruption described above stems, as shown by Bardhan (1997), from government weakness and political instability, resulting in the decentralization of corruption, rather than being inevitable consequences of democracy or the decentralization of governance. Coolidge and Rose-Ackerman (1997) show, for instance, that in Somalia the dictatorial rule of Barre was characterized by anarchic bribery because Barre was too weak to control local fiefdoms, while they attribute Botswana's relatively favourable record on corruption to its political stability.

A more relevant similarity among the examples given above is that they all represent periods of transition – that is, periods prior to the establishment of functioning, decentralized democracies – after the overthrow of authoritarian rule. Also, notably, all share a long history of kleptocracy by a clique of elites, which induces those who received relatively few benefits in the past to maximize personal benefits during what they fear are small windows of opportunity.

Illegal logging, corruption and the political environment in Indonesia during the Suharto regime

During the first two decades of Suharto's rule, the government could be characterized as a strong regime. Suharto centralized control over natural resources, such as oil and timber, and exploited them both for political patronage and for projects such as transmigration programmes to the outer islands in order to consolidate his power throughout the country (Ascher, 1999). Forests long used by local communities under informal rights were declared state forests. Large-scale logging concessions were granted to forestry conglomerates controlled by Indonesian-Chinese entrepreneurs, with government officials and the military as partners (McCarthy, 2000; Barber and Talbott, 2003). In return for timber rents the military enforced internal obedience to Suharto's policies throughout the country. Control over the provinces was maintained through military officers who were appointed to head provincial and district governments (Barber and Talbott, 2003).

Non-collusive corruption flourished during the Suharto regime. In return for granting privileged access to forests, Indonesian-Chinese entrepreneurs granted shares in timber enterprises to Suharto's family and contributed massive funds, which Suharto used for off-budget spending to further political objectives (Ascher, 1999; Brown, 1999). The military also benefited from non-collusive corruption by selling its influence to secure favoured access to forests for business entrepreneurs (Barber and Talbott, 2003).

During the last decade of Suharto's rule, some degree of disunity arose within the government (Ascher, 1999), which, in turn, facilitated collusive corruption. With his political power well established, Suharto distanced himself from the military's timber interests (Ascher, 1999; Barber and Talbott, 2003). The military now turned to supplementing its income through collusive corruption. Timber processing capacity by now far exceeded sustainable timber supplies, and concessionaires met the shortfall by harvesting above their annual allowable cut, repeat harvesting before the approved cutting cycle and logging outside approved areas both within and outside their concessions (Barr and Resosudarmo, 2002). Timber brokers also provided logs from unauthorized areas to processors who had inadequate supplies (Obidzinski, 2001). The military benefited substantially from collusive corruption during this period by extorting fees from illegal operators.

Thus, for most of the Suharto regime, non-collusive corruption was the dominant form of corruption. Collusive corruption became widespread in the last decade of his rule. However, one organization – the military – was the main beneficiary of bribes from collusive corruption and it still maintained strong political ties with Suharto, although economic ties were now considerably weaker. Thus, collusive corruption was far more coordinated than the archetypal collusive corruption that occurs under weak, fragmented governments.

Illegal logging, corruption and the political environment after the fall of Suharto

After the fall of Suharto, the government became weak, fragmented and politically unstable. Indonesia had three heads of state within three years, East Timor Province broke away and separatist movements in Aceh and Irian Jaya provinces experienced a resurgence. The aftermath of the Asian economic crisis added to the problems, as the currency depreciated steeply, leading to banks being saddled with a high level of non-performing loans caused by the technical bankruptcy of prominent manufacturing organizations with foreign currency debt service obligations. Unemployment soared and petty crime became widespread.

It was in this chaotic environment that administrative and regulatory authority was decentralized, primarily to the district level, with district heads reporting to locally elected legislative assemblies. Under decentralization, regional governments were entitled to a larger share of resource revenues and were given authority to oversee management of community forests. Customary rights to forests were restored to local communities. In order to generate revenue from local sources after decentralization, district governments issued numerous short-term, small-scale forest conversion permits, (in the case considered here, these are known as timber extraction and utilization permits, or IPPKs), largely to companies that were joint ventures between Indonesian regional entrepreneurs, locally known as 'contractors', and Indonesian or Malaysian timber buyers, locally known as 'investors' (Obidzinski, 2001). Before obtaining a permit, companies secure a timber-harvesting agreement with the community under which they pay a small royalty to the community in exchange for harvesting rights. IPPKs are supposed to be granted in community forests that lie outside areas defined by the national government as permanent forest estate. In practice, however, many IPPKs have been granted within the boundaries of logging concessions awarded during the Suharto period (Barr et al, 2001).

In effect, IPPKs provide a means by which much of the illegal logging that was going on before could be legalized (see Chapter 3 in this volume; Obidzinski, 2001; Tacconi et al, 2004). Nevertheless illegal logging is widely believed to have exploded after the fall of the Suharto regime, based on reports by researchers and NGOs (see Chapter 3; EIA/Telapak, 1999; Scotland et al, 2000; Obidzinski, 2001), as well as on public acknowledgement of the problem by the Ministry of Forestry and Estate Crops (quoted in Scotland et al, 2000). Corruption is also perceived to have worsened. In Transparency International's Corruption Perception Index, the score for Indonesia has fallen from 2.65 (out of 10 for a highly clean country) in 1996 to 1.9 in 2002 (Transparency International, 2003).

The study area

The study area consists of three districts in North-East Kalimantan: Bulungan, Malinau and Nunukan. Kalimantan is estimated to contain about 30 per cent of Indonesia's forest area and around 50 per cent of Indonesia's production forests – that is, forests designated by the government for timber extraction (Ismael, 2000). Deforestation in Kalimantan is estimated to be around 706,000ha per year (World Bank, 2000).

IPPKs have rapidly been issued in Bulungan and Malinau and at a somewhat slower rate in Nunukan (see Table 5.1). Permit holders are liable for an area-based tax, known as the Third Party Tax, equivalent to about US$20 per hectare and a volume-based production tax equivalent to about US$1.5 per cubic metre. A royalty, usually around US$3 per cubic metre is negotiated with local communities. Exports require a permit and a fee equivalent to about US$12 per cubic metre for Meranti species. In addition, transport permits are required, the fees for which vary by species.

Table 5.1 Tax revenues[a] and estimated informal payments[b] from local logging permits: Bulungan, Malinau and Nunukan districts, North-East Kalimantan, Indonesia, August 2001

District	Active local logging permits[c] (thousand hectares)		Taxes payable (US$,000)		Realized tax revenue (US$,000)		Realized as percentage of payable		Estimated informal payments (US$,000)		Realized tax revenues as percentage of informal payments	
	2000	2001 (January–June)	2000	2001	2000	2001	2000	2001	2000	2001	2000	2001
Bulungan	10	23	200	240	15	50.4	8	21	142	311	11	16
Malinau	32	16.5	640	330	329.4	292	51	88	451.2	683.9	73	43
Nunukan	3.7	5.1	No tax	No tax	No tax	3.6	No tax	4	52.2	124.1	No tax	3

Notes: [a]Data source: Economics Section, District offices: Bulungan, Malinau, Nunukan, North-East Kalimantan. Figures include Third Party Tax (US$20 per hectare) only.
[b]Three payments of US$25,000 per average permit size of 1766ha are made to obtain approval of logging permits. Data for year 2000 assumes only first payment of US$25,000 per 1766ha is made. Data for year 2001 assume first payment of US$25,000 per 1766ha for new permits plus second payment of US$25,000 per 1766ha for permits issued in 2000. Data source: average figures from three confidential interviews with Indonesian timber contractors in Malinau and Bulungan (North-East Kalimantan).
[c]Data source: Economics Section, District offices: Bulungan, Malinau, Nunukan, North-East Kalimantan.

Much of the timber harvested from East Kalimantan is exported to Sabah, Malaysia, where plywood companies face an acute shortage of raw materials due to the depletion of local timber stocks and increased enforcement of regulations on timber extraction. Data on the timber trade were therefore also collected from Tawau, Sandakan and Kota Kinabalu in Sabah, Malaysia.

Data collection methods

Data were collected by first using rapid rural appraisal methods, consisting of semi-structured interviews with key informants, including government officials, timber industry actors and members of local communities. IPPK industry operators had to be identified through informal contacts, given their shadowy nature and their unwillingness to be formally interviewed. Data on corruption were collected through informal confidential interviews with anonymous IPPK operators. Rapport was established by requesting their cooperation in understanding the IPPK system and the problems that its operators face. We emphasized that we were not requesting information on specific companies or officials. Given the sensitive nature of the data and the time-consuming process of establishing rapport, information was collected opportunistically from very small samples, often a single case, to three or four cases. In addition, primary and secondary documents on government statistics and laws at the district level were also reviewed.

Empirical evidence

Although IPPKs provide a cover of legality for previously illegal logging activities, our study reveals that, in practice, a high degree of irregularity exists in the IPPK system. The maximum volume of timber authorized to be harvested from the IPPK area is usually well in excess of planned harvests, thus enabling companies to harvest areas substantially larger than the authorized area. Inspection of a random sample of 10 per cent of IPPK records in the study area showed the average authorized volume as 49 cubic metres per hectare. According to confidential interviews, however, volumes actually harvested averaged approximately 20 cubic metres per hectare. The implication is that the area actually harvested may be 2.5 times greater than the authorized area. This figure is consistent with reports documented by Barr et al (2001) of the influx of logging equipment into East Kalimantan well in excess of requirements for logging authorized areas. District heads openly acknowledged logging outside authorized areas and blamed provincial forestry officials for not checking the feasibility of authorized volumes in the field. A member of the district legislative assembly claimed that provincial forestry officials and district government officials had stakes in IPPK companies. Some district officials claimed that they were unable to convict offenders because of the intervention of the police and military, which were under the control of the central government.

Although IPPKs were granted to generate revenue for district governments, tax evasion is widespread. Official data (see Table 5.1) on the area-based tax of US$20 per hectare show that tax receipts were only 8 and 21 per cent of taxes payable in 2000 and 2001 in Bulungan, and 4 per cent of taxes payable in 2001 in Nunukan. Comparable figures from Malinau are significantly better for 2001, but indicate that considerable tax evasion occurred in 2000.

While data on tax evasion in the timber industry during the Suharto regime are not available, we do know that the government obtained substantial revenues from the timber industry in spite of keeping timber royalties at low levels. The Reforestation Fund, for instance, into which the reforestation fee was deposited, was estimated to hold US$800 million. These data indicate that Suharto was able to exact a higher degree of tax compliance from the timber industry than district governments after his fall and had better control over the proceeds.

Substantial illegal exports from East Kalimantan to Sabah, Malaysia, also appear to be occurring, given the significant volume of unaccounted logs – that is, the difference between the official log supply in Sabah (domestic plus net imports) and the volume of logs processed there. For the purposes of this analysis, official log supply in Sabah (LS) is estimated as:

$$LS = M + LP - X \qquad [5.1]$$

where:

- M = Malaysian data on official imports into Sabah;
- LP = domestic log production in Sabah; and
- X = official exports of logs from Sabah.

The volume of unaccounted logs (UNL) is estimated as:

$$UNL = LP - LS \qquad [5.2]$$

where:

- LP = input of logs into the timber processing industry in Sabah.

Table 5.2 shows that the estimated volume of unaccounted logs was 1.86 million cubic metres in 2000 and 1.59 million cubic metres in the first seven months of 2001. If total imports are taken as the sum of official imports (M) and unaccounted logs (UNL), then official Malaysian imports are estimated to be only approximately 10 per cent of estimated total imports (see Table 5.2). If M in Equation 5.1 is substituted by Indonesian data on official exports from North-East Kalimantan to Sabah, Malaysia, official Indonesian exports are estimated to be only approximately 3 per cent of estimated total imports (see Table 5.2). These figures should be taken as only rough estimates. We assume that all imports into Sabah are from Kalimantan. While this need not be the case, it is arguably not very inaccurate because industry representatives say that virtually all imports are from Kalimantan. Data on LP may also under-represent actual log production in Sabah. Arguably, however, under-reporting in Sabah is relatively small because industry representatives claimed that improvements in enforcement of logging restrictions had left them with little choice but to import logs from Indonesia. Both the above qualifications imply that unaccounted logs may have been overestimated to some extent. Notwithstanding these caveats, the data indicate the existence of substantial illegal exports from Kalimantan to Sabah. This is consistent with statements by the Indonesian Ministry of Forestry and Estate Crops (quoted in Scotland et al, 2000) and observations by field researchers (see Chapter 3 in this volume; Obidzinski, 2001).

Data on illegal exports during the Suharto regime are unavailable. However, enforcement of a log export ban in the 1980s was clearly sufficiently effective to force Indonesian timber entrepreneurs into domestic log processing in spite

Table 5.2 *Estimates of unaccounted log exports from North-East Kalimantan to Sabah, Malaysia, 2000 and 2001*

		2000	2001 (January–July)
		Million m^3	
1	Indonesian data on official exports from North-East Kalimantan[a]	0.06[b]	0.01[c]
2	Malaysian data on official imports into Sabah[d]	0.21	0.13
3	Domestic log production in Sabah[e]	3.7	1.7
4	Official exports of logs from Sabah[f]	0.37	0.62
5	Input of logs into timber industry in Sabah[e]	5.4	2.8
6	Unaccounted logs (based on Malaysian imports) (5 – (2 + 3 – 4))	1.86	1.59
7	Unaccounted logs (based on Indonesian exports) (5 – (1 + 3 – 4))	2.01	1.71
8	Official imports into Sabah as percentage of estimated total imports (2/(6 + 2))	10	8
9	Official exports from North-East Kalimantan as percentage of estimated total exports (1/(6 + 1))	3	0.6

Notes: [a] Office of Trade and Industry and Customs Office, Tarakan and Customs Office, Nunukan.
[b] Exports to all countries.
[c] Exports to Sabah.
[d] Sabah Timber Industries Association, Kota Kinabalu and Sabah Department of Forestry, Sandakan. The above organizations claim that almost all imports are from North-East Kalimantan.
[e] Sabah Department of Forestry, Sandakan.
[f] Malaysian Timber Industries Board, Kota Kinabalu.

of processing being less profitable than raw log exports (Ascher, 1999). Thus, it is likely that illegal exports were lower during the Suharto regime.

Confidential interviews with IPPK operators showed that the above illegal activities were made possible by collusive corruption. Timber contractors are able to specify authorized timber volumes according to their convenience, subject to informal payments to provincial forestry officials. Local communities are aware of harvesting outside authorized areas but have little incentive in reporting irregularities since the royalty they receive is volume based. Village heads also connive with IPPK operators because of informal payments that they receive in return for harvesting agreements in customary forests, as well as for the employment opportunities provided by IPPK operations. IPPK operators make payments to the police and military to overlook logging outside authorized areas and the transport of logs to the border without transport permits. Border patrol personnel are paid off to permit exports without export permits. In addition, informal payments are made to Malaysian officials in order to 'legalize' illegal exports from Indonesia. Although we were not told of informal payments made specifically to overlook tax evasion, it is likely that informal payments were made in lieu of taxes. Alternatively, it is possible, as one member of the district legislative assembly claimed, that tax revenues were being siphoned off by district and province officials. Unofficial payments are also made to district government officials to obtain approval for IPPK permits, thus indicating the existence of non-collusive corruption. Table 5.1 (columns 6 and 7) shows that estimates of informal payments for approval of IPPK permits far exceed realized revenues from area-based logging taxes.

In addition to illegal logging and exports in connection with the IPPK system, logging within national parks exploded during this period. National parks are attractive because of their commercially valuable stands. At the same time, they became particularly vulnerable after decentralization because they remained the responsibility of the central government, whose weakness erodes its capacity to secure these remote areas. District governments offer little cooperation on enforcement since illegal activities within national parks are perceived as the product of unjust treatment by the Suharto government towards the regions (Barr and Resosudarmo, 2002). During the Suharto regime, by contrast, while enforcement in national parks was not entirely effective, it was tougher than now because of policing activities by the military (Barr and Resosudarmo, 2002).

Illegal logging in national parks is also facilitated by collusive corruption, with government officials, military and police receiving bribes for overlooking these activities (McCarthy, 2000). The judiciary has also reportedly been paid off to prevent conviction of high-profile figures behind illegal logging gangs (EIA/Telapak, 1999).

Thus, in addition to non-collusive corruption remaining widespread after the fall of Suharto, the evidence also points to an explosion of collusive corruption. As we argue below, government weakness makes this type of corruption and the illegality associated with it particularly difficult to eradicate.

Impact of a weak, fragmented government

Decentralization in Indonesia was largely a political manoeuvre in response to separatist movements and the dissatisfaction of resource-rich regions with the centralization of resource revenues under the Suharto regime (Barr and Resosudarmo, 2002). As a result, decentralization was rushed through before strong institutions necessary for a stable, functioning democracy could be established. The whole process has also been *ad hoc* in nature, with little coordination among national, provincial and district governments. On the contrary, it has been characterized by a power struggle among the different levels of government and conflicts among different categories of forest stakeholders.

The power struggle between different levels of government has blurred the lines between legality and illegality and made illegal activities easier. Provincial and district officials have instituted reforms that extend well beyond the authority granted to them under central government regulations. As a result, laws have often been contradictory and unclear. For example, IPPKs were issued in the study area before the regulations implementing the central government's transfer of authority to the regions had been finalized (Barr et al, 2001). Allocation of IPPKs by district governments continued in spite of central government instructions to suspend the granting of IPPKs because of the high level of irregularities. Barr and Resosudarmo (2002) report that this is because district governments recognize that the national government has little capacity to block their allocation of new permits. Contrary to national-level laws, IPPKs have been issued within logging concessions granted by Suharto. Barr et al (2001) contend that this is to emphasize to concessionaires that access to timber profits depends upon the support of the district government and cannot be guaranteed, as in the past, by connections with the central government.

Weakness of the government in enforcing the rule of law has also resulted in anarchy and widespread conflict among forest stakeholders. Conflict rages between concessionaires and IPPK operators over harvesting rights, among different groups of local communities over claims to community forests, and between provincial and district governments over forest administration and control. This anarchic situation has further blurred the lines between legality and illegality.

Because the lines between legality and illegality are blurred, IPPK operators have preferred to remain as anonymous shadowy characters. For example, in our study we were unable to openly interview any IPPK operator. IPPK offices refused interviews and district offices could not supply us with the names of IPPK operators. Interviews had to be carried out anonymously and confidentially after identifying IPPK operators through local networking – even though one of the authors (Suramenggala) hails from the study area and another (Obidzinski) has spent two years doing fieldwork in the area. This secrecy surrounding IPPK operations makes the corruption associated with it more insidious and therefore more difficult to detect and root out. In contrast to IPPK operators, we were able to openly interview holders of concessions granted under the Suharto regime. While many concessionaires are guilty of illegal activities and corruption, as well, it has also been possible to openly involve them in multi-stakeholder discussions on improving forest management in Indonesia, thus increasing the chances of addressing the underlying causes that made them opt for illegal activities and corruption.

Government weakness has made it possible for a wide variety of agents to now benefit from illegal activities and the corruption associated with it, ranging from the police and military to local government officials at the district and provincial levels, local communities, and timber contractors and investors, as well as customs officials and the wood-processing industry in Malaysia. This is pernicious because as the number of corrupt people increases, the gains from corruption also increase. This occurs because expected punishment, if detected, declines when more people are corrupt since it is cheaper to be detected by a corrupt person than a non-corrupt one (Cadot, 1987; Andvig, 1991). Many of those now involved in illegal activities and corruption are new actors (Barr and Resosudarmo, 2002) who were deprived of forest rents during the Suharto regime. Most significantly, these actors are now independent agents, motivated by diverse agendas. Thus, the nature of corruption has changed from being controlled and driven by Suharto and his cronies to an anarchic explosion of attempts to grab a slice of the pie, while the governance vacuum provides a window of opportunity.

Towards strategies for controlling corruption and illegal logging

Where corruption is widespread, as in Indonesia, straightforward strategies are unlikely to be effective. Oh (1995) shows, for example, that increasing the level of penalties for corrupt activities is unlikely to work unless surveillance is also improved. But how does one improve surveillance if the police are corrupt? Other measures, such as anti-corruption investigative units, are unlikely to be effective in highly corrupt countries because corruption units are likely to become corrupt themselves. Nor is an increase in the salaries of civil servants likely to work. It is more likely to increase the number and rate of attractive positions that are bought (Andvig and Fjeldstad, 2000). Bardhan (1997) argues that, where corruption is widespread, a critical mass of opportunistic individuals will have to be convinced over a long enough period of time that corruption no longer pays. Thus, strategies have to be sustained over the long term and are likely to require broader reform. Here we evaluate, in the Indonesian context, a few such strategies, some of which reduce the gain from corruption and others that increase the probability of corrupt activities being detected and punished.

Economic competition

One strategy advocated for reducing the gain from corruption is economic liberalization through measures such as removal of subsidies, opening of the economy to foreign competition and breaking up monopolies. When economic competition is restricted, industries protected from competition make 'excessive' profits or rents – that is, profits higher than the market rate of return on investments. Thus, they seek to corner access to resources in protected sectors by bribing officials in charge of allocating these resources. When economic competition is introduced, rents are dissipated and, thus, the returns to corruption are reduced. In particular, non-collusive corruption, which increases costs for the private sector, is likely to become less attractive when protected industries are exposed to competition. Shleifer and Vishny (1993) show, however, that economic competition causes collusive corruption to spread more widely. If one company lowers costs through collusive corruption, those that do not will find it increasingly difficult to survive. What is more, Bardhan (1997) points out, even non-collusive corruption may be exacerbated by economic competition. As restrictions on trade, for instance, are relaxed and economic activity increases, government officials may see more opportunities for non-collusive corruption. Under intense competition, the private sector, too, may resort to 'grease payments' to speed up transactions. These qualifying factors probably explain why econometric analysis shows that government trade policies are only weakly related to levels of corruption (Andvig and Fjeldstad, 2000).

In the Indonesian case, non-collusive corruption during the Suharto regime was supported by subsidies, including low timber royalties, monopolies in the export of processed wood products and loans at subsidized rates. Arguably, the financial contributions made by the timber industry for Suharto's benefit may not have been economically viable without the subsidies that artificially inflated its profits.

The situation in Indonesia after the fall of Suharto, however, appears to support Shleifer and Vishny's (1993) contention that economic competition may increase collusive corruption. Economic competition in the plywood industry increased markedly after Suharto's fall because plywood from Indonesia, which is produced from tropical hardwoods, began to be undercut by plywood with cheap timber cores from China. At the same time, demand from Japan declined because of the downturn in the Japanese economy (ITTO, 2002). These factors are reflected in a substantial fall in the price of Indonesian plywood, whose price index (with January 1997 as 100) declined during the course of 2001 from 60 to 45 (ITTO, 2002). Confidential interviews with timber investors and contractors in East Kalimantan revealed that, faced with the price decline, companies used collusive corruption to cut costs and thus maintain returns to investments in machinery and distribution infrastructure. In fact, one claimed that it was no longer possible to make profits without resorting to such means.

The implication is that while economic competition may, arguably, reduce non-collusive corruption, it also runs the risk of exacerbating collusive corruption. Economic liberalization therefore needs to be accompanied by dramatically improved law enforcement in order to contain the risk of increased collusive corruption.

Political competition

Some aspects of political competition reduce the gains from corruption, while others increase the probability of detection and punishment.

Reducing the gain from corruption

Shleifer and Vishny (1993) argue that in a federal or decentralized government, competition among states or districts in the provision of government supplied goods (say, logging permits) could drive bribes for non-collusive corruption down to zero. Non-collusive corruption increases costs for the private sector. Companies would therefore choose to invest in the state or district where bribes are lowest. This would reduce the gains from corruption for government officials and lessen incentives for non-collusive corruption. Shleifer and Vishny (1993) caution, however, that competition among districts may increase collusive corruption. Collusive corruption drives down costs for the private sector and companies may choose to invest where collusive corruption enables them to keep costs down. Thus, as in the case of economic competition, political competition needs to be accompanied by significant improvements in law enforcement in order to control collusive corruption.

Increasing the probability of detection and punishment

An increase in the probability of detection and punishment increases the amount of bribe that has to be paid for overlooking an illegal activity and thus reduces the private sector's willingness to participate in either collusive or non-collusive corruption. Political competition has built-in mechanisms for increasing the probability of detection and punishment due to competition among officials of the ruling party, as well as from opposition parties. Corrupt officials may be voted out of office. Competitors for office may reveal corrupt activities by their competitors. Democracies also provide more space for public pressure against corruption through laws, democratic elections, parliamentary oversight, and independent judiciaries, press and watchdog bodies (Shleifer and Vishny, 1993).

Democracy, decentralization and corruption

While the above arguments appear to be eminently sensible, Andvig and Fjeldstad (2000) report several econometric studies that show that the effect of democracy and decentralization on corruption is dubious. Their analysis shows that corruption is highest in situations of political transition from authoritarian to democratic rule – that is, before a fully functioning democracy with checks and balances and legitimate and accountable institutions has been established. This view is consistent with Huntington (1968), who attributes the high incidence of corruption in political transitions to underdeveloped institutions. The concept also fits neatly into the Indonesian experience, where the incidence of corruption appears to have increased during a political transition that was hasty and *ad hoc* in nature. This circumstance highlights the importance of building strong public institutions at all levels of government to speed up the transition to a strong and fully functioning democracy.

Strengthening public institutions

Bardhan (1997) has argued that a strong government, capable of enforcing laws and property rights, is more important for reducing corruption than economic or political competition. In countries moving towards democracy, strength and political stability come from the legitimacy of the government in the eyes of the people. This, in turn, requires the government to address the grievances and aspirations of the people. This process is particularly important for the forestry sector in Indonesia, given its long history of appropriation of forest rents by a select few. In countries undergoing political transitions, institutions to make local legislative assemblies accountable to their constituents need to be established, such as free and fair elections, public meetings, and democratic and transparent organizations for debating policy issues at the village level and between villages and the district and provincial governments. Ideally, participants in such activities should include all forest stakeholders, including local communities, the private sector, government officials and members of civil society groups. However, involving the private sector may be difficult to achieve, at present, in Indonesia, given the shadowy nature of IPPK operators.

In Indonesia, improving law enforcement and establishing the rule of law will require, at the very minimum, first, a clarification of the law through clear demarcation of national, provincial and district jurisdictions and the elimination of contradictions among laws passed at different levels of government. Second, a thorough reform of the judicial system will be needed in order to make law enforcement fair and equitable. In Indonesia today, active civil society groups have successfully revealed and identified illegal logging operators (see EIA/Telapak, 1999, for instance). Offenders have been brought to justice only selectively, however, because of judicial corruption (Scotland et al, 2000). In this context, it is worth noting that judges and the Office of the Prosecutor were among the public institutions receiving the lowest scores for integrity from the public (Partnership for Governance Reform, 2001). Third, public oversight needs to be encouraged by further strengthening civil society organizations and the freedom of the press. Barber and Talbott (2003) argue that strengthening counter-balancing institutions of government and civil society is also the most promising route to reforming the military.

Conclusions

This analysis illustrates the importance of distinguishing between collusive and non-collusive corruption and understanding their interface with the political and institutional environment. While non-collusive corruption was widespread during the autocratic Suharto regime, a more insidious type of corruption – collusive corruption – exploded after his fall as the country moved towards a decentralized and more democratic regime. The explosion of collusive corruption cannot, however, be attributed to democracy and decentralization. On the contrary, fully functioning democracies create public pressure against corruption and bring about institutional changes to control corruption.

It was the weak, fragmented nature of government – characterized by power struggles, anarchy, conflict and contradictory laws – that blurred the lines between legality and illegality and therefore made it easier for illegal logging, supported by collusive corruption, to flourish. Periods of transition from autocracy to democracy are particularly vulnerable to burgeoning collusive corruption because, during transitions, institutions essential for fully functioning democracies are still underdeveloped, leading to a governance vacuum.

Because collusive corruption reduces costs for the private sector, it is more difficult to root out. While some strategies, such as economic competition and competition among officials in the provision of government goods, may be effective in controlling non-collusive corruption, these measures actually exacerbate collusive corruption.

This chapter shows that a strong government capable of enforcing the rule of law is required for controlling widespread collusive corruption. In Indonesia, this will require wider sustained reform and institutional strengthening. The results suggest that political stability, mechanisms to make governments accountable to their constituencies, removal of inconsistencies in the legal framework, judicial reform and encouragement of public oversight could be useful, if daunting, cornerstones in the fight against illegal logging and corruption, particularly during political transitions.

Acknowledgements

The UK Department for International Development provided partial funding for this study. The authors thank David Kaimowitz, Gill Shepherd, Adrian Wells and Petrus Gunarso for useful comments on a seminar upon which this chapter is based. This chapter is reprinted from Smith, J., Obidzinski, K., Subarudi, S. and Suramenggala, I. (2003) 'Illegal Logging, collusive corruption and fragmented governments in Kalimantan, Indonesia', *International Forestry Review*, vol 5, no 3, pp293–302, with permission.

References

Andvig, J. C. (1991) 'The economics of corruption: A survey', *Studi Economici*, vol 43, pp57–94

Andvig, J. C. and Fjeldstad, O.-H. (2000) *Research on Corruption: A Policy Oriented Survey*, Report commissioned by NORAD (Norwegian Agency for Development Cooperation), Bergen and Oslo, Chr Michelsen Institute and Norwegian Institute of International Affairs

Ascher, W. (1999) *Why Governments Waste Natural Resources*, Baltimore and London, Johns Hopkins University Press

Barber, C. V. and Talbott, K. (2003) 'The chainsaw and the gun: The role of the military in deforesting Indonesia', in Price, S. V. (ed) *War and Tropical Forests: Conservation in Areas of Armed Conflict*, Binghamton, Haworth Press

Bardhan, P. (1997) 'Corruption and development: A review of issues', *Journal of Economic Literature*, vol 35, pp1320–1346

Barr, C. and Resosudarmo, I. A. P. (2002) *Decentralization of Forest Administration in Indonesia: Implications for Forest Sustainability, Community Livelihoods, and Economic Development*, Bogor, Indonesia, Centre for International Forestry Research

Barr, C., Wollenberg, E., Limberg, G., Anau, N., Iwan, R., Sudana, I. M., Moeliono, M. and Djogo, T. (2001) *The Impacts of Decentralization on Forests and Forest-Dependent Communities in Malinau District, East Kalimantan*, Case Studies on Decentralization and Forests in Indonesia, Bogor, Indonesia, Centre for International Forestry Research

Brown, D. (1999) *Addicted to Rent: Corporate and Spatial Distribution of Forest Resources in Indonesia: Implications for Forest Sustainability and Government Policy*, Report No PFM/EC/99/06, Jakarta, DFID UK, Tropical Forest Management Programme

Cadot, O. (1987) 'Corruption as a gamble', *Journal of Public Economy*, vol 22, pp223–244

Callister, D. J. (1999) *Corrupt and Illegal Activities in the Forestry Sector: Current Understandings and Implications for World Bank Forestry Policy*, Forest Policy Implementation Review and Strategy Development: Analytical Studies, Washington, DC, World Bank

Cohen, M. A. (1999) 'Monitoring and enforcement of environmental policy', in Tietenberg, T. and Folmer, H. (eds) *International Yearbook of Environmental and Resource Economics*, vol 3, Cheltenham, UK, Edward Elgar Publishers

Contreras-Hermosilla, A. (2001) *Forest Law Enforcement*, Washington, DC, World Bank

Coolidge, J. and Rose-Ackerman, S. (1997) *High-Level Rent Seeking and Corruption in African Regimes: Theory and Cases*, Washington, DC, World Bank

EIA (Environmental Investigation Agency)/Telapak (1999) *The Final Cut: Illegal Logging in Indonesia's Orang-Utan Parks*, London and Indonesia, Environmental Investigation Agency and Telapak

FIN (Forest Integrity Network) (2003) *FIN Newsletter*, no 5, June

Frye, T. and Shleifer, A. (1997) 'The Invisible hand and the grabbing hand', *AEA Papers and Proceedings*, vol 87 no 2, pp354–358

Huntington, S. P. (1968) *Political Order in Changing Societies*, New Haven, Yale University Press

Ismael, N. M. (2000) *Rencana Stratejik. Rakernas 2000*, Ministry of Forestry, 26–29 June, Jakarta, Indonesia

ITTO (International Tropical Timber Organization) (2002) *Tropical Timber Market Report 16–31 December 2001*, www.itto.or.jp/market/
McCarthy, J. F. (2000) 'The changing regime: Forest property and *Reformasi* in Indonesia', *Development and Change*, vol 31, pp91–129
NRM Headline News (2003) 'Asia pulse', 18 June, quoted in *NRM Headline News*, vol 11, 26 June 2003, Natural Resources Management Program Indonesia, USAID
Obidzinski, K. (2001) *Illegal Logging in East Kalimantan, Indonesia and Its Transformation in Times of Regional Autonomy*, ASIT workshop paper, 27 March
Oh, Y. (1995) 'Surveillance or punishment? A second-best theory of pollution regulation', *International Economic Journal*, vol 9, pp89–101
Palmer, C. E. (2001) *The Extent and Causes of Illegal Logging: An Analysis of a Major Cause of Deforestation in Indonesia*, London, Centre for Social and Economic Research on the Global Environment
Partnership for Governance Reform in Indonesia (2001) *A National Survey of Corruption in Indonesia*, Jakarta, Indonesia, Executive Office of the Partnership for Governance Reform in Indonesia
Scotland, N. et al (2000) *Indonesia Country Paper on Illegal Logging*, Prepared for the World Bank–World Wide Fund for Nature Workshop on Control of Illegal Logging in East Asia, 28 August, Jakarta, Indonesia
Shleifer, A. and Vishny, R. W. (1993) 'Corruption', *The Quarterly Journal of Economics*, vol 108, pp599–617
Tacconi, L., Obidzinski, K., Smith, J., Subarudi, Suramenggala, I. (2004) 'Can "legalization" of illegal forest activities reduce illegal logging? Lessons from East Kalimantan', *Journal of Sustainable Forestry*, vol 19, pp137-151
Transparency International (2003) *Transparency International's Corruption Perceptions Index 2002*, www.transparency.org/
World Bank (2000) *Deforestation in Indonesia: A Preliminary View of the Situation in 1999*, Draft report, Jakarta, Indonesia, World Bank

6
Forest Law Enforcement and Rural Livelihoods

David Kaimowitz

Introduction

International concern about illegal forestry activities has grown markedly (see Chapter 1). There are good reasons for concern. Illegal forestry activities deprive governments of billions of dollars in tax revenues. They also cause environmental damage and threaten forests upon which many people depend. Forest-related corruption and widespread violation of forestry laws undermine the rule of law, discourage legitimate investment, and give the wealthy and powerful unfair advantages due to their contacts and ability to pay large bribes. Money generated from illegal forestry activities has even been used to finance armed conflict.

Nonetheless, greater enforcement of forestry and conservation laws also has the potential to negatively affect rural livelihoods. That is because:

- Existing legislation often prohibits forestry activities such as small-scale timber production, fuelwood collection and hunting, upon which millions of poor rural households depend.
- Most small farmers, indigenous peoples and local communities are ill equipped to do the paperwork required to engage in forestry activities legally or to obtain the technical assistance needed to prepare management plans.
- Millions of rural households live on lands that governments have classified as state-owned forestland or protected areas, and existing laws often consider them encroachers even though their families may have lived there for generations.

- Forestry and wildlife departments generally enforce forestry and protected area legislation more vigorously and with less respect for due process and human rights when poor people are involved.
- In some countries, forestry and wildlife officials engage in illegal activities that harm the poor. Measures that empower these officials and give them more resources can make it easier for them to act with impunity.

The magnitude of these risks varies greatly from country to country. Some countries have little interest or capacity to enforce their forestry and conservation laws, and the increased international attention to forest law enforcement will probably not change that. Others focus their regulatory efforts almost exclusively on curtailing abuses by large logging companies, which is less likely to have major negative effects on rural livelihoods. But there are many countries where existing efforts to enforce forestry and conservation laws already have significant negative impacts on rural livelihoods. In those cases, greater law enforcement efforts may make the problem worse.

This chapter examines the opportunities and threats that increasing forest law enforcement efforts pose for rural livelihoods. It first defines rural livelihoods and identifies eight relevant aspects. Then it looks at potential negative livelihood impacts of both illegal forestry activities and law enforcement efforts. Next, it discusses the links between legality and the sustainability of the natural resource base. This topic is pertinent since proponents of strict forestry law enforcement usually justify their positions by arguing that enforcing the laws will encourage people to manage their forests sustainably. The chapter then looks at how distinct law enforcement scenarios may affect livelihoods and how the results may vary depending upon the context. Finally, it examines options for increasing the positive livelihood impacts of forest law enforcement and for reducing the negative ones.

The chapter addresses complex and difficult problems. In many cases, attempts to solve one set of problems will create others. Policies that work well in one location may have unanticipated or disastrous consequences in others. Clearly, there are situations where the positive benefits from the enforcement of forestry and conservation laws outweigh the negative impact it may have on livelihoods; therefore, governments and communities sometimes need to take measures that restrict the options of poor rural households. Similarly, it would be unwise to be naive about how easy it is to get communities themselves to regulate the use of forests effectively. Still, there are good reasons to question many of the existing and proposed efforts to regulate forests and to take steps to ensure that regulations do not simply justify wealthy and powerful groups gaining a monopoly on access to forest resources, rather than protecting the resource.

The chapter represents an initial exploration of a topic about which there is still surprisingly little empirical information. Hopefully it will be followed up by in-depth empirical studies and analysis that can provide the basis for seriously assessing various policy options.

Although the chapter illustrates its arguments with examples, the author has no desire to single out or criticize specific countries or individuals. The examples used happen to be ones that the author had information on. Some are out of date. In other cases, the author cannot fully verify the accuracy of the information taken from published sources. Thus, it is important that the reader focus on the broad issues that the chapter raises and not just the examples. The case studies have been included only to demonstrate that the arguments have an empirical basis and are not simply conjectural.

Forestry and rural livelihoods

To seriously analyse how forest law enforcement affects rural livelihoods, one must define livelihoods and identify criteria and indicators for assessing how policies affect them. According to the UK Department for International Development, 'A livelihood comprises the capabilities, assets and activities required for a means of living' (DFID, 2001). The same publication outlines criteria for assessing livelihood outcomes. They imply that persons have a better livelihood if they:

- have higher incomes;
- receive more government services;
- have their physical security respected;
- have better health;
- have adequate food;
- are less vulnerable to changes in markets or their own environment;
- rely on natural resources that are managed sustainably;
- can participate in political processes; and
- can maintain their cultural heritage and self-esteem.

These criteria link up with the discussion about forest law enforcement and rural livelihoods in the following ways:

- *Income*: in the context of forestry activities, higher incomes for low-income rural households can come from small-scale forest-based activities or wage labour.
- *Government services*: taxes and revenues from forestry can finance government services for the rural poor.
- *Physical security*: to respect forest users' physical security implies not physically mistreating them or imprisoning them without adequate due process.
- *Food, health and vulnerability*: poor rural households with access to wild meat, vegetables, fruits, medicinal plants and animals are likely to have better food security and health. This access is especially crucial in situations where families have already exhausted the food from their last harvest and in periods of economic crisis, war or crop failure.

- *Sustainable natural resource management*: if people use forest resources sustainably, rural families should be able to maintain the benefits that forests provide over time.
- *Participation and cultural heritage*: governments and other groups can enfranchise low-income forest-dependent people politically, protect their cultural heritage and legal rights, and encourage their self-esteem by providing institutional mechanisms for participation in decision-making and respecting their rights, cultures and opinions.

The DFID's livelihood approach also posits that people with more natural, physical, financial, human and social capital generally have better livelihoods. In the context of this chapter, natural and social capital merit the greatest attention. One may expect poor rural households to live better if they have secure access to forest resources and if they have effective and efficient social mechanisms to regulate forest use, to manage their forests and to distribute the benefits.[1]

Given the above, this chapter assumes that forest law enforcement policies favour forests' contributions to rural livelihoods if they:[2]

- increase the amount of forest products that poor rural households can sell and the prices they receive;
- increase wage labour in forestry and the salaries that forestry workers earn;
- increase tax revenues from forestry companies;
- decrease the number of poor rural people who are physically mistreated, forced to pay bribes, or inappropriately arrested or fined;[3]
- increase poor rural households' access to forest resources and make it more secure, including access by women;
- help to maintain the long-term supply of forest products and services that poor households use;
- promote poor people's participation in decision-making and collective action; and
- respect poor households' rights, cultures and traditions.

The next two sections look at how illegal forestry activities and inappropriate forest law enforcement can negatively affect these eight aspects.

Negative impacts of illegal forestry activities on rural livelihoods

Illegal forestry activities often negatively affect rural livelihoods. Indeed, this is the main reason why development agencies, whose primary mission is poverty alleviation, have become increasingly concerned about the problem.

Lost income from forest products

Situations of widespread corruption and disrespect for the rule of law typically favour groups who have sufficient resources to pay bribes, develop informal links with government officials and hire armed guards (World Bank, 1997). This conduct puts households who engage in small-scale forestry activities at a clear disadvantage. Often, they can operate only if they agree to sell their products to wealthier 'patrons' who protect them from forestry officials and provide credit.[4, 5] The assistance of these patrons comes at a high price. They pay producers much less than they would receive if they could borrow from formal lending agencies and sell their products legally to whomever they wanted to.

Lost job opportunities

Illegal logging may generate employment in the short run; but in the longer run it can contribute to the depletion of timber resources and the subsequent collapse of forest industries. This has already happened in several West African and South-East Asian nations and could well happen in others, such as Cambodia and Indonesia.

Less government revenue

Every year developing country governments lose billions of dollars in revenues because of illegal tax evasion in the forestry sector and unauthorized timber harvesting in publicly owned forests.[6] This situation leaves governments less money to spend on services such as health, education, roads, electricity and agricultural extension. Lack of transparency in government budgets in countries with widespread corruption makes it less likely that whatever funds governments do receive will go to services for the poor. Weak rule of law and corruption also limit long-term economic growth, which further reduces tax revenues (Thomas, 2000).

Threats to physical security

When local people complain about illegal forestry activities, the implicated parties often respond with threats or even violence. In addition, corrupt government officials sometimes take action against local people to protect their interests or those of illegal loggers and poachers.

Loss of access to forest resources

In heavily corrupted systems, the only way to get access to forest resources may be through bribes and connections. People who are unwilling or unable to use those mechanisms cannot access forest resources, or risk fine and arrest by accessing them illegally. Without transparency in decision-making and a functioning system of legal due process, poor rural households have little recourse when government officials or private companies and individuals illegally deny them access to forest resources.

Forest loss and degradation

Illegal forest clearing, poaching and failure to respect timber harvesting regulations can deplete the natural resources upon which poor rural households rely, such as wild fruits and vegetables, bush meat, medicinal plants, fuelwood and timber. For example, illegal commercial hunting for bush meat in Central Africa poses a serious threat to the medium-term availability of animal protein for millions of rural people (Bennett, 2002). Illegal forestry activities can also negatively affect environmental services that are important to poor rural households, such as the provision of clean water, pest and disease control, pollination, and regulation of the climate, stream flow and groundwater levels.

Political disenfranchisement and loss of social capital

Corruption and lack of respect for the rule of law subvert the democratic process. Elected public officials lose influence or fail to represent the interests of those who elected them, while small elite groups can use bribery and private business associations with government officials to influence policies in their favour (Contreras-Hermosilla, 2002). Corrupt forestry officials discourage democratic participation and independent oversight out of concern that it might undermine their authority and income sources.

When individuals or groups within a community engage in illegal forestry activities or support others involved in such activities, it may create discord and undermine pre-existing mechanisms for regulating the use of forest resources. Formal community forestry initiatives have difficulty competing with groups who operate illegally since the latter can sell their products more cheaply because they do not pay taxes, prepare management plans or devote resources to paperwork. Ironically, formal community forestry groups often find it more difficult to get government permits than illegal operators do because they are less able or willing to pay bribes to obtain them.

Negative livelihood impacts of forest law enforcement

Although illegal forestry activities can be bad for rural livelihoods, so can enforcement of existing forestry laws – and doing it more effectively may make the problem worse. This observation applies particularly to situations where legislation and/or law enforcement practices discriminate against poor rural households.

Lost income from forest and agricultural products

Most small-scale commercial forestry activities in developing countries are illegal or have unclear status under existing laws. Those involved generally do not have permits or formal management plans and do not pay taxes, and they often work without permission in forests claimed by governments or large landholders. Small-scale forest producers may violate regulations that oblige them to sell their products to government marketing agencies or prohibit them from engaging in specific practices, such as using chainsaws to process timber, hunting or engaging in swidden cultivation. Frequently, they have to pay bribes to get permits or to operate without them.

A large but unknown number of people engage in informal forestry activities. Poschen (1997) calculated that fuelwood and charcoal activities employed the equivalent of some 13.3 million full-time workers during the early 1990s, the vast majority of them outside the legal framework. The United Nations Food and Agriculture Organization (FAO) estimated that during the early 1990s, India, Indonesia, the Philippines and Thailand had between 175 million and 200 million people living on land classified as government forest, a large proportion of whom could be classified as encroachers under national laws (FAO, 1998). In the mid 1980s, an estimated 70 per cent of the world's protected areas were inhabited and many, if not most, of those inhabitants were officially 'illegal' (Dixon and Sherman, 1991). No one knows how many of those villagers practise small-scale informal timber harvesting, but the number is probably in the millions.

The high transactions costs associated with operating legally are a major factor that typically confines small-scale commercial forestry to the informal sector. Existing laws and regulations require extensive paperwork, payments and visits to government offices. Professional foresters must sign certain papers and the offices that process those papers are typically far away. When low-income people go there, the officials whom they need to talk to may be away or unwilling to receive them. It frequently takes a long time to get any response, and officials may send papers back several times for corrections.

Enforcement of existing forestry laws sometimes reduces small-scale producers' incomes by discouraging them from engaging in forestry activities or forcing them to sell their products illegally for lower prices. It also increases their costs associated with avoiding detection and paying fines and bribes. The literature on agro-forestry and small-scale woodlots is full of references to small farmers not planting trees because of the permits and bribes required to sell the wood once those trees come to maturity. Enforcement of forestry and protected area legislation that prohibits households from clearing forests for agriculture reduces their real or potential income from agriculture.

Lost job opportunities

Restricting the activities of larger commercial forestry and agricultural operations may negatively affect rural livelihoods to the extent that it limits employment opportunities for low-income rural people. This is most apparent in the case of logging bans by which formal-sector timber production and its associated jobs completely disappear (see the section on 'Logging bans and moratoriums').

Threats to physical security

In some countries, forestry officials and police arrest low-income people for violating forestry and protected area legislation without due process, forcibly expel them from their houses and fields, hit them, rape them or even kill them. Unless governments take measures to prevent this violence, attempts to encourage forest law enforcement could easily worsen this problem.

Loss of access to forest resources

Under existing legislation, governments and large private landowners officially own about three-quarters of the forests in developing countries (White and Martin, 2002). Poor rural households have no legal right to use most of that forest or have limited rights restricted to collecting certain products for a household's own use.[7] The extent to which governments actually restrict poor rural households from accessing these resources varies widely, in practice. Many governments essentially tolerate poor families living on forestlands and protected areas claimed by the government; but there are also numerous cases where families have been evicted from such areas, often forcibly.[8] If governments were to strictly apply existing forestry and conservation laws limiting poor rural households' access to forest resources, this could have dramatically negative impacts upon them.

Forest loss and degradation

In some instances, enforcing or attempting to enforce existing forestry legislation may encourage forest destruction. This may occur because forestry laws undermine previously existing community-based mechanisms for regulating forest use, or because the law promotes tenure regimes and forestry practices that negatively affect forest conditions. This may be the case, for example, where the law favours large-scale industrial logging and the conversion of forests to agro-industrial plantations, but discourages small-scale, low-impact forestry activities.

Political disenfranchisement and loss of social capital

Considering the residency or livelihood activities of large numbers of rural people as illegal essentially 'criminalizes' these people and makes it easier to deny them their political and legal rights, and the opportunity to participate in decisions related to natural resource management. As noted above, trying to enforce laws that fail to recognize and build upon pre-existing 'informal' mechanisms to collectively regulate the use of forest resources may also undercut those mechanisms and make it more likely that forest resources will essentially become open access.

Existing forestry laws often discourage the development of formal community forestry activities, even when government officials support such activities in principle. They create an additional paperwork and tax burden that informal sector activities do not face, and forestry officials often process the papers for community forestry groups more slowly because they are unwilling or unable to pay bribes. As a result, such groups can usually only compete with large-scale and informal forestry operations when they receive donor support. This support, in essence, allows them to cover the additional cost of operating legally and to get forestry officials to take action without receiving bribes. The result, however, is that such donor-supported community forestry efforts are almost never sustainable or replicable because without external support to cover the costs of paperwork and not paying bribes, formal community-sector initiatives could not operate.

In principle, one may say that this problem is the result of weak law enforcement, rather than strong law enforcement, since if the laws were strictly enforced formal community forestry efforts would not face 'unfair competition' from the informal sector. For this reason, the problem was also mentioned in the section on 'Negative impacts of illegal forestry activities on rural livelihoods'. Nonetheless, in practice, efforts to improve law enforcement often focus on formal operations that claim to operate legally, rather than on clandestine activities. That can make it all the more difficult for the former to operate.

Lack of respect for local culture and traditions

Many forestry and conservation laws fail to recognize indigenous and nomadic peoples' rights over the territories that they have historically occupied, and to take into account their traditional farming, hunting, fishing, grazing and gathering practices. That makes it more difficult for many local people to maintain their traditional diets, health practices and ways of life. One common example of this, particularly in Asia, is seen in laws that prohibit swidden cultivation (also known as shifting cultivation or slash-and-burn cultivation). Swidden cultivation forms an integral part of the traditional practices of many peoples and, in many cases, is the main livelihood option that people have available to them. Table 6.1 summarizes the negative livelihood impacts discussed above.

Does legal mean sustainable?

Discussions about forest law enforcement sometimes practically equate sustainable forest management with complying with forestry laws, but the two differ markedly. A large portion of forestry legislation focuses on administrative requirements, fees, taxes and property rights, rather than on how forests are really

Table 6.1 *Threats to rural livelihoods from illegal forestry activities and from forest law enforcement*

	Illegal forest activities	**Forest law enforcement**
Forest product income	Small foresters earn less because they pay bribes and depend upon patrons.	Small foresters earn less because governments stop their 'illegal' activities or greater law enforcement leads them to have to pay higher bribes or depend more upon patrons.
Wages from forestry	Over-harvesting makes the forestry sector collapse.	Government actions reduce logging.
Government revenues	Tax evasion and illegal logging in public forests lower revenue.	Reduced logging due to law enforcement lowers revenue.
Physical security	Illegal loggers and corrupt officials threaten and attack villagers.	Officials inappropriately threaten, attack, arrest or expel villagers and destroy their crops and houses.
Access to forest resources	Wealthy groups and officials illegally deny access to forests and due process.	Wealthy groups and officials 'legally' deny access to forests and due process.
Long-term supply of forest goods and services that poor households use	Damage to forests by illegal logging reduces forest product supply and disrupts environmental services.	Damage to forests due to loss of social capital reduces forest product supply and disrupts environmental services.
Collective action and participation	Illegal logging undermines local forest management institutions, and bribes and influence peddling replace democratic process.	Laws that fail to recognize local forest management institutions undermine them, and police action substitutes for dialogue.
Respect for cultures and tradition	Illegal logging undermines traditional institutions.	Laws that disallow traditional practices or fail to respect local institutions undermine peoples' traditional cultures.
Economic growth	Widespread failure to respect rule of law reduces investment and growth.	Limiting forestry activities reduces (short-term?) economic growth.

managed. Some regulations actually encourage unsustainable management and some people who violate forestry laws manage forests sustainably.

This section discusses the relation between forest law enforcement and sustainable forest management for two reasons. First, it may be possible to justify enforcing forestry laws that negatively affect rural livelihoods if that makes forest production more sustainable or protects the environment. Improving rural livelihoods is certainly not the only relevant policy goal, and clearly governments have to make trade-offs between objectives. Second, if enforcing forestry laws helps to maintain forests in good condition, this could benefit rural livelihoods.

Successfully enforcing laws that prohibit forest clearing, logging, hunting and collecting vegetable products usually, although not always, directly helps to protect the forest resources involved, at least in the short run.[9] The situation is less straightforward when it comes to laws and regulations that specify annual allowable cuts, harvesting rotations and minimum harvesting diameters. Enforcing these regulations usually will not suffice to sustain commercial timber production and environmental services over the long term and may even make things worse.[10] Many existing prescriptions for tropical forest management have a surprisingly weak scientific basis (Fredericksen, 1998; Putz et al, 2000; Sist, P., Fimbel, R., Nasi, R., Sheil, D. and Chevallier, M. H. (2001) *Towards Sustainable Management of Mixed Dipterocarp Forests: Moving beyond Minimum Diameter Cutting Limits*, Unpublished manuscript). Frequently, they fail to take into account the regeneration requirements of commercial timber species and the role of animals in seed dispersal, pollination, and pest and disease control (Sheil and Van Heist, 2000). Allowable cuts usually reflect political, economic and administrative concerns as much as the biological capacity of a forest to sustain timber production. Most legally sanctioned approaches to designing forest management plans assume that forest ecosystems are in a steady state, rather than being path-dependent outcomes of episodic disturbances. In principle, foresters are supposed to adapt management plans to the dynamic of each forest; but most foresters in developing countries lack the training and information required to do that, and forestry officials often will not accept the plans if they do.

Many forestry laws require logging companies to plant trees to compensate for the trees that they have logged. This is unlikely to be effective. In many contexts, forests will regenerate naturally without enrichment planting, and unless companies plant trees because it is profitable to do so they almost always take poor care of them.

Some countries prohibit people from processing timber with chainsaws, a common practice among small informal timber producers, on the grounds that it is inefficient. They argue that less efficient processing increases pressure on forests since people have to cut more logs to produce the same amount of sawn wood. In reality, using chainsaws to process timber is not always less efficient than using sawmills. Even though one gets lower recovery rates from the portions of the tree trunk used, processing with chainsaws often allows one to use a higher portion of the trunk and branches.[11] Besides, the total amount of sawn wood produced may increase if greater efficiency makes it more profitable to produce sawn wood. This could lead to more logs being harvested (Barr, 2001).

Similarly, many countries justify the prohibition of swidden cultivation partially on environmental grounds. Swidden cultivation undoubtedly alters the local ecosystem; but in many places such disturbances began hundreds, if not thousands, of years ago. Whether continued swidden cultivation will further degrade the ecosystem varies widely depending upon the location (Thrupp et al, 1997).

Forestry laws focus heavily on defining who has property over and access to forest resources. Some lay out the conditions for allocating publicly owned forests to large companies in the form of forest concessions. Others assign rights over 'forestland' to forestry agencies – even if it has no forest on it. Still others give rights over forest to indigenous peoples, communities or private forest owners.

If the group who receives rights over forests manages them better than those who do not, enforcing those rights should lead to more sustainable forest management. However, there is surprisingly little serious information available about which groups manage forests better. Many, if not most, forests that governments and large forestry companies have managed have ended up in poor condition. The same can be said of many forests that communities or individual households have managed. Indigenous peoples and extractivists have historically tended to maintain their forests in better condition than other groups; but that does not necessarily mean they would continue to do so. As a general rule, those with more limited access to markets and less capital to invest in forestry and agriculture destroy forests much less than wealthier groups do, at least on a per capita basis, although they can still do substantial damage.[12] This is so not necessarily because they are more environmentally conscious, but rather because they have fewer incentives and means to destroy the forest.

Some people argue that secure tenure rights encourage sustainable forest management; but several studies raise serious doubts about this (Angelsen, 1996; Boscolo and Vincent, 1998; Walker and Smith, 1999). Under current conditions, even companies with secure tenure over forest concessions in the tropics have little economic incentive to think beyond the first rotation, given the slow growth rates of commercially valuable trees in most natural tropical forest and the companies' desire for immediate profits. Having secure tenure may also make it more profitable for farmers to invest in clearing forests for crops and pasture and, hence, may increase deforestation.

Many forestry regulations require paperwork that only indirectly relates to forest management, if at all. Loggers, hunters, truck drivers, traders and processors must prepare and submit the appropriate forms, plans, inventories and impact assessments, and get the right signatures. They must obtain all the right permits and receipts and make them available upon request. Fulfilling these administrative requirements, however, often has little to do with how one actually manages a forest. One can manage a forest well and still fail to comply with the burdensome paperwork requirements. Similarly, in most countries, one can have most or all of the paperwork in order but still manage the forests quite poorly.[13]

In theory, having a formal management plan, getting it approved, implementing it and tracking the timber harvested in accordance with what it prescribes form one coherent system. In reality, there is often little connection among what the plan says, having the required permits and what happens in the forest. Having the paperwork in order, *per se*, says little about how a forest is managed, especially where forestry officials rarely visit the forest and/or sign the papers in return for bribes.

The same argument applies to taxes and fees. In principle, the structure and rate of taxation can influence how forests are managed (Karsenty, 2000). For example, one might expect area-based taxes to induce loggers to exploit forests more intensively and harvest more low-value timbers. Similarly, how much governments tax each timber species should influence what species are harvested. Some economists claim that higher taxes on timber volumes encourage more efficient harvesting and processing. Higher taxes, in general, should make forestry operations less profitable and discourage logging in natural forests (as well as forest management and plantation development).

In fact, there is little hard evidence on how tax regimes affect forest management in practice. Tax regimes change frequently, it has proven difficult to get enough data to analyse the issue, many factors confound the results, and there are usually wide discrepancies between the theoretical tax structures and what people really pay.

In summary, there is little doubt that enforcing some forest laws could encourage sustainable forest management. Nevertheless, the relation is less clear and direct than most people may think. While effectively enforcing some forestry laws and regulations may have a positive impact, enforcing others may make things worse. In many instances, enforcing the laws is unlikely to affect how forests are managed. Many forestry laws and regulations that discriminate against small-scale farmers and foresters and local communities have no scientific basis for doing so. Nonetheless, proponents of such regulations typically justify such inequitable rules on environmental grounds. These rules need to be reviewed and reformed.

The effects on rural livelihoods of enforcing different forestry and conservation laws and regulations

The previous discussion showed that enforcing some forestry laws and regulations affects forest management differently than when enforcing others. The same applies to rural livelihoods. This section examines the livelihood impacts of enforcing different forestry and conservation laws and regulations. It first looks at the impacts of laws and regulations that prohibit or sharply restrict logging, hunting and other human activities in forests. It then looks at measures designed to improve forest management, but not necessarily to limit logging.

Logging bans and moratoriums

Logging bans are simple. Once one bans all logging in a region or a country the authorities can safely assume that any logging that continues must be illegal. Nonetheless, such bans have only had mixed success at reducing environmental destruction (Boyer, C. R. (2000) *Conservation by Fiat: Mexican Forests and the Politics of Logging Bans, 1926–1979*, Unpublished manuscript; FAO, 2001). Few countries have the political will and capacity to stop all logging in the designated areas, in part because of political pressure from the people whom the bans affect and the local governments that represent them.

Where logging bans have been implemented in places where many people depend upon forestry activities for their livelihoods, great hardship has resulted. China represents the clearest case, even though its government has made major efforts to compensate those most affected. Similar problems have arisen in several South-East Asian countries, although the governments there have generally implemented the bans less effectively and the commercially valuable timber was already largely exhausted when governments imposed the bans.

Partial logging bans sometimes deny access to timber to small-scale loggers and loggers who lack political connections, while giving access to others. In fact, some partial logging bans end up becoming little more than an excuse for ensuring that only those favoured by key individuals within the government can have access to the resource.

Strict enforcement of all timber harvesting laws and regulations

In many countries, attempting to strictly enforce all existing forestry laws and regulations affecting timber harvesting would be tantamount to imposing a logging ban. The laws and regulations are so demanding that loggers would find it practically impossible to comply with them and still earn a profit, if, indeed, they could comply at all. There are simply too many requirements, they are too difficult and costly to meet, and some even contradict each other. Without bribes to avoid inspections and speed up the paperwork, the approval of plans and permits would slow down significantly. This projection implies that truly rigorous forest law enforcement would put out of business practically everyone engaged in forestry, in both the formal and informal sectors.

Like logging bans, strict enforcement of all the existing forestry laws would have decidedly mixed effects on rural livelihoods. Under the unlikely assumption that governments were able to achieve complete enforcement, there would be much less forest loss and degradation; but forestry workers and people engaged in small-scale forest-based activities would lose jobs and income.

There would be very little forestry tax revenue. To get everyone to strictly obey all of the laws and regulations may require repression, which could threaten households' physical security and undermine traditional mechanisms of forest management. Communities and poor rural households would lose access to forests where they currently live and work without legal recognition.

Strict protection of conservation areas

Creating strictly protected conservation areas is similar to establishing a logging ban that applies only to one particular area. Completely prohibiting activities makes it easier to detect when the law has been violated. In this case, the prohibitions may include clearing forest for agriculture, hunting, fishing, cattle grazing and harvesting forest plants, as well as logging.[14] Focusing law enforcement efforts on protected areas has the advantage of allowing officials to concentrate on a limited number of compact geographic locations and permits them to devote their attention to laws that link directly to what happens in the forest, rather than to administrative requirements.

Poor rural households, many of whom are indigenous peoples, inhabit the great majority of protected areas in the tropics.[15] Some protected area management plans are the products of extensive consultation with local people and adequately take into account their needs. Often, this is not the case. Many households were expelled from their land and forest when governments established the protected areas.[16] Throughout the tropics, park officials regularly face conflicts with communities over the use of park resources and try to evict villagers from the protected areas. Even when existing regulations permit indigenous peoples to remain in those areas, they are often not allowed to harvest timber or to make agreements with timber companies for the latter to harvest timber in their territories. Greater enforcement of existing restrictions associated with protected areas could easily deny many poor rural households their incomes, access to forest resources and the ability to maintain their traditional customs and lifestyles, and would probably lead to large numbers of arrests and human rights violations.

Prohibitions on shifting cultivation and upland agriculture

Most countries in South and South-East Asia have laws prohibiting shifting cultivation and/or agriculture on steep slopes, and governments there have forcibly resettled people engaged in these activities. Although such laws have generally been justified on environmental grounds, they seem to be motivated, in part, by a desire to keep villagers from competing with logging companies for forest resources, cultural prejudices against indigenous peoples and the desire to concentrate rural populations in order to make it easier to control them and provide services (Thrupp et al, 1997).

The results have frequently been expensive failures. There is little evidence that such prohibitions have led to less deforestation or more sustainable forest management (Sadeque, 2000). But they have often lowered households' income, threatened their physical security, limited their access to forest resources, destroyed their local cultures and undermined their social capital.

Restrictions on commercial bush-meat hunting

Few issues pose as many stark dilemmas as the complex trade-offs between environmental and livelihood objectives, and between short- and long-term goals associated with commercial hunting for bush meat. The issue is particularly important in Central and West Africa. Wild meat provides much of the protein that both urban and rural households consume there, and many rural families earn substantial income from commercial hunting. Nevertheless, in many areas the current levels of hunting are clearly unsustainable (Wilkie and Carpenter, 1999; Fa and Peres, 2001; FAO, 2001; Bennett, 2002; Kaimowitz, 2002).

In most Central and West African countries, commercial hunting of wild meat is illegal but widely tolerated. International environmental NGOs have been pressing governments and logging companies to stop these activities. This could have a significant negative impact on urban and rural diets, although there is some debate about how easily families would find alternative sources of protein (Bennett, 2002). It could also deprive a large number of families of an important source of cash income.[17] At the same time, it must be recognized that failure to regulate commercial hunting of bush meat will not only reduce biodiversity, but also deplete the resource that people rely on as a source of food and income. In some places, this may happen soon.

Outcome-oriented approaches to using forestry laws to improve commercial logging

As noted above, many forestry regulations focus on aspects that have little direct relation to how people manage forests and what happens as a result (Bennett, 1998). Recently, however, some international groups have emphasized enforcement of those laws that most influence forest management and tax revenue. Rather than concentrating on whether logging companies meet all of the multiple administrative requirements, they focus on whether companies:

- have management plans based on serious forest inventories and only harvest logs specified in those plans;
- follow government restrictions concerning annual allowable cuts, minimum diameters, rotation periods and conservation areas;
- monitor and track each log from when it is harvested until it reaches its final destination; and
- pay all the mandated taxes and fees.

This approach has been designed largely for industrial logging companies, particularly those involved in export markets. There is still little practical experience with its implementation and limited data on its cost and effectiveness.

A priori, it is difficult to predict how such an approach may affect rural livelihoods. This would largely depend upon:

- how effectively the initiative was implemented and with what degree of fairness and transparency;
- whether the initiative substantially improved how forests were managed or simply made sure that management was well documented;
- how the initiative affected the sector's profitability and harvest levels;
- the characteristics of the informal forestry sector and how law enforcement efforts affected it; and
- the extent to which the new system re-enforced and legitimized control over forest resources by large-scale logging companies, rather than local communities, indigenous peoples, and small-scale foresters and farmers.

Cracking down on informal timber and fuelwood harvesting

As noted previously, millions of poor rural households engage in fuelwood, charcoal and timber activities that are officially illegal or of uncertain legality, although many countries allow families to harvest small amounts of forest products for their own consumption. Under normal circumstances, most countries make little effort to regulate these informal forestry activities. Local officials may sporadically make their presence felt, particularly when looking for bribes; but otherwise they turn a blind eye.[18]

To the extent that forestry officials *do* occasionally enforce some laws, however, it usually harms rural livelihoods. Local people have to pay bribes, sell their products for lower prices, and face problems of intimidation and threats to their physical security. Formal government structures that contradict traditional mechanisms regulating forestry activities undermine the latter.

Anecdotal evidence suggests that officials make greater efforts to regulate the informal timber and fuelwood sectors in two types of situation. First, when the level of informal timber production seriously competes with industrial loggers, the companies often pressure governments to restrict informal forestry activities, which they consider unfair competition. Ironically, the same companies that call on governments to stop informal forestry activities often engage in illegal practices themselves. Second, official attempts to promote community forestry, such as joint forest management in India and community forestry in Nepal, often cause elements within communities to restrict informal fuelwood and timber collection by people from other villages and by the poorest groups within the village themselves (Sundar et al, 2002). In some cases, this can have serious negative impacts on poor households' livelihoods.

Differences across countries and contexts

The previous discussion has made it clear that forest law enforcement efforts will affect rural livelihoods differently depending upon the context. Key variables that influence these outcomes are the characteristics of the forestry:

- *sector* – for example, the size and characteristics of the forest itself; forest tenure; the type of producers involved; product composition; market orientation; and the types of links between harvesters, processors, traders, lenders and investors;
- *legislation* – for example, to whom it assigns property rights; how large a technical and administrative burden it presents; and to what extent it restricts small-scale forestry activities; and
- *institutions* responsible for law enforcement – for instance, their territorial presence; technical capacity; level of decentralization; and degree of transparency, corruption and respect for human rights.

Places rich in commercially valuable timber with large processing facilities attract regulators' interest since they are potential sources of tax revenues and informal payments. In such contexts, forestry legislation and institutions often help certain groups to grab the resources at the expense of others (Ross, 2001). Governments usually allocate large forestry concessions to private companies with strong political connections and with little regard for rural livelihoods. This applies particularly to Central Africa and South-East Asia. Forestry laws in these contexts typically legitimize the more powerful and wealthier groups' monopoly over forest resources. Since governments already enforce these laws sufficiently to ensure that poor rural households have only limited access to commercial forest resources, greater law enforcement in the more commercially valuable areas would pose a real threat to the rural poor only if the government went after informal-sector activities that are relatively marginal in terms of commercial timber production.

Prominent protected areas with great international support have also been associated with significant regulatory efforts. Here the concern for rural livelihoods has varied. Some locations have put a strong emphasis on integrating conservation with local development. Others have taken more coercive approaches.

Indonesia represents a special case as a country with a large forestry sector in which national authorities and substantial forestry conglomerates have recently lost some of their control over forest resources (see Chapters 3 and 4). This development has opened many new opportunities for regional and local elites and, in some cases, local communities and poorer households who operate in violation of what national authorities consider the law. To a certain extent, the call for the 'restoration of law and order' in such circumstances represents a call for restoring the monopoly over forest resources by national public and private elites – and has uncertain impacts on rural livelihoods.

In China, India, Nepal and several other Asian countries, governments have traditionally given substantial attention to regulating forest use. These nations have large, powerful and deeply entrenched state bureaucracies with strong historical traditions and limited transparency and accountability. The countries are forest poor; but large numbers of people rely heavily on forest resources. The potential risk to rural livelihoods from increased forest law enforcement may be greatest in such contexts since people depend greatly upon forests, and at times the governments have demonstrated both the will and capacity to take measures that limit access to those forests by poor households and ethnic minorities.

In most other countries, efforts to regulate forests have been more limited and sporadic. Forestry departments are heavily under-staffed and have little political power or influence. Forestry law enforcement efforts are unlikely to be effective, but by the same token are also less likely to have major negative impacts on rural livelihoods – although they may still make some producers' lives more difficult and increase the costs of engaging in small-scale forestry activities.

Policy options

The previous discussion has focused on potential negative livelihood impacts of both illegal forestry activities and forest law enforcement efforts. This section discusses options for addressing those problems. Table 6.2 summarizes some of these as they relate to each of the livelihood concerns raised above.

Forestry law reform

One key element of ensuring that enforcement of laws and regulations relating to forests does not harm rural livelihoods is to reform the laws and regulations so that they discriminate less against low-income households, ethnic minorities and women. Key elements include the following:

- Establish simple low-cost mechanisms to formally recognize the rights of local communities and smallholders over forest resources that they already manage and allocate additional resources to them. This must include, among other things, appropriate mechanisms for resolving competing claims.
- Reduce the number of administrative and technical requirements, simplifying them and allowing decisions about them to be made at the local level.
- Establish clear and accessible legal mechanisms to allow people to seek redress for government decisions and actions that may have harmed them illegally.
- Empower local community organizations to monitor compliance of forestry laws with support from government authorities.
- Guarantee full public availability and transparency of government information related to forest regulation.
- End prohibitions on swidden cultivation and on processing timber with chainsaws and permit rural households to engage in activities that form part of their cultural heritage.
- Formally recognize and implement international laws, treaties and agreements that support the rights of indigenous peoples, ethnic minorities and women.

Institutional reform

Reform of laws and regulations will have limited impact unless one also reforms the institutions charged with implementing them.[19] These institutions have to become more efficient and outcome oriented, less corrupt, more transparent, more accountable and more responsive to the needs of smallholders and local communities.

Experience shows that it is not easy to achieve this. Many government officials depend economically upon formal and informal payments associated with the existing regulatory regimes. They may be reluctant to give up the authority associated with their discretionary ability to enforce or not enforce existing legislation and the status associated with their supposed scientific understanding of how to manage forests. Working for wealthy and powerful forestry companies and farmers provides greater status and benefits than working for small farmers and foresters and indigenous peoples. Forestry officials have been trained and socialized under existing paradigms and many aspects of their institutional cultures reinforce them.

Donor support for agencies and officials who take into account livelihood concerns can encourage reform in government forestry agencies. So can the appointment of reform-minded officials, the implementation of training programmes, the sanctioning of corrupt officials and the recruitment of younger and more idealistic forestry officials. As much as possible, it is important to make government officials feel that they are part of the reform process and benefit from it, rather than feel threatened.

Besides working with government agencies, it is also important to strengthen civil society organizations that independently monitor government agencies and forestry companies, provide legal and technical assistance to communities and smallholders, and promote multi-stakeholder dialogues and informal mechanisms for resolving conflicts. These organizations include NGOs, the mass media, professional associations and grassroots organizations. In addition to providing services directly, these organizations can encourage government agencies to become more accountable and transparent.

Focusing on the biggest violators

One obvious suggestion for reducing the potential harm to rural livelihoods from forest law enforcement would be to concentrate enforcement efforts on the largest violators – especially those that provide limited employment. In some, but certainly not all, contexts these are also the groups responsible for the greatest amounts of forest destruction and most of the tax evasion.

Certainly, one aspect of forest law enforcement that stands out compared with law enforcement in other fields is the limited effort to target the greatest offenders and the locations where most offences occur. Forest law enforcement activities often give the impression of being random, at best, or of even discriminating against poorer or less well-protected groups. It makes sense to change this.

Table 6.2 Options to address threats to rural livelihoods from illegal forestry activities and from forest law enforcement

	Illegal forestry activities	Forest law enforcement	Options
Forest product income	Small foresters earn less because they pay bribes and depend upon patrons.	Small foresters earn less because governments stop their 'illegal' activities or greater law enforcement forces them to pay higher bribes or depend more upon patrons.	Reduce and simplify forestry regulations. Exempt small-scale activities from some regulations. Focus regulatory efforts where problems are greatest. Achieve greater transparency in regulation.
Wages from forestry	Over-harvesting makes the forestry sector collapse.	Government actions reduce logging.	Regulators give preference to labour-intensive activities.
Government revenues	Tax evasion and illegal logging in public forests lower revenue.	Reduced logging because of law enforcement lowers revenue.	Implement progressive and transparent tax collection.
Physical security	Illegal loggers and corrupt officials threaten and attack villagers.	Officials inappropriately threaten, attack, arrest or expel villagers and destroy their crops and houses.	Strengthen human rights institutions, grassroots organizations and independent judiciary and oversight. Provide legal assistance and promote legal literacy.
Access to forest resources	Wealthy groups and officials illegally deny access to forests and due process.	Wealthy groups and officials 'legally' deny access to forests and due process.	Implement tenure policies that increase community and smallholder access to forests. Recognize indigenous territories and increase efforts to protect them from encroachment. Use multi-stakeholder dialogues to resolve conflicts and increase access to forests by the poorest groups.

Long-term supply of forest goods and services that poor households use	Damage to forests by illegal logging reduces forest product supply and disrupts environmental services.	Damage to forests due to loss of social capital reduces forest product supply and disrupts environmental services.	Focus regulatory efforts on maintaining forest resources of value to poor families. Support local efforts to protect forest resources legally, politically and financially.
Collective action and participation	Illegal logging undermines local forest management institutions, and bribes and influence-peddling replace democratic process.	Laws that fail to recognize local forest management institutions undermine them, and police action substitutes for dialogue.	Recognize and support community efforts to protect forests. Employ multi-stakeholder dialogue and informal mechanisms to resolve conflicts. Compensate communities for environmental services.
Respect for cultures and tradition	Illegal logging undermines traditional institutions.	Laws that disallow traditional practices or fail to respect local institutions undermine peoples' traditional cultures.	Avoid regulations that unduly restrict peoples' traditional activities. Implement international agreements concerning indigenous peoples.
Economic growth	Widespread failure to respect rule of law reduces investment and growth.	Limiting forestry activities reduces (short-term?) economic growth.	Make regulatory systems more transparent, democratic and equitable. Promote small-scale forestry activities and partnerships between companies and communities.

To restrict money laundering and international trade linked to illegal forestry activities has the advantage of affecting mostly formal-sector operations that make substantial financial transactions and/or export their products. On average, these tend to be wealthier than those that sell to domestic markets and/or form part of the informal sector. It must be noted, however, that in some countries industrial logging companies purchase substantial amounts of timber from small-scale producers and provide them with operating capital. Hence, the separation between large- and small-scale activities may be less straightforward than it first appears.[20] In addition, some workers may lose their jobs if the companies that employ them are sanctioned.

Enforcing laws that favour rural livelihoods

Some forest-related laws specifically favour poor rural households and ethnic minorities, so those groups should benefit from their enforcement. For example, over the last few decades many governments in Latin America have recognized indigenous peoples' rights over large territories, but indigenous peoples often find it difficult to protect these territories from encroachment by loggers, miners and farmers. Greater efforts to protect indigenous peoples' rights could improve their situation and help to guarantee their continued access to the forest products upon which they depend. The same applies to the legal rights of people living in extractive reserves in Brazil, indigenous peoples and community-based forestry organizations in the Philippines and similar groups.

One problem that rural communities and small farmers frequently suffer in many countries is the failure of logging companies to fulfil their promises to construct roads, fund social services and scholarships, pay fees and provide other benefits in return for permission to log their forests. Establishment and enforcement of legal contracts could potentially go a long way towards solving this problem.

Community-based law enforcement

In many countries, rural communities play an increasing role in monitoring and reporting forestry law violations to government officials, confronting law violators themselves and regulating forest use among community members. In some cases, they collaborate closely with government officials; in other cases, the two conflict. One example of the former comes from the Philippines, where the forestry department helped to organize multi-stakeholder forest protection committees that brought together various public agencies and civil society organizations to work on forest law enforcement issues. Similarly, in India, the forestry departments have sponsored forest protection committees as part of their joint forest management programme. In both cases, the groups have worked well in some areas, less so in others, and the government and communities have shown varying degrees of initiative and dynamism.

One issue that has arisen in the case of joint forest management has been the tendency for more powerful groups within the villages to exclude poorer groups within the village from collecting fuelwood and other products from the local forests. In contrast, in Honduras, Mexico and other countries, local communities have organized to expel outside logging companies accused of illegal forestry activities without the support of government departments, often with success. Generally, this thrust has involved communities who depended upon the forests and were not involved in the logging operations themselves.

In principle, organizing communities to defend their own interests in relation to forest law enforcement should be an important element of any strategy in order to make sure that law enforcement efforts do not negatively affect rural livelihoods. Nonetheless, it would be important to synthesize the lessons from existing experiences before drawing any definitive conclusion. As the joint forest management experience reminds us, communities are not homogeneous entities, and some community law enforcement efforts may negatively affect the livelihoods of poorer and weaker groups.

Sequencing

Clearly, something needs to be done about illegal forestry activities and the weak rule of law in forested regions. Just as clearly, many existing forestry and conservation laws discriminate against small-scale farmers and foresters and indigenous peoples, and enforcing those laws more effectively would only make the problem worse.

In principle, the logical thing to do would be to reform the laws and the institutions that implement them and then have the reformed institutions enforce the new laws. But illegal forestry activities are causing major damage today, and reforming the forestry legislation and the institutions that implement it could easily take years, or fail completely. This raises a serious sequencing dilemma. Is it better to wait until the laws and institutions are improved before pressing for greater law enforcement, or would it be better to push existing institutions to enforce the current laws now, even though this could have a negative impact on rural livelihoods? The question has no easy answer. One probably has to work on both simultaneously; but it is important not to lose sight of the fact that enforcement of many existing laws can have negative consequences. Forestry agencies and civil society organizations must work hard to focus on those law enforcement activities that have the greatest potential for improving forest management and tax revenues with the least negative impact on livelihoods.

Adaptive management and learning

There is still much to learn about how efforts to regulate forest use affect rural livelihoods and what can be done to get people to manage their forests more sustainably without making life harder for groups whose lives are already difficult enough. This learning needs to happen. We need much more information about how forestry and conservation laws and regulations are currently enforced and what the impacts have been, as well as to learn from interesting experiences and examples of best practice. At present, it is practically impossible to answer questions such as how do most forestry officials and park guards spend their time? How many people do they fine or arrest? How common and large are the bribes people pay? How likely is it that forestry violations are detected and prosecuted and result in punishment? How common are human rights abuses linked to forest law enforcement? How do these dynamics affect small-scale forestry producers' costs and incomes? How many people depend upon forestry activities that are currently illegal and in what ways? How much deforestation and forest degradation that negatively affect rural livelihoods result from illegal forestry activities? Until there are more answers to such questions, it will be hard to design appropriate forest law enforcement strategies that take into account the implications for rural livelihoods.

Most of the existing information about illegal forestry activities is anecdotal or speculative. While it has been extremely useful for increasing public awareness about the problem and for stimulating action in particular cases, it is less helpful in coming up with appropriate policy responses. To answer the more systematic questions about the links between forest law enforcement and rural livelihoods will require formal research. But it will also require well-organized multi-stakeholder study tours, improved data collection in forest law enforcement agencies, and workshops and visits through which people learn from each other's experiences, among other things.

Conclusion

Governments and communities must regulate the management and use of forests in order to ensure that their useful functions are maintained over time, that benefits are shared equitably, that conflicts are resolved in a fair and transparent manner, and that sufficient tax revenues are obtained to pay for necessary public expenses. The widespread violation of existing forest laws and regulations has major negative impacts on forests, livelihoods, public revenues and the rule of law. Something must be done about this.

The problem is that many existing forests and conservation laws themselves have unacceptable negative impacts on poor people, ethnic minorities and women, and in many places they are enforced in a fashion that is discriminatory and abusive. Ways must be found to address the problems associated with illegal forestry activities that at least do not aggravate the negative impacts of existing regulatory efforts on the rural poor. This will not be easy; but it will certainly be utterly impossible unless the challenge is recognized from the outset. If this chapter contributes to this recognition, it will have served its purpose.

Acknowledgement

This chapter is reprinted, with slight editing, from Kaimowitz, D. (2003) 'Forest law enforcement and rural livelihoods', *International Forestry Review*, vol 5, no 3, pp199–210, with permission.

Notes

1 The concept of social capital is justifiably controversial. As used in the context of this chapter, the concept is synonymous with the local institutional capacity to regulate and manage forest resources effectively, efficiently and equitably.
2 One could also argue that policies favouring economic growth contribute to rural livelihoods. This chapter occasionally refers to the impact of policies on growth.
3 For the purposes of this chapter, an arrest or fine is considered 'inappropriate' if it:

- does not contribute to sustainable forest management;
- does not follow due process; or
- results from laws or law enforcement practices that discriminate in favour of wealthier or more powerful groups.

4 Private banks, NGOs and government credit agencies generally will not lend money to independent small-scale foresters, in part because of concerns about the legality of such activities.
5 Obidzinski (2003) documents in great detail how such patronage networks operate in East Kalimantan, Indonesia, and how they affect the distribution of benefits.
6 Contreras-Hermosilla (2002) estimates the total annual loss from such illegal activities as being at least US$10 billion each year. The figure includes developed and transition countries, as well as developing countries.
7 Contreras-Hermosilla (2002, p13) notes: 'In other cases local communities and indigenous people have traditional rights over forestlands; but these are not recognized by the state and thus their use of resources is technically illegal.'
8 One particularly complex and difficult example of this displacement is that of India. The country's supreme court had ordered that, by 31 May 2003, government officials should evict all of the families encroaching upon the country's reserve forestland. The author was unable to locate any reliable estimate of the number of people this may have affected; but the number is clearly in the hundreds of thousands – and may be even higher (Sharma, 2003).
9 Nonetheless, there are situations where failure to log, hunt or harvest plants can lead to ecological imbalances, fire hazards or other problems. One should also remember that forests are not static. With or without further human disturbance, they change over time.
10 For example, regulations that encourage companies to avoid large canopy gaps may impede regeneration of major commercial timber species, such as mahogany. There are also situations in which it is less profitable to manage forests sustainably than to manage them unsustainably because of the high cost of preparing the required management plans and similar documents. For example, Davies (Davies, J. (1998) 'Secondary Forests', Unpublished manuscript) has shown that the main reason why it is commercially unviable for landowners in northern Costa Rica to promote the natural regeneration of secondary forest for timber production on abandoned pastures is the high cost of the associated paperwork that the law requires.
11 Jan Laarman, pers comm, 1997, based on a thesis that presented measurements of recovery rates from the two systems using data from Ecuador.
12 Small-scale logging along rivers using human labour or animal traction almost invariably has a much smaller environmental impact than large-scale logging using heavy machinery and the establishment of large numbers of logging roads.

13 During the late 1990s, the Brazil Forestry Department carefully reviewed a sample of approved forest management plans. The agency concluded that only one third of them were of acceptable quality, and cancelled or suspended the remainder. A subsequent review several years later found that the situation had only marginally improved (Barreto and Souza, 2002).
14 Not all categories of protected areas prohibit all of these activities.
15 During the early 1990s, it was estimated that 86 per cent of Latin America's protected areas were inhabited (Amend and Amend, 1992, and Kempf, 1993, cited by Colchester, 1994).
16 Dangwal (1999) estimates that the Indian government relocated 600,000 forest dwellers in the process of establishing its existing network of national parks and sanctuaries.
17 Studies have shown that commercial bush-meat hunting offers higher incomes than most other activities for rural people in several Central African countries (Anadu et al, 1988; Dei et al, 1989; Dethier, 1995; Muchal and Ngandjui, 1995; Nguegueu and Fotso, 1996).
18 This should be considered a hypothesis to be verified. There are very few data on the extent or consequences of regulating the informal forestry sector – but see Chapter 8.
19 This may be only partially true when it comes to legal reforms that limit the institutions' functions and authority.
20 In several countries, including Cameroon and Indonesia, industrial logging companies have even created their own 'community' organizations to take advantage of government policies favouring community forests and cooperatives.

References

Anadu, P. A., Elamah, P. O. and Oates, J. F. (1988) 'The bushmeat trade in south-western Nigeria: A Case study', *Human Ecology*, vol 16, no 2, pp199–208

Angelsen, A. (1996) *Deforestation: Population or Market Driven? Different Approaches in Modeling of Agricultural Expansion*, Working Paper no 9, Bergen, Norway, Chr Michelsen Institute

Barr, C. (2001) *Banking on Sustainability: Structural Adjustment and Forestry Reform in Post-Suharto Indonesia*, Bogor, Indonesia, Centre for International Forestry Research and Macroeconomics for Sustainable Development Programme Office, World Wide Fund for Nature

Barreto, P. and Souza Jr., C. (2002) *Controle do desmatamento e da exploração da madeira na Amazônia, Diagnóstico e sugestoes*, Belem, Brazil, Imazon

Bennett, C. (1998) 'Output based policies for sustainable forestry in community forestry: Reducing forestry bureaucracy', in Wollenberg, L. and Ingles, A. (eds) *Incomes from the Forest, Methods for the Development and Conservation of Forest Products for Local Communities*, Bogor, Indonesia, CIFOR and IUCN

Bennett, E. I. (2002) 'Is there a link between wild meat and food security?', *Conservation Biology*, vol 16, no 3, pp590–592

Boscolo, M. and Vincent, J. R. (1998) *Promoting Better Logging Practices in Tropical Forests: A Simulation Analysis of Alternative Regulations*, World Bank Working Paper Series 1971, Washington, DC, World Bank

Colchester, M. (1994) *Salvaging Nature, Indigenous Peoples, Protected Areas, and Biodiversity Conservation*, Discussion Paper 55, Geneva, Switzerland, United

Nations Research Institute for Social Development, World Rainforest Movement and World Wide Fund for Nature

Contreras-Hermosilla, A. (2002) *Law Compliance in the Forestry Sector, An Overview*, WBI Working Papers, World Bank Institute, Washington, DC, World Bank

Dangwal, P. (1999) 'Whose forests are they anyway? A case study of the Rajaji National Park in northwest India', in Colchester, M. and Erni, C. (eds) *Indigenous Peoples and Protected Areas in South and Southeast Asia*, Proceedings of the Conference at Kundasang, Sabah, Malaysia, IWGIA Document no 97, Copenhagen

Dei, G. J. S., Sedgely, M. and Gardner, J. A. (1989) 'Hunting and gathering in a Ghanaian rainforest community', *Ecology of Food and Nutrition*, vol 22, pp225–243

Dethier, M. (1995) *Etude Chasse Report*, ECOFAC Programme, Yaoundé, Cameroon, Ministry of Environment and Forests

DFID (UK Department for International Development) (2001) *DFID Sustainable Livelihoods Guidance Sheets*, London, DFID

Dixon, J. A. and Sherman, P. B. (1991) *Economics of Protected Areas: A New Look at Benefits and Costs*, London, Earthscan

Fa, J. E. and Peres, C. (2001) 'Hunting in tropical forests', in Reynolds, J. D., Mace, G. M., Redford, K. H. and Robinson, J. G. (eds) *Conservation of Exploited Species*, Cambridge, UK, Cambridge University Press

FAO (United Nations Food and Agriculture Organization) (1998) *Asia-Pacific Forestry Towards 2010*, Report of the Asia-Pacific Forestry Sector Outlook Study, Rome, FAO

FAO (2001) *Forestry Outlook Study for Africa: A Regional Overview of Opportunities and Challenges Towards 2020*, Rome, FAO

Fredericksen, T. S. (1998) 'Limitations of low-intensity selective and selection logging for sustainable tropical forestry', *Commonwealth Forestry Review*, vol 77, pp262–266

Kaimowitz, D. (2002) *Not by Bread Alone, Forests and Rural Livelihoods in Sub-Saharan Africa*, Bogor, Indonesia, CIFOR

Karsenty, A. (2000) *Economic Instruments for Forests: The Congo Basin Base*, London, IIED, CIRAD and CIFOR

Muchal, P. K. and Ngandjui, G. (1995) *Wildlife Populations in the Dja Reserve (Cameroon): An Assessment of the Impact of Village Hunting and Alternatives for Sustainable Utilization, Report*, Yaoundé, Cameroon, Ministry of Environment and Forests

Ngnegueu, P. R. and Fotso, R. C. (1996) *Chasse Villageoise et Conséquences pour la Conservation de la Biodiversité dans la Réserve de Biosphère du Dja*, Yaoundé, Cameroon, Progamme de Conservation et Utilisation Rationelle des Forestiers d'Afrique Central

Obidzinski, K. (2003) *Logging in East Kalimantan, Indonesia: The Historical Expedience of Illegality*, PhD thesis, Amsterdam, Universiteit van Amsterdam

Poschen, P. (1997) *Forests and Employment, Much More than Meets the Eye*, World Forestry Congress, Antalya, Turkey, 13–27 October, vol 4, Topic 20, Rome, FAO, www.fao.org/montes/foda/wforcong/PUBLI/V4/T20E/1-5.HTM#TOP, accessed May 2003

Putz, F. E., Redford, K. H., Robinson, J. G., Fimbel, R. and Blate, G. M. (2000) 'Biodiversity conservation in the context of tropical forest management', *Conservation Biology*, vol 15, pp7–20

Ross, M. (2001) *Timber Booms and Institutional Breakdown in Southeast Asia*. Cambridge, UK, Cambridge University Press

Sadeque, S. Z. (2000) 'Shifting cultivation in Eastern Himalayas: Regulatory regime and erosion of common pool resources', Paper presented at the Eighth Conference of the International Association for the Study of Common Property, Bloomington, Indiana, 31 May–4 June

Sharma, R. (2003) 'The states, eating up forestland', *Frontline*, vol 20, no 6, www.flonnet.com/fl2006/stories/20030328001505000.htm, accessed December 2003

Sheil, D., and Van Heist, M. (2000) 'Ecology for tropical forest management', *International Forestry Review*, vol 2, pp261–270

Sundar, N., Jeffry, R. and Thin, N. (2002) *Branching Out, Joint Forest Management in India*, Oxford, UK, Oxford University Press

Thomas, V. (ed) (2000) *The Quality of Growth*, Oxford, UK, Oxford University Press

Thrupp, L. A., Hecht, S. and Browder, J. with Lynch, O., Megateli, J. N. and O'Brien, W. (1997) *Diversity and Dynamics of Shifting Cultivation, Myths, Realities and Policy Implications*, Washington, DC, World Resources Institute

Walker, R. T. and Smith, T. E. (1999) 'Tropical deforestation and forest management under the system of concession logging: A decision theoretic analysis', *Journal of Regional Science*, vol 33, pp387–419

White, A. and Martin, A. (2002) *Who Owns the World's Forests? Forest Tenure and Public Forests in Transition*, Washington, DC, Forest Trends

Wilkie, D. S. and Carpenter, J. (1999) 'Wild-meat hunting in the Congo Basin: An assessment of impacts and options for mitigation', *Biodiversity and Conservation*, vol 8, pp927–955

World Bank (1997) *Helping Countries to Combat Corruption: The Role of the World Bank*, Washington, DC, World Bank

7

Rural Livelihoods, Forest Law and the Illegal Timber Trade in Honduras and Nicaragua[1]

Adrian Wells, Filippo del Gatto, Michael Richards, Denis Pommier and Arnoldo Contreras-Hermosilla

Introduction

Honduras and Nicaragua qualify as heavily indebted poor countries (HIPCs) and are among the poorest in the Western hemisphere. Both countries have extensive natural forest cover,[2] and economic and livelihood dependence upon forests is significant. In 1997, the forest sector in Honduras contributed some 9 per cent of gross domestic product (GDP) and 7 per cent of national foreign currency earnings (Lazo, 2001), while in Nicaragua it contributed 3.2 per cent of GDP (Friends of the Earth, 1999). However, levels of deforestation are also high.[3] In remoter areas, where dependence upon forests is highest, deforestation is partly the result of widespread illegal logging. Yet, the links between illegal logging and rural livelihoods, and, in particular, poverty, remain poorly understood.

This chapter presents evidence generated by one of the first attempts at a systematic policy, legal and institutional assessment of illegal logging in Honduras and Nicaragua, conducted by a team of local and international researchers during 2002.[4] Reports from this study are available at www.talailegal-centroamerica.org.

The case study outlines the scale and dynamics of the illegal timber trade, focusing on the involvement of local communities, the legal, institutional and economic factors that drive it, and the livelihood and poverty implications (positive and negative), both directly and in terms of impacts on state revenue. The case study examines how the existing legal and institutional framework, within and beyond the forest sector, presents community-based forest producers with significant barriers to legal compliance. These span denial of secure tenurial and resource rights, over-complex regulation and associated corruption.

Such barriers to legality criminalize local forest users. The transaction costs of complying with complex regulation, combined with low timber prices (itself a result of illegal logging), renders legal forest production uneconomic for small-scale producers. This situation leaves them vulnerable to economic capture by powerful illegal timber traders, who in many cases operate in collusion with forest governmental employees and law enforcement officials. This scenario is especially the case in remote rural areas with weak state presence and high levels of organized crime as a result of the drug trade.

The chapter shows that involvement in illegal logging is a source of valuable short-term income for the poor, where few other livelihood options exist. But it also comes at a considerable cost to natural and social capital, reducing the livelihood resilience of the poor in remote areas. The case study argues for a combination of measures to tackle illegality, while also securing the livelihood benefits of forests for the rural poor. These include regulatory and institutional reforms to reduce barriers to legality, as well as better targeted law enforcement to address corruption and organized crime – within and beyond the forest sector.

The chapter is divided into ten sections, the next of which discusses the relationship between forests and rural livelihoods in Honduras and Nicaragua. Successive sections present a working definition of illegal forest production; outline the scale of the illegal timber trade in Central America; describe key actors and, in particular, the role of local communities; address legal and policy barriers to legality and how these provide an economic incentive for illegal logging; focus on failures in public administration, including institutional corruption, and how these compound the effects of regulatory barriers to legality; address the positive and negative impacts of the illegal timber trade on rural livelihoods; and, finally, recommend a range of policy prescriptions.

Forests and rural livelihoods in Honduras and Nicaragua

In both Honduras and Nicaragua, around 40 per cent of the population live in and near forest areas. The majority are poor or extremely poor.[5] The heavily forested Atlantic regions examined in this case study are home to some of the most persistent poverty, due to isolation and lack of investment. In Nicaragua, extreme rural poverty in these areas actually increased between 1993 and 1998, whereas other regions saw significant reductions.[6]

Forests users in Honduras and Nicaragua include indigenous peoples and ethnic minorities,[7] as well as migrant *mestizo* settlers. They also include increasing numbers of timber traders and landowners. Forest incomes have an important role in livelihood diversification. In poorly integrated areas with limited investment and little access to urban employment opportunities, forests provide important supplementary income to farming. This situation partly reflects policy and market failures affecting the agricultural sector. The sale of forest products and employment in the timber trade help to finance accumulation in agriculture and to pay off debts accrued during stress periods (McSweeney, 2003).

The existing mix of forest users, especially on the Atlantic coasts of Honduras and Nicaragua, is a product of the advancing agricultural frontier. The agricultural frontier acts as an escape valve for the effects of policy and market failures in more densely populated drier areas to the south and west. These include highly skewed land distribution (in Honduras, almost half the farms are less than 2ha); export-oriented agriculture that has left *campesinos* with the poorest soils; declining terms of trade for agriculture (in mid 2002, the coffee price paid to farmers was about one third of its peak 1997 value); lack of credit and rural marketing infrastructure; and declining domestic market opportunities due to more competitive (subsidized) grain imports. These pressures, compounded by natural disasters and land degradation, have led to declining productivity in many areas. Even the rural non-farm economy is struggling to maintain the rural poor (Richards et al, 2003b). This creates strong pressure for out-migration, mainly to urban areas, but also to the forest frontier.

Many observers fear that, with the coming into effect of the Central American Free Trade Agreement (CAFTA) in 2006, the pressure for migration will increase even more. Augmented imports of subsidized grains from the US will further reduce the market opportunities for local small-scale grain growers, while the expected investments in urban-based industries should attract new waves of job-seeking people to the country's main cities. On the other hand, agro-industrial investors looking to purchase pasture land will likely exert added pressure on cattle ranchers and poor farmers to move into remoter forested areas, encouraging even greater deforestation.

Illegal forest production: A working definition[8]

This case study focuses on the role of local communities in activities considered illegal under national law. It provides a lens through which to identify legal, institutional and economic barriers to legality that work to criminalize local forests users' livelihoods – whether such barriers are a denial of tenurial and resource rights or over-complex regulation. For these purposes, illegal forest production is defined as that which fails to conform to national laws and standards regulating forest resource allocation, forest management and extraction, processing, transport and trade.

That said, 'illegality' is a problematic concept. Societal objectives vary and also shift with time. The legitimacy of laws and regulations is therefore likely to be perceived differently by different sets of stakeholders. Local forest users

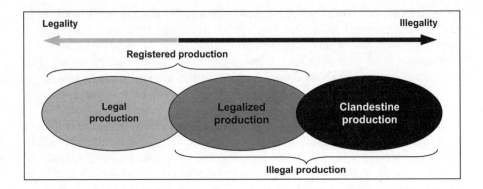

Figure 7.1 *Legality and illegality in forest production in Central America*

Source: del Gatto (2003a)

may perceive national forest law as illegitimate if, for example, it ignores their customary rights in favour of assigning harvesting rights to third parties.

It is also hard to distinguish between legal and illegal production. Much apparently legal forest production is fraudulently 'legalized' at some point along the production chain. In the Central American context, it is useful to group forest production into three (albeit overlapping) categories of legal, legalized and clandestine production (see Figure 7.1). The latter two categories conform to illegal logging. Poor people are involved in and affected by both.

Legalized production is timber fraudulently legalized at the stump (e.g. by adding timber from an unauthorized area), in transport (in Nicaragua, it is common for a transport permit to be used up to ten times) or in processing. Legalized production is accompanied by official documents, forest fees are paid, and it enters official statistics. Clandestine timber, by contrast, evades all documentation and fees and remains unregistered.

Bribing the state forestry authority, local government and law enforcement (police) officials to secure permits or to turn a blind eye forms an integral part of both legalized and clandestine timber. In official data, there is, therefore, an overlap between legal and illegal logging, as shown in Figure 7.1, making it difficult to estimate the scale of illegal production and its impacts on national revenue.

The scale of the illegal timber trade in Central America[9]

The illegal timber trade in Honduras and Nicaragua

Based on a combination of secondary data,[10] key informant estimates and supply and demand analysis,[11] clandestine production in Honduras is estimated at 75 to 85 per cent (about 125,000 to 145,000 cubic metres) of total hardwood production and 30 to 50 per cent (350,000 to 600,000 cubic metres) of softwood production. For Nicaragua, it represents about half (30,000 to 50,000 cubic metres) and 40 to 45 per cent (110,000 to 135,000 cubic metres) of total hardwood and softwood production, respectively. Anecdotal evidence suggests that most of the remaining, or official, timber production is fraudulently legalized. Levels of clandestine hardwood production may be higher given weak state presence in remoter source areas on the Atlantic coast, as well as the higher value of hardwood trees compared to softwood pine. Furthermore, a greater proportion of pine is in private or communal hands, whereas most broadleaf forests remain under state control.

Legalized and clandestine timber production chains feed a variety of markets. Processing and export industries generally located around the main cities consume pine and higher-value timber, including mahogany for export as furniture or furniture parts (mainly to the US and Caribbean markets).[12] But production chains also feed national markets (for urban consumption) and local demand (for mines and construction in local towns and villages). National demand focuses on a broader spectrum of species, from traditional high-value timbers such as mahogany to pine and lower-value hardwoods.

Transboundary movement of illegal timber

Across Central America as a whole, an estimated 20 to 30 per cent of the regional timber trade is undocumented (clandestine). Notwithstanding the limitations of the data, cross-country comparisons made on the basis of the FAOSTAT (2002) *Bilateral Trade Matrices*, as well as anecdotal reports, suggest considerable under-declaration of exports of Nicaraguan and Honduran softwoods and hardwoods, including mahogany, especially to the Dominican Republic and the US. In 1999, the US International Trade Commission registered the import of 2222 cubic metres of Honduran mahogany (sawn wood), revealing the violation of Honduras's 1998 hardwood log and sawn-wood export ban.

Anecdotal evidence also suggests that Honduras exports significant quantities of illicit timber to Nicaragua, only to re-import it as 'legal' timber. El Salvador, the least forested country in Central America, has been a significant consumer of both Honduran and Nicaraguan timber, especially in recent years given the need for reconstruction following Hurricane Mitch and two earthquakes. On the Honduras–El Salvador border, the illegal timber trade is exacerbated by old territorial disputes – resulting in timber conflicts involving the local authorities, as well as local armed groups that include Salvadorans living in Honduras. Costa Rica has been much more successful than its neighbours in controlling illegal logging (Campos Arce et al, 2001); but Costa Rica partly makes up for its national supply shortfall by importing illegal Nicaraguan timber across the highly permeable San Juan River border (del Gatto, 2003c).

Key actors in the illegal timber trade: The role of local communities[13]

Illegal timber production chains in Honduras and Nicaragua involve a wide range of actors, including forest owners, forest squatters, migrants, community leaders, forest professionals, timber truckers, timber industrialists and public officials. Arrangements between these actors enable access to forest resources; provision of upfront capital and equipment; transportation, processing and marketing; as well as accompanying formal and informal transactions to 'legalize' production or circumvent the legal and fiscal system.

In both countries, timber traders and intermediaries are instrumental in facilitating supply chains. The latter usually operate by advancing funds and equipment to local communities, including forest owners, local timber producer associations and individual sawyers.

The role of local communities in illegal timber production chains varies depending upon existing institutional structures and tenurial arrangements. In Nicaragua, some processing companies use intermediaries to systematically buy up community and/or non-commercial use permits as a means of accessing the resource and 'legalizing' illegal cutting. The role of indigenous communities and small-scale private forest owners is therefore limited to giving permission for their land to be logged. In Honduras, intermediaries have infiltrated and even established forest producer organizations under the Social Forestry System (SSF). This scheme allows intermediaries to obtain different types of cutting permits, which most of the time just serve as means of 'legalizing' otherwise illicit extraction. Intermediaries take control of decision-making, using local sawyers as little more than hired labour (see Box 7.1). In return, well-connected intermediaries offer community producers the means of securing permits through patronage, as well as access to credit and markets.

The fraudulent use of community permits to legalize production would not be possible without the collusion of technicians and senior government officials in the state forestry agencies. In Honduras, the State Forestry Administration–Honduran Forestry Development Corporation (AFE–COHDEFOR) introduced a policy of issuing 'deadwood' licences following Hurricane Mitch in 1998. These were used to fraudulently cut standing trees (see Box 7.1). In Nicaragua, informal schemes are negotiated between officials of the National Forestry Institute (INAFOR), local government (municipalities), community leaders (*síndicos*)[14] and other interest groups to legalize production. These plots include issuing permits to cut larger volumes than are obtainable from an authorized area.

Barriers to legality: Legal and policy constraints[15]

A variety of legal and institutional constraints leave local community organizations highly vulnerable to capture by timber traders. Actors may perceive the

Box 7.1 The Social Forestry System and illegal mahogany exploitation in the Sico-Paulaya Valley, Honduras

The Honduran Social Forestry System (SSF) was introduced in 1974, granting management and harvesting permits to groups of farmers or indigenous peoples organized as agro-forestry co-operatives. In the years following Hurricane Mitch, the State Forestry Administration–Honduran Forestry Development Corporation (AFE–COHDEFOR) began issuing licences to community-based producer groups under the SSF, authorizing the exploitation of *deadwood* felled as a result of natural causes (such as Hurricane Mitch) or due to changes of land use to agriculture or animal husbandry. In the Sico-Paulaya Valley, this practice led to widespread forest exploitation during 2000 and 2001. In all, 8696 cubic metres of mahogany timber was released for commercial exploitation, 93 per cent of it being ostensibly deadwood. Evidence suggests that around 80 per cent of the timber extracted, in fact, came from other sources. In other words, the deadwood licences that were issued served as a cloak for trade in illegally logged timber.

Officially, forest exploitation is carried out by community-based groups organized under the SSF of the state forestry agency AFE–COHDEFOR. Implementation of the SSF in the Paulaya Valley, however, has been strongly influenced by the interests of timber merchants (a common pattern in high-value forests). Middlemen infiltrate local producer organizations under the SSF through a combination of backhanders, patronage systems and, where necessary, intimidation. This provides intermediaries with a front to obtain timber use permits under the SSF system, transacting directly with AFE–COHDEFOR with little or no local participation.

Intermediaries advance money to local communities to cover the costs of production, and fix the price and place for the timber purchase. Cutting is carried out by dozens of sawyers working in small independent teams of three to six people with one or two chainsaws. In spite of the SSF, local sawyers are frequently just daily workers. Intermediaries may also hire workers from outside the valley.

Source: del Gatto (2003e)

law as unfair and resist its application where it fails to recognize local tenurial rights, restricts access to forest resources and/or places unjustifiable volume constraints on production. Other key constraints include regulations that are inconsistent and difficult to comply with, as well as overlapping or conflicting government responsibilities. These constraints criminalize livelihoods and increase the transaction costs associated with staying legal for community-based enterprises.

Tenurial constraints

Forest land tenure is undergoing a profound change in Honduras, and there is a shortage of reliable, up-to-date data. According to most observers, however, around 50 per cent of forestland is private (including community collective titles), while the remainder is either state owned (30 per cent) or *ejidal* land owned by municipalities (20 per cent). The tenurial distribution of forest land in Nicaragua is currently unclear, given ongoing disputes with indigenous peoples and rural land titling processes.

The tenurial or use rights of people living on state forestland are also unclear – except for the few who have benefited from policy initiatives on co-management, for example under social forestry schemes or in protected areas. For example, Article 41 of the 2003 Nicaraguan Forest Law defines national forestland as that without owners, which is problematic given that 75 per cent of land in Nicaragua lacks clear title (FAO, 2000).

Without legal recognition, migrants living on state forestlands are instead subject to a system of *de facto* rights, based on economic power. Individual *campesinos* may secure informal property rights depending upon whether they are the first to claim and clear a piece of forestland or have acquired such property rights from a previous owner. Informal rights may provide sufficient tenurial security in some areas. In more recently settled areas, however, tenurial insecurity is often greater (Broegaard et al, 2002). Evidence from Honduras shows that informal property rights can easily be appropriated by other members of the community, including groups of legal and illegal loggers. Poor people are often pushed farther and farther into remote forests to find land not already claimed, as the cattle ranches expand.

In Nicaragua, a series of rural land titling schemes[16] has worked to formalize property rights, granting individual households title deeds to plots of land. In Honduras, the 1992 Law of Agricultural Modernization returned forest property rights to land owners and municipalities, reversing 1974 legislation that nationalized forests on private and municipal (*ejidal*) lands. It also conferred ownership rights to *de facto* owners of national forest that had been under agricultural use for at least three years before 1992 (Lazo, 2001).

Land titling can, however, increase tenurial insecurity for the poor due to delayed, partial, corrupt, improper or reversed implementation. Land registration is long and expensive, and total costs may equal half the initial land price. In Nicaragua, lack of formal documents due to failures in the registration system mean that households are sometimes forced off their land by more powerful interests (Broegaard et al, 2002). In Honduras, by allowing titling of state forestland under agricultural use for at least the past three years, the 1992 law appears to have resulted in widespread speculative land clearance by wealthy locals and outsiders since they had the resources and lawyers to take advantage of the situation (Suazo et al, 1997). Titling has also been slower in remoter areas, meaning that many settlers and indigenous peoples continue to live without tenurial security.

Land reform and rural titling schemes have also yet to fully address indigenous land claims. In Honduras, indigenous land rights are recognized under Article 346 of the Constitution of the Republic. Groups such as the Tawahka have had some success in securing title and protection for their lands. Claims by other groups living on national land remain unresolved, however. Such groups find themselves increasingly in conflict with settlers and extractive industries. In Nicaragua, indigenous land rights are recognized under the 1987 Constitution, Law 28 on Autonomy of the Caribbean Coastal Regions and, more recently, Law 445 on Indigenous Communal Property Regimes in the Atlantic regions. Nevertheless, the case of *Mayagna (Sumo) Awas Tingni Community v Nicaragua* in the Inter-American Court demonstrates the difficulties faced by indigenous communities in securing demarcation of their territories even where this is provided for in the law. These hindrances include failures in public administration, as well as difficulties in accessing justice.

Limits on rights of access

Access to national forestland in Honduras

In the last three decades, access to state forestland in Honduras has been shaped by two main acts. The 1974 COHDEFOR Act nationalized all forests, placing even private forest areas under state control. The 1992 Law of Agricultural Modernization reversed elements of the COHDEFOR Act by returning control of private forests to landowners. It also devolved ownership of *ejidal* forests to the municipalities. The 1992 law restricted AFE-COHDEFOR's[17] responsibilities to managing national forests, including natural protected areas, and controlling and setting standards for operations in private and municipal forests.

For forestlands that remain under national control, the 1996 Regulation Governing the Rights of the People on National Lands with Forest Potential limits access by people to so-called traditional uses of the land that do not involve logging. It only permits them to harvest fuelwood from trees that have died a natural death or to extract timber after thinning-out and culling operations, while it consents to resin, oil, latex and seed extraction, as well as grazing, recreation, harvesting medicinal plants, hunting and fishing.

The only means for local people to legally participate in timber harvesting is under the SSF if they are able to organize themselves in order to prepare a management plan. Even then, official support for the SSF has declined and the system no longer provides a source of secure access rights for people living on state forestlands (see Box 7.2). Beneficiary families only represent some 60,000 persons; many others have been excluded (Lazo, 2001).

Nor do communities gain significantly from commercial cutting operations under the auction system to sell standing timber in national forests. Companies tend to cut and run, leaving communities with few benefits. The system has met stiff local opposition. In particular, in the department of Olancho, there have been significant popular protests over logging by companies under the auction system, attracting huge media attention (see Box 7.3).

> **Box 7.2 The Social Forestry System: Benefits but no secure rights**
>
> Community-based groups that organize themselves in order to prepare a management plan may secure legal rights to harvest timber, resins or other products under the Social Forestry System (SSF). The SSF was introduced in 1974, granting management and harvesting permits to groups of farmers or indigenous peoples organized as agro-forestry co-operatives. Initially, it focused on resin-tapping co-operatives, but then provided a platform for communities to secure a much wider range of benefits, including sale of roundwood and stakes, as well as sawing of timber, boards and poles.
>
> During the mid 1990s, SFF worked on the basis of long-term forest usufruct agreements between the State Forestry Administration–Honduran Forestry Development Corporation (AFE-COHDEFOR) and community-based forest organizations. Community usufruct agreements were later suspended, however, because of opposition by the private sector, which claimed that these agreements were, in reality, concessions. Concession contracts had previously been abolished under the 1992 Law of Agricultural Modernization, to be replaced by an auction system.
>
> In response to growing concern over the legal status of usufruct agreements, the subsequent political administration (January 1998–January 2002) instead developed forest management agreements. These new agreements have been signed with a significant number of community producer groups under SSF; but they expire at the end of each four-year political cycle, and their renewal is delayed and uncertain.
>
> Consequently, community groups currently have practically no secure legal rights of access to the areas that they manage, apart from a few indigenous groups with recognized land rights. The SSF has also suffered because of the decline in pine resin prices, over-complex regulation, the lack of administrative experience of the co-operatives, and the limited capacity of AFE-COHDEFOR to provide technical assistance and other support services to community groups.
>
> *Source*: Lazo (2001); Colindres (2002a); del Gatto (2003d)

Access to state forestland In Nicaragua

Access to state forestlands in Nicaragua is now subject to the 2003 Law for the Conservation, Growth and Sustainable Development of the Forest Sector, which seeks to clarify the existing mass of often contradictory legislation. This includes the 1993 Forest Regulation, which was the primary instrument governing the sector and the first attempt to rationalize forest law.

Under Chapter VII of the 2003 law, natural or juridical persons (save some exceptions defined in the constitution) may obtain commercial forest

> **Box 7.3 Grassroots protests against logging in Olancho, Honduras**
>
> The north-eastern Honduran department of Olancho has long been the country's main source of timber. For decades, licensed logging companies have carried out intensive timber extraction with few, if any, benefits for local communities and little or no concern for the environment. Because of the alleged collusion of local and national authorities, residents have been unable to influence the actions of timber companies and large local landowners, who have also resorted to intimidation and violence to defend their interests.
>
> Given increasing awareness about land degradation and its impacts on water supplies, mounting discontent surfaced over the last five to six years in the form of the grassroots environmental protest movement Movimiento Ambientalista de Olancho (MAO), organized by a coalition of religious leaders, environmental activists, community members and others concerned with the degradation of Olancho's forests.
>
> In June 2003 and again in June 2004, two Marches for Life were organized: six-day walks from Olancho and other parts of the country to the capital Tegucigalpa, joined by thousands of Honduran citizens. The marches were supported by dozens of religious, human rights, *campesino*, student and worker organizations. Their demands included an immediate halt to commercial logging in Olancho and other critical forest regions, an independent evaluation of the actual status of forest resources in these areas and the creation of an independent commission to monitor the cutting suspension and to facilitate its evaluation.
>
> *Source:* La Prensa (2003a, 2003b); Swedish (2003)

concessions on state land in accordance with the provisions of the new law and its implementing regulations (Article 43). Concessions are approved by the Ministry of Industrial Growth and Commerce. Under Article 21 of the Forest Law, it is also possible to obtain a permit to exploit natural forests on state land from the state forest agency INAFOR. This process is subject to the approval of a management plan, including prior consultation with local government representatives.

The ability of local communities to secure access through exploitation permits will depend upon the design of regulations and administrative procedures to implement the new law. This includes the development of simplified procedures for small-scale 10ha permits under Article 23. To date, local communities have had difficulty in securing permits and have, instead, resorted to innovative schemes to obtain a greater share of benefits. For example, in Bilwi, local community leaders (*síndicos*)[18] colluded with INAFOR officials and the municipality to operate a system of 'community permits' for Caribbean pine (*Pinus caribaea* var *hondurensis*) extraction. While strictly illegal, the community permit facilitated access by indigenous community members to their forest resources and raised their employment and income levels (Ampié Bustos, 2002).

Access to private and *ejidal* forests in Nicaragua and Honduras

On private or *ejidal* land, the major constraint is not access rights, but the transaction costs of complying with regulations governing the production and transport of timber. For example, under the 2003 Forest Law in Nicaragua, landowners are required to prepare management plans for exploitation of natural forests on their own land, as well as to obtain permits to transport timber (not a requirement for agricultural produce).[19] In municipal or *ejidal* areas of Honduras, the requirement for a management plan constitutes a significant constraint given limited means to meet technical specifications. Some municipalities have established municipal forestry offices with the responsibility for devising forest management and protection plans, as well as providing capacity-building and support to forest users and community enterprises (Vallejo Larios, 2003). Municipalities with limited technical capacity, however, have had difficulty in securing AFE–COHDEFOR approval of management plans and continue to suffer rampant illegal logging.

Volume constraints

Even when local communities are assigned national forestland, strict harvest limits may be imposed. Under Nicaragua's forest law, farmers have limited rights to harvest planted trees on their own land. In Honduras, community timber co-operatives are not entitled to the full annual permitted cut stipulated in management plans for their area (see Box 7.4). Levels of harvesting permitted under management plans may also not reflect how farmers, in reality, draw on forest resources to cope in drier years or economic crises (Nitlapán– University of Central America, 2002).

Unclear and complex regulations

The legal framework governing the forest sector comprises a complex mass of rules and regulations as new legislation has progressively altered the ideas underlying earlier legislation. When a new law is enacted, little effort is made to ensure that it is complete, and previous legislation is only partially repealed. The resulting body of law therefore contains contradictions and duplications, leading to arbitrary interpretation and confusion in law enforcement (Lazo, 2001).

Forest regulations in both countries also require actors to comply with unrealistic rules. Communities are asked to prepare highly technical forest management plans, even though they lack the financial resources and technical know-how to submit them. Compliance can take up several days of work, often

> **Box 7.4 Volume constraints in the approval process in Honduras**
>
> In accordance with Article 10 of the regulations attached to Chapter VI (Forestry) of the Law for Agricultural Modernization, peasant farmer collectives, acting under the umbrella of the Social Forestry System (SSF) and managing land in public forest areas, may not avail themselves of the full 100 per cent of the annual allowable cut (AAC) envisaged in the forest management plan for their area. Instead, they are restricted to only 200 cubic metres per organization per year in the case of broadleaf forest and to 1000 cubic metres per organization per year in the case of conifers. The unused portion of the AAC should then be auctioned off to the highest bidder.
>
> Two hundred cubic metres divided amongst 20 or 30 members in a collective leaves an insufficient profit margin for members to stay with the collective. There is little or no incentive for them to protect the forest from timber theft or clearance pressures if they are entitled to only a small portion (often less than 20 per cent) of the sustainable production of the forest in question.
>
> Since 1998, attempts have been made to apply this regulation more rigorously (even though there has not been a single broadleaf forest auction since then). The implications of this are serious:
>
> - a loss of motivation and cohesion among the groups;
> - a drastic fall in membership as many ex-members became involved in illegal extraction (see, for example, the decline in co-operative membership noted in Box 7.6);
> - growth in the influence and power wielded by local and outside intermediaries; and
> - the difficulty of producer groups building up capital, as well as a fall in household incomes.
>
> *Source:* del Gatto (2003d)

including wasted visits to distant offices (see Box 7.5). Compliance can appear meaningless when forest administrations obviously lack the capacity to control thousands of farmers, each logging and marketing a few trees. But as legal requirements proliferate, so too does the custom of demanding and paying backhanders, creating a confusing and frequently damaging bureaucratic jungle.

Overlapping or conflicting government responsibilities

Overlapping or conflicting forest governance responsibilities compound legal uncertainty and regulatory complexity. In Nicaragua, there are often

> **Box 7.5 Complex management plans as a regulatory barrier to legality for community-based forest producers in Honduras**
>
> The law in Honduras requires a management plan, duly approved by the State Forestry Administration–Honduran Forestry Development Corporation (AFE–COHDEFOR), for 'all market-oriented cut[s] or harvesting of forests'. It also makes AFE–COHDEFOR the legal body responsible for drawing up and implementing management plans in national forests. But AFE–COHDEFOR does not have the staff or the money to fill this role. It is no surprise that the same can be said of the communities entrusted, under the aegis of the Social Forestry System (SSF), with managing and exploiting a national forest, especially in view of the technical complexity involved.
>
> This situation has generally led to one of three scenarios:
>
> 1 A community draws up its management plan under the aegis of a project and, when the project comes to an end, drifts back into the same situation it was in before (i.e. a mixture of dependence upon intermediaries and illegal activity).
> 2 Not having access to any external help, a community fails in its bid to be admitted as part of the SSF and the local forest continues to be exploited in a thoroughly clandestine manner, usually on the orders of one or more intermediaries, until the supply of the most valuable timber is exhausted.
> 3 The management plan is financed by an outside intermediary (against an undertaking to repay the cost in timber). The intermediary then uses the local organization and its management plan as a front to mask his or her own direct access to forest resources.
>
> *Source:* del Gatto (2003d)

conflicts between central and regional 'autonomous' governments, such as the Autonomous Region of the North Atlantic, over the rights to assign harvesting permits (Ampié Bustos, 2002). The confusion is heightened by an unclear distribution of law enforcement responsibilities between forestry agencies and other state institutions, such as the police and judiciary. Such conditions create opportunities for corrupt public officials and corporate interests to 'breach the law, either unintentionally due to confusion of roles or unclear procedures, or intentionally by exploiting the incoherencies' (Global Witness, 2002). In Honduras, the 2002 Administration Simplification Act presents an opportunity to correct the effects of administrative and regulatory complexity; but further research is needed to understand how this law can be practically applied to the forest sector.

High transaction costs and low prices: The economic incentive to operate illegally[20]

Over-complex rules greatly increase the transaction costs of compliance, leaving legal producers vulnerable to market competition from lower-cost illegal timber. Although it is difficult to obtain completely credible data, evidence from Honduras suggests that the costs of producing and transporting illegal wood may be 75 per cent that of legally harvested wood (del Gatto, 2003a). Studies of Nicaragua show similar results and that the financial incentive to operate illegally is higher for precious woods. If a mahogany logger evades 60 per cent of taxes and 50 per cent of the cost of forest management plans and transport permits, the profit per cubic metre of sawn wood increases from about US$119 to US$200 or $250. This compares to an increase in profits by just US$12 per cubic metre of pine from around US$29 (Ampié Bustos, 2002). The case of the Cooperativa Agroforestal Colón, Atlántida Honduras Limitada (COATLAHL) in northern Honduras (see Box 7.6) illustrates how high transaction costs and 'unfair competition' from illegal loggers make legal production unattractive.

Box 7.6 Economic and social problems of the Cooperativa Agroforestal Colón, Atlántida Honduras Limitada (COATLAHL)

COATLAHL, a timber co-operative on the north coast of Honduras, has been supplied by small community groups in the surrounding mountains since 1977. Establishing legal and equitable forest management has been a costly foreign aid endeavour. Administrative, processing and storage inefficiencies, and increased regulatory and stumpage charges, as well as excessive transaction costs, have been compounded by 'unfair competition' from illegal logging. COATLAHL-affiliated groups have found it difficult to survive given their higher production and transaction costs, and only marginally higher sale prices, than illegal loggers. Increasing conflict and insecurity is also discouraging long-term forest management since timber is vulnerable to theft by armed bands. Ex-COATLAHL members have now switched to illegal logging as a livelihood option. They have effectively been criminalized by the economic consequences of illegal logging, as well as over-complex regulations. Since the early 1990s, the number of affiliated groups has halved and grassroots membership has fallen by 75 per cent.

Source: del Gatto (2003b)

The substantial financial incentive to operate illegally will continue so long as the costs of legal timber production are higher and such timber does not fetch a significantly higher price than illegal timber. Since legal wood intrinsically carries higher costs due to the need to comply with forest management plans, obtain transport certificates and so on, the cost of illegal wood needs to be increased through better detection and stiffer penalties, financial or otherwise. Such actions are also likely to improve the market price for legal operators by reducing the price-dampening effect of a market flooded with illegal wood. If, however, stiffer enforcement is not to harm the rural poor, it needs to be accompanied by efforts to reduce barriers to entry into legal forest management, including more secure rights in forest resources, a more realistic and predictable regulatory framework and simplified administrative procedures.

Failures in public administration: Compounding legal and policy barriers to legality[21]

Institutional corruption and distortion of decision-making

Surveys in both countries have identified corruption as a major problem. Transparency International (2002) ranked Nicaragua as the 21st and Honduras the 31st most corrupt countries globally. Recent World Bank-sponsored surveys in Honduras show that corruption is particularly severe in the judiciary and police (World Bank Institute, 2001). Measures to tackle corruption include the National Anticorruption Council and Anticorruption Strategy in Honduras (CNA, 2002).

Forestry-related corruption in Honduras and Nicaragua occurs at both political ('state capture') and bureaucratic levels. The latter includes facilitation payments to speed up regulatory and bureaucratic processes, as well as payments for officials to step outside of their legal mandate to ignore illegal acts.

State forestry agencies and local governments are more susceptible to distortion by illegal logging where forest fees constitute a major source of revenue. Logging companies and timber merchants work to disproportionately influence decision-making, cut through red tape and distort environmental monitoring procedures through a combination of credit, bribes and threats. Forestry officials and field technicians become party to elaborate informal arrangements with traders and community leaders in order to fraudulently 'legalize' production. This has knock-on effects for the broader network of institutions, regulations and values, including police force standards and public respect for the law.

Figure 7.2 depicts the informal arrangements, and associated power relations between intermediaries, officials and local elites governing illegal forest exploitation in the Sico-Paulaya Valley in Colón Province, Honduras, over the two-year period of 2000 to 2001 (see also Box 7.1).

Three sets of power relations are depicted in Figure 7.2, in which intermediary timber traders are instrumental in distorting decision-making:

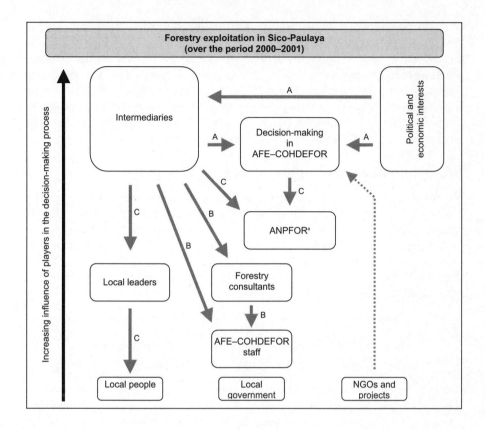

Figure 7.2 *Power relations governing forestry exploitation in the Sico-Paulaya Valley, Honduras*

Note: ᵃ National Association of Broadleaf Forest Producers.

Source: authors

1 *influencing public decision-making on law and policy ('state capture')* (arrows marked 'A') – in this case, to retain a policy granting community groups the right to harvest mahogany deadwood, when there was already ample evidence that this provided a front for legalizing illegal harvesting;
2 *influencing implementation by 'street-level bureaucrats'* ('B') – through a combination of bribes and threats; and
3 *influencing community institutions and their representative institutions* ('C') – to take control of the production chain and to obtain the necessary political backing to influence public policy.

Links to organized crime

In remote rural areas, a combination of timber, drugs, unemployed youth and arms appears to have led to the collapse of civil governance. Anecdotal evidence suggests that illegal logging is often linked to such criminal activities. The Sico-Paulaya Valley in Honduras is part of a drug trafficking route and has become a refuge for people involved in criminal activities in urban centres (such as kidnappings, armed assaults and car robberies). This has generated significant sums of ready cash, which can be conveniently invested in cutting and selling mahogany. One raid on an unregistered sawmill uncovered illegal timber, hijacked lorries, stolen goods and firearms – *prima facie* evidence of the links between illegal logging and organized crime.

Weak penalties and enforcement

No law is applied in full, especially where it threatens the interests of powerful groups. Most fines are substantially lower than the potential benefits of operating illegally. Weak inspection and enforcement mean that there is a low probability of being caught: 'crime pays'. Where law enforcement occurs, it usually takes the form of crackdowns with little, if any, regard to the conflicts that this action may generate at the local level or other poverty impacts that it may have (see Box 7.7).

> ### Box 7.7 Crackdowns on illegal logging in the Rio Platano Biosphere Reserve
>
> During recent years, illegal logging in the Rio Platano Biosphere Reserve has attracted intense public and media attention. The response has commonly been the organization of field policing operations aimed at detecting and suppressing illegal activities, usually with the participation of different public institutions and involving relatively large numbers of forest law enforcement officers, police and army soldiers. The results have rarely met expectations, especially when compared with the costs of such operations.
>
> The impacts on poor local people may be significant. Typically, undocumented timber is held by logging crews or small community-level intermediaries, while more powerful intermediaries usually secure at least some 'legalizing' paperwork before taking charge of timber. Thus, most of the confiscated timber belongs to poorer stakeholders in the production chain, who lack equivalent income-generating opportunities. In contrast, influential intermediaries suffer little, if any, of the consequences, apart from delays in the scheduled supplies of timber.

Weak penalties and poorly targeted enforcement reflect a variety of failures in public administration, including lack of financial and human resources. Box 7.8 describes a typical situation. Loggers and companies have sometimes paid forestry officials for the logistical and subsistence costs of performing their official duties, casting doubts on official impartiality. Where forestry agencies depend upon forest revenues, revenue collection rather than forest law enforcement tends to become the main priority. The presence of armed gangs and drug traffickers, especially in agricultural frontier areas, makes law enforcement a dangerous and undesirable task for badly paid government staff. Enforcement officers in remote areas are highly vulnerable to pressure and bias when they come to report on cutting operations.

Box 7.8 Law enforcement in a Nicaraguan frontier region

The National Forestry Institute (INAFOR) office of Puerto Cabezas municipality in Nicaragua has one forestry officer, two assistants and a secretary. There is a single motorbike for transport. This team has to enforce the forest law in this and in the neighbouring Waspam municipality, a total of more than 1.5 million hectares. Attempts to enforce the law are subject to death threats; salary levels do not justify taking such risks.

Source: Ampié Bustos (2002)

Lack of information, transparency and accountability

The lack of information, transparency and accountability also compounds corruption and organized crime. Public knowledge about forest resources and their management is scarce in both countries. Forest inventories are incomplete or out of date, and the public have limited access to them. Forestry authorities lack the resources and data to control the implementation of forest management plans and to monitor forest conditions. Lack of information and limited public access to the information that exists mean that NGOs find it hard to adopt a public watchdog role.

Impacts on rural livelihoods[22]

The importance of illegal logging as a livelihood option

Illegal logging constitutes a livelihood diversification strategy. As explained in the section on 'Forests and rural livelihoods in Honduras and Nicaragua', the non-farm economy is becoming increasingly important for rural households across Latin America, in part due to the decline in the small-farm economy. Timber extraction presents an important livelihood opportunity in remote areas, where opportunities to access credit for agricultural accumulation and other forms of non-farm employment are limited, and where mobility constraints (such as lack of skills) prevent migration to urban areas (Richards et al, 2003b).

Income and employment opportunities

Timber companies and traders provide employment opportunities for poorer rural households. Remuneration is inadequate, however. In Nicaragua, local co-operatives and forest owners receive only 5 to 10 per cent of the timber's gross value, whether extracted legally or illegally. Most of the benefits go to the *síndicos*, loggers, truckers, traders and export companies.

Furthermore, while illegal logging may inject large amounts of cash into an area, the livelihood benefits are thin and temporary. In the Sico-Paulaya Valley, Honduras, legalized and clandestine production of about 11,000 cubic metres of mahogany over 2000 and 2001 brought local people about US$1.2 million in wages and profits. Over half the income was grabbed by powerful community members, leaving only about one quarter for poorer groups. Table 7.1 sets out the profits, and how they were ultimately used, for one of the main co-operatives involved. Misappropriation of funds by elites and the purchase of chainsaws for the cutting operation itself meant that only 6 per cent of the value of the timber produced could ultimately have been invested in local livelihoods. The poorest households captured very little of this, mainly as wage labour sawing and running timber down rivers.

Delayed and sporadic payments by timber traders further reduced the possibility of investing in capital assets, such as tools and equipment. After two years of intensive activity, little or nothing was left of the money saved either in cash or liquid assets (livestock, jewellery, etc.). Several chainsaw operators and local organizations ended the period significantly in debt.

This study did not specifically look at the livelihood impacts on women and children. But evidence from Bilwi, Nicaragua, suggests that they benefit little from timber income, with the men spending most of the money on such items as alcohol and cigarettes. It is estimated that only 30 to 40 per cent of their total income is spent on family welfare (Ampié Bustos, 2002).

Physical security, collective action and participation

Illegal logging is contributing to an increasingly 'uncivil' society in remote areas, where state presence is already weak. In Rosita, Nicaragua, armed gangs make their living by threatening loggers and traders. Violence is compounded by insecure tenure and conflicts over land. In the Sico-Paulaya area of Honduras, many villagers invested in arms for family protection during the illegal logging boom. Because it is difficult and dangerous for forestry officers or the police to indict armed and powerful actors, the poor are often the first to be accused of illegal logging.

Case studies from both Nicaragua and Honduras also show how infiltration of community organizations by timber merchants can divide and alienate a community. In the Sico-Paulaya Valley in Honduras, local people accused each other of 'grassing' to the authorities, and rival chainsaw gangs emerged. Grassroots membership of the COATLAHL co-operative in Honduras has fallen by 75 per cent since the early 1990s as more and more find themselves

Table 7.1 *Gross and net income and its use by the Collective Society of Romero Barahona and Associates, Sico-Paulaya, Honduras, 2000 and 2001*

Description	Value in lempiras	Value in US$	Percentage of gross income*	Percentage of net income	Comments
Gross income	4,142,250	257,680	100		Assuming lempiras of 315,000 per 100m^3
Total cost of production and commercialization	3,350,500	220,870	86		
Net income	591,750	36,810	14	100	
Misappropriation of funds by community leaders	220,000	13,685		37	Reported misappropriation by three community leaders for personal use
Remainder distributed within the group	371,750	23,125			
Purchase of chainsaws	118,000	7340		20	Nine new chainsaws bought at 12,000 lempiras each, and two second-hand chainsaws bought at 5000 lempiras each
Remainder for other uses	253,750	15,785		43	

Notes: US$1 = 16,075 lempiras (February 2002).

*The gross income column shows the division of gross income between costs and net income. The net income column indicates how this net income (US$36,810) was spent: 37 per cent misappropriation, 20 per cent chainsaws and 43 per cent other uses.

forced into illegality. In Nicaragua, illegal logging has corrupted indigenous leaders and eroded traditional institutions. Erosion of community social capital reduces resilience to outside shocks.

Long-term supply of forest goods and services that poor households use

Illegal logging is often the first step in a downward ecological spiral, leading to the degradation of forest ecosystems and, eventually, deforestation. As forest ecosystems degrade and farm productivity gradually declines due to the loss of local ecosystem benefits, it will be the poorest farmers who cannot afford cattle and who are forced to move farther afield, into remoter areas where health and education facilities are weakest or non-existent. Illegal logging also creams off the most valuable species, removing future sources of income generation for the rural poor (Campos et al, 2001; del Gatto, 2003e).

Conclusions and policy recommendations

This case study found that barriers to legality within the legal, regulatory and institutional framework governing forests leave small-scale rural forest producers with few options but to break or avoid the law. The transaction costs of complying with complex and unrealistic regulations mean that legal forest production is often uneconomic – especially in markets flooded with cheaper illegal timber. Weak public institutions, ineffective state enforcement capacity, administrative confusion and low penalties create opportunities for corruption and organized crime. This compounds the barriers to entry for small-scale forest producers by distorting decision-making processes and increases their reliance on powerful patrons in order to secure access to resources, credit and markets.

The results of this study were discussed in a series of stakeholder consultations in Honduras and Nicaragua between August 2002 and February 2003. Discussions prioritized the need for tackling discriminatory and over-complex regulation, for better targeted and more effective law enforcement, and for measures to reduce institutional corruption.

Measures to reduce legal barriers to entry for small-scale forest producers need to focus on:

- Strengthening rights in land and resources, including demarcation of indigenous land boundaries, land titling for migrant families and allocation of secure community concessions. Evidence from Nicaragua suggests, however, that small-scale forest owners with title are equally vulnerable to capture by the illegal timber sector if other legal and institutional constraints to production are not addressed simultaneously.
- Removing volume constraints where these cap harvesting by small-scale producers at a level below the annual allowable cut, as well as limitations on the rights of farmers to plant and harvest trees on their own land.
- Simplifying administrative procedures, including approvals for forest management plans, to reduce transaction costs and to increase the returns from legal forest management for small-scale producers.

Measures are also required to aim law enforcement at actors with genuinely corrupt and criminal intent, given that they compound existing legal and institutional barriers by capturing and distorting decision-making structures. In particular, measures are required to increase the costs and risks of illegal logging through better detection and stiffer penalties. Stakeholders suggested:

- increasing transparency, data collection and access to information in order to make illegal logging and other forest crimes more difficult to hide;
- strengthening monitoring, in partnership with NGOs and local community groups;
- reducing the financial dependence of the state forestry agencies upon forest fees, currently an incentive for forest officials to collude with illegal timber traders; and
- depoliticizing the selection of senior forestry officials in order to prevent state capture of public decision-making processes.[23]

Stakeholder consultations also stressed that reforms are contingent upon:

- enhancing civil society participation in decision-making, including through decentralized policy fora; and
- adequate incentives for sustainable forest management, including payments for environmental services and certification (controlling illegal logging can increase market prices and therefore the returns to legal operators).[24]

Yet, the challenges to improved law enforcement in the forest sector are so great in these countries that it is doubtful whether the governments alone can accomplish much, and support from international donors, based on a sound understanding of the political power relationships, is essential. The reform agenda requires political mobilization and action by actors both inside and outside the government, including the industrial sector, the general public, forest communities, NGOs and consumer groups in importing countries.

Finally, a regional trade flow study (del Gatto, 2003c) and other literature (Richards, 2003) show that policy prescriptions also require coordination at the regional level, given that better control of illegal logging in one country (e.g. Costa Rica) increases demand pressures on illegal logging in normally poorer neighbouring countries with weaker regulations (e.g. Nicaragua).

Acknowledgements

This chapter has been possible due to the efforts of the field research staff, often working in dangerous social environments, including Gilberto Alcocer; Eduardo Ampié, Oscar Castillo, Ibis Colindres; Danilo Dávila; Jaime Guillén; Arístides Jiménez; Arnoldo Paniagua; and Abelardo Rivas. We are also grateful for comments by Nalin Kishor and Bill Hyde on an early draft of the economic calculations. The case study is based on a research study coordinated by the

Overseas Development Institute, and funded by the UK Department for International Development, the World Bank and the Canadian International Development Agency. The study was implemented in Honduras mainly by the Honduran Network for Broadleaf Forest Management, with official backing from the State Forestry Administration and the Honduran Federation of Agroforestry Co-operatives. In Nicaragua, the main collaborator was the Nicaraguan NGO NICAMBIENTAL, with official backing from the National Forestry Institute. The study received technical assistance from the United Nations Food and Agriculture Organization, as well as from Global Witness and the Royal Institute of International Affairs.

Notes

1. This chapter is based on research carried out in collaboration with Gilberto Alcocer, Eduardo Ampié, Oscar Castillo, Ibis Colindres, Danilo Dávila, Jaime Guillén, Arístides Jiménez, Arnoldo Paniagua and Abelardo Rivas.
2. Covering about 47 per cent of Honduras and 48 per cent of Nicaragua (Harcourt and Sayer, 1996).
3. About 3.5 per cent (80,000ha to 100,000ha) in Honduras and 2.3 per cent (70,000ha to 75,000ha) in Nicaragua.
4. See the 'Acknowledgements'.
5. In Nicaragua, 70 per cent of rural residents are poor, 29 per cent of whom are extremely poor (see www.poverty.worldbank.org/files/Nicaragua_PRSP.pdf). In Honduras, 75 per cent of rural households are poor and 61 per cent of these are extremely poor (see www.poverty.worldbank.org/files/Honduras_PRSP.pdf).
6. See www.poverty.worldbank.org/files/Nicaragua_PRSP.pdf.
7. Such as the Tawahka, Miskito, Pech and Garifuna in northern and eastern Honduras, and the Miskito, Mayagna and Rama in the Atlantic Coast regions of Nicaragua.
8. This section is based on del Gatto (2003a) and Richards et al (2003a).
9. This section is based on del Gatto (2003a) and Alcocer López (2003).
10. For Honduras, data from a 1987 inspection survey of timber quantities in processing plants were available.
11. The clandestine cut can be crudely estimated as the residual of demand (composed of national consumption and exports) less the sum of official supply and imports. One problem of this calculation is the difficulty of estimating clandestine imports and exports.
12. According to a recent investigation carried out by the Environmental Investigation Agency (2005), in Honduras two of the main buyers of 'legalized' or clandestine mahogany, often sourced from protected areas at the Rio Platano Biosphere Reserve, are Milworks International and Caoba de Honduras. Both companies export luxury mahogany products to the US.
13. This section is based on Ampié Bustos (2002); Colindres (2002b); del Gatto (2003e); Paniagua (2003); NICAMBIENTAL (2003); REMBLAH (2003).
14. A *síndico* is an indigenous community leader in the northern Atlantic region of Nicaragua.
15. This section is based on Colindres (2002a); Contreras-Hermosilla (2003); del Gatto (2003d); Pommier (2003).

16 Land titling is currently the responsibility of the Office of Rural Titling, formerly INRA under Law No 14, 'Amendment to the Agrarian Reform Law' 1986, and before that MIDINRA under the Agrarian Reform Law 1963.
17 The acronym AFE was added after the approval of the Law of Agricultural Modernization to reflect the new title of State Forest Administration (Administración Forestal del Estado) applied by that law to COHDEFOR.
18 Leader of a community or group of indigenous communities in the Atlantic north of Nicaragua.
19 Perversely, if a farmer were to replace natural trees with planted trees, this would count as a 'plantation' – exploitation of which is not subject to permits and fees (although a transportation permit is still required) (Nitlapán–University of Central America, 2002).
20 This section is based on Ampié Bustos (2002); del Gatto (2003a, 2003b); Paniagua (2003).
21 This section is based on Contreras-Hermosilla (2002); del Gatto (2003b); Pommier (2003).
22 This section is based on Ampié Bustos (2002); Alcocer López (2003); del Gatto (2003b); Pommier (2003); Richards et al (2003a, 2003b).
23 This has been achieved in Bolivia, where the head of the Superintendencia Forestal is selected through a transparent process and appointed for six years, straddling the presidential term of five years. This partly protects the agency from political interference and corruption (see also Chapter 9). In Honduras, similar reforms have been achieved outside the forest sector, as in the case of the Supreme Court of Justice. Such experiences should inform forest institutional reform.
24 This will also require parallel investments in support of the wider rural economy, including human capital development, low-technology agricultural inputs (as promoted by the *campesino-a-campesino* movement), agricultural extension, niche markets and pro-poor tourism (Richards et al, 2003b).

References

Alcocer López, G. (2003) *Estimación de los costos económicos de la tala y el comercio ilegal para la economía nacional de Nicaragua*, Consultancy report for the project Illegal Logging in Central America – Tackling Its Impacts on Governance and Poverty, London, REMBLAH–FEHCAFOR–NICAMBIENTAL–FAO–ODI

Ampié Bustos, E. (2002) *La producción forestal no controlada en el Municipio de Puerto Cabezas, Región Atlántico Norte*, Consultancy report for the project Illegal Logging in Central America – Tackling Its Impacts on Governance and Poverty, London, REMBLAH–FEHCAFOR–NICAMBIENTAL–FAO–ODI, www.talailegal-centroamerica.org

Broegaard, R., Heltberg, R. and Malchow-Møller, N. (2002) *Property Rights and Land Tenure Security in Nicaragua*, Copenhagen, Centre for Economic and Business Research, www.econ.ku.dk/heltberg/Papers/landtenureNicaragua.pdf

Campos Arce, J. J., Camacho Calvo, M., Villalobos Soto, R., Rodriguez, C. and Gomez Flores, M. (2001) *La Tala Ilegal en Costa Rica. Un análisis para la discusión*, Turrialba, Costa Rica, Informe elaborado por el Centro Agronómico Tropical de Investigación y Enseñanza (CATIE) para la Comisión de Seguimiento del Plan Nacional de Desarrollo Forestal

CNA (Consejo Nacional Anticorrupción) (2002) *Estrategia Nacional Anticorrupción*, Tegucigalpa, Honduras, www.worldbank.org/wbi/governance/honduras/pdf/hon_estrategia-ac.pdf

Colindres, I. (2002a) *Contexto social, institucional y político del subsector forestal en Honduras*, Consultancy report for the project Illegal Logging in Central America – Tackling Its Impacts on Governance and Poverty, REMBLAH–FEHCAFOR–NICAMBIENTAL–FAO–ODI, www.talailegal-centroamerica.org/

Colindres, I. (2002b) *La Zona Sur de la Biosfera del Río Plátano: La Madera de Caoba un Recurso en Disputa*, Case study for the project Illegal Logging in Central America – Tackling Its Impacts on Governance and Poverty, REMBLAH–FEHCAFOR–NICAMBIENTAL–FAO–ODI, www.talailegal-centroamerica.org/

Contreras-Hermosilla, A. (2002) 'Policy and legal options to improve law compliance in the forest sector', in *Proceedings, Reforming Government Policies and the Fight Against Forest Crime*, Rome, 14–16 January 2002, United Nations Food and Agriculture Organization, pp43–91

Contreras-Hermosilla, A. (2003) *Barriers to Legality in the Forest Sectors of Honduras and Nicaragua*, Consultancy report for the project Illegal Logging in Central America – Tackling Its Impacts on Governance and Poverty, REMBLAH–FEHCAFOR–NICAMBIENTAL–FAO–ODI, www.talailegal-centroamerica.org/

del Gatto, F. (2003a) *La producción forestal no controlada en Honduras. ¿Qué es? ¿Cuánta es? ¿Y cuánto cuesta? Unas respuestas preliminares*, Consultancy report for the project Illegal Logging in Central America – Tackling Its Impacts on Governance and Poverty, REMBLAH–FEHCAFOR–NICAMBIENTAL–FAO–ODI, www.talailegal-centroamerica.org/

del Gatto, F. (2003b) 'The impacts of unregulated forestry production in Honduras', Briefing paper for the project Illegal Logging in Central America – Tackling Its Impacts on Governance and Poverty, REMBLAH–FEHCAFOR–NICAMBIENTAL–FAO–ODI, www.talailegal-centroamerica.org/

del Gatto, F. (2003c) *El Comercio No Documentado de Madera en Centro América: Comparación de Datos Estadísticos y Evidencia Anecdótica*, Consultancy report for the project Illegal Logging in Central America – Tackling Its Impacts on Governance and Poverty, REMBLAH–FEHCAFOR–NICAMBIENTAL–FAO–ODI, www.talailegal-centroamerica.org/

del Gatto, F. (2003d) 'The forestry sector in Honduras: the legal barriers', Briefing paper for the project Illegal Logging in Central America – Tackling Its Impacts on Governance and Poverty, REMBLAH–FEHCAFOR–NICAMBIENTAL–FAO–ODI, www.talailegal-centroamerica.org/

del Gatto, F. (2003e) *'¡El Magnate de la Maderiada Soy Yo!' Defraudando el Sistema Social Forestal en el Valle del Río Paulaya*, Case study for the project Illegal Logging in Central America – Tackling Its Impacts on Governance and Poverty, REMBLAH–FEHCAFOR–NICAMBIENTAL–FAO–ODI, www.talailegal-centroamerica.org/

Environmental Investigation Agency (2005) *The Illegal Logging Crisis in Honduras*, Washington, DC, Environmental Investigation Agency

FAO (United Nations Food and Agriculture Organization) (2000) *Bibliografía comentada – Cambios en la cobertura forestal: Nicaragua*, Rome, FAO, www.fao.org/forestry/fo/fra/docs/Wp34_spa.pdf

FAOSTAT (2002) *Bilateral Trade Matrices: FAOSTAT Forestry Data, Food and Agriculture Organization of the United Nations*, Rome, United Nations Food and Agriculture Organization, http://apps.fao.org/cgi-bin/nph-db.pl?subset=forestry

Friends of the Earth (1999) *The IMF: Selling the Environment Short*, Washington, DC, Friends of the Earth
Global Witness (2002) *Independent Forest Monitoring and Support to Forest Law Enforcement*, London, Global Witness
Harcourt, C. S. and Sayer, J. A. (eds) (1996) *The Conservation Atlas of Tropical Forests: the Americas*, New York, Simon and Schuster
La Prensa (2003a) 'Si el padre Tamayo no se va hoy del país lo matarán', *La Prensa*, Tegucigalpa, 30 May
La Prensa (2003b) 'Garantizan vida del padre Tamayo', *La Prensa*, Tegucigalpa, 4 June
Lazo, F. (2001) *Honduras Country Profile for the 'Forum on the Role of Forestry in Poverty Alleviation'*, Facilitated by the Forestry Department, Rome, United Nations Food and Agriculture Organization, September
McSweeney, K. (2003) 'Tropical forests as safety nets? The relative importance of forest product sale as smallholder insurance, eastern Honduras', Paper presented at International Conference on Rural Livelihoods, Forests and Biodiversity, May, Bogor, Indonesia, Centre for International Forestry Research
NICAMBIENTAL (Nicaraguan Society for the Conservation of Nature and Environmental Restoration) (2003) *La Producción Forestal No Controlada en El Municipio De Rosita*, Case study for the project Illegal Logging in Central America – Tackling Its Impacts on Governance and Poverty, REMBLAH–FEHCAFOR–NICAMBIENTAL–FAO–ODI, www.talailegal-centroamerica.org
Nitlapán–University of Central America (2002) 'El Bosque como ahorro para tiempos difícíles y como inversión para dejar la pobreza', Equipo Recursos Naturales, Managua, *El Nuevo Diario*, 24 May
Paniagua, A. (2003) *La Producción Forestal No Controlada en Rio San Juan, Nicaragua*, Case study for the project Illegal Logging in Central America – Tackling Its Impacts on Governance and Poverty, London, Overseas Development Institute, www.talailegal-centroamerica.org/
Pommier, D. (2003) 'Barriers to legal compliance and good governance in the forest sector, and impacts on the poor in Nicaragua', Briefing paper for the project Illegal Logging in Central America – Tackling Its Impacts on Governance and Poverty, London, Overseas Development Institute, www.talailegal-centroamerica.org/
REMBLAH (Honduran Network for Broadleaf Forest Management) (2003) *Diagnóstico de la Producción Forestal Ilícita en El Departamento De Atlántida*, Case study for the project Illegal Logging in Central America – Tackling Its Impacts on Governance and Poverty, London, Overseas Development Institute, www.talailegal-centroamerica.org/
Richards, M. (2003) 'Higher international standards or rent-seeking race to the bottom? The impacts of forest product trade liberalisation on forest governance', Paper presented at FAO Expert Consultation on Trade and Sustainable Forest Management – Impacts and Interactions, 3–5 February, Rome, United Nations Food and Agriculture Organization
Richards, M., del Gatto, F. and Alcocer López, G. (2003a) *The Cost of Illegal Logging in Central America: How Much Are the Honduran and Nicaraguan Governments Losing?*, Consultancy report for the project Illegal Logging in Central America – Tackling Its Impacts on Governance and Poverty, London, Overseas Development Institute, www.talailegal-centroamerica.org/
Richards, M., Maxwell, S., Wadsworth, J., Baumeister, E., Colindres, I., Laforge, M., López, M., Noé Pino, H., Sauma, P. and Walker, I. (2003b) *Options for Rural Poverty*

Reduction in Central America, ODI Briefing Paper, London, Overseas Development Institute

Suazo, J., Walker, I., Ramos, M. and Zelaya, S. (1997) 'Políticas Forestales en Honduras: Análisis de las Restricciones para el Desarrollo del Sector Forestal', in Seguro, O., Kaimowitz, D. and Rodríguez, J. (eds) *Políticas Forestales en Centro América*, Bogor, IICA-Holanda/LADERAS-CA/CCAB-AP/Programa Frontera Agrícola/CIFOR, pp230–267

Swedish, M. (2003) 'Environmental activist murdered, priests threatened', in *Central America and Mexico Report*, Washington, DC, Religious Task Force on Central America and Mexico

Transparency International (2002) *Corruption Perceptions Index 2002*, Berlin, Transparency International

Vallejo Larios, M. (2003) 'Gestión forestal municipal; una nueva alternativa para Honduras', in Ferroukhi, L., Larson, A. and Pacheco, P. (eds) *La Gestión Local Forestal Municipal en América Latina*, Bogor, Indonesia, Centre for International Forestry Research, and Ottawa, International Development Research Centre, www.cifor.cgiar.org/publications/pdf_files/Books/La_gestion27.pdf

World Bank Institute (2001) *Honduras Governance Diagnostics*, Washington, DC, World Bank Institute, www.worldbank.org/wbi/governance/honduras/

8
Livelihoods and the Adaptive Application of the Law in the Forests of Cameroon

Guillaume Lescuyer

Introduction

Cameroon is generally seen as a country with abundant natural resources, especially with its dense rain forest covering around 17.5 million hectares. The deforestation rate is estimated at around 0.9 per cent per year, with an annual loss of roughly 220,000ha over the last decade. In spite of the range of resources, however, most Cameroonians are impoverished. According to the United Nations Development Programme (UNDP) *Report on Human Development*, Cameroon is ranked 142nd, at the very beginning of the low human development countries category. Three indicators are given to justify this ranking: life expectancy, 48 years; adult literacy rate, 72.4 per cent; gross domestic product per inhabitant, US$559. The situation is all the more worrying in that the UNDP index has decreased over the last decade, mainly because of the negative evolution of the gross domestic product per inhabitant ratio of around −0.6 per cent per year.

Natural resources therefore have a specific place on the political agenda with a view to restarting economic growth in rural areas, which are the most stricken by poverty. Of the total population of around 16 million Cameroonians, rural populations form the majority. An important characteristic of the livelihood systems of rural dwellers in the humid forest zone is their direct (hunting, non-timber forest product gathering, etc.) and indirect (employment, ecological functions, etc.) reliance on forest products. A wide range of products are utilized either directly for their daily lives or to generate important income, especially for minority groups such as women or pygmies. For the rural poor of Cameroon, the forest resources constitute an essential basis for any socio-economic development. This stake is taken into account in the new forestry and environmental regulations of Cameroon.

During the mid 1990s, a new legal and regulatory framework was introduced to improve forest resource utilization, with the dual objective of economic development and sustainable management. The involvement of local populations in decisions concerning forestry management was also highly recommended.

All of these objectives are ratified by two major documents: Law No 94/01 of 20 January 1994, which lays down forestry, wildlife and fisheries regulations, and its Decree of Implementation No 95/531 of 23 August 1995. These documents were complemented by the Master Law 96/12 of 5 August 1996 relating to environmental management. These ambitious objectives and the tentative application of an innovative legislation have put Cameroon in the limelight. It is viewed by international agencies as something of a test bed for interventions in the forestry sector, with the implication that these may be replicable, at least in other Congo Basin countries. This potential model status means that it is worth investigating how these forestry regulations are being applied, in practice, in Cameroon and their impacts on rural livelihoods.

This chapter analyses the impacts arising from the enforcement, as well as the lack of enforcement, of the forestry regulatory framework on rural livelihoods at the local scale, and more precisely on a local council scale. Other forest-related laws, such as protected area laws, land tenure laws, and human rights laws, are not considered. Given space limitations, the detailed focus on the application of the law precludes the analysis of the 'macro' players (e.g. central services at ministries, World Bank, aid agencies and logging industry lobbies), who were influential in the design of the law (Ekoko, 2000) and, obviously, influence its implementation. The chapter also does not consider how forest regulations influence forest practices.

The focus chosen for the chapter presents the advantage of circumscribing our attention around the main forest stakeholders according to the law and to ignore other secondary actors (e.g. planters and industrial projects). But the approach also has the drawback of favouring a normative view of reality, with the danger of understating the tactics and ways of working of the different actors in their use of the law. The field surveys carried out for this study attempt to minimize this risk; but they probably do not reflect well enough the capacities of the stakeholders to adapt themselves to new legal prescriptions.

The chapter is organized around the five main law-related issues:

1   classification of the permanent forest estate;
2   rights related to hunting and gathering of non-timber forest products (NTFPs);
3   commercial timber production;
4   small-scale timber production; and
5   community forests.

Each of these topics is the subject of one section of the chapter. Finally, the concluding section summarizes the key lessons, makes policy recommendations and identifies four research issues.

Methods

Several illustrative sites were selected in southern Cameroon in order to meet stakeholders and to gather data (see Table 8.1). Because of time constraints, only eight days were spent in each study area. These field meetings were complemented by discussions in Yaoundé, the capital of Cameroon, with several (technical, international and research) institutions concerned with application of the forest law.

Table 8.1 *Study areas*

Site name	Main characteristics
Campo Ma'an (Sud Province)	National park surrounded by timber production forests and many villages Community forest projects
Djoum (Sud Province)	Buffer zone of the Dja natural reserve One council forest, 15 community forest projects Small-scale timber production
Lokoundjé-Nyong (Sud Province)	One timber production forest that is classified and used according to a management plan Disputes with local populations
Lomié (Est Province)	Functioning community forests Buffer zone of the Dja reserve Presence of Baka pygmies
Ngambe-Tikar (Centre Province)	Three functioning community forests, five planned Small-scale timber production No zoning plan

A common field methodology was applied in four steps:

1. review of the literature regarding the site;
2. selection of key informers in each category of players;
3. semi-structured interviews (one to one and a half hours); and
4. visits to logging areas, sawmills and village infrastructures.

At each site, personal interviews were conducted with different stakeholders: farmers, village elites, association chairmen, illegal pit sawyers, ministry decentralized services, NGOs, elected representatives, administrative authorities and timber companies.

Table 8.2 *One common tool to assess law contributions to livelihoods*

		Contribution to rural livelihoods	
		Positive	**Negative**
Legal instrument	Implemented	Appropriate implementation of the law	How should the law and/or local practices be changed?
	Non-implemented	How should the law and/or local practices be changed?	How should the law be applied?

The following sections are more or less similarly structured, describing

1. the specific legal and regulatory mechanisms;
2. their enforcement (or not) in the field; and
3. their impact on rural livelihoods, with the information summarized in a table whose generic format is presented in Table 8.2.

The ideal situation for the each legal instrument would be that its appropriate application leads to positive impacts on livelihoods. However, two other situations may occur. First, the implementation of the law has negative impacts on rural livelihoods or its non-implementation has positive impacts. In both cases, measures need to be found to change the law and/or local practices. Second, non-implementation of the law produces harmful effects on rural livelihoods. The question then becomes how to apply the law.

Of course, the framework summarized in Table 8.2 offers a simplistic view of reality since a legal instrument is never fully implemented or fully unimplemented. Likewise, the contribution of one forest law prescription hardly has one single clear influence; most often, it produces secondary impacts or undesirable effects that reduce its (positive or negative) scope. Nevertheless, the framework adopted here is useful in depicting the most pronounced impacts of the forest law on rural livelihoods. Its purpose is to display the major trends observed in the field, even if reality is more complex.

The difficult steps of zoning and classifying permanent forest

The letter of the law

The forest law provides for two types of status for forested areas, with different types of management, as depicted in Table 8.3.

Table 8.3 *Forest classification in Cameroon*

National forest area				
Permanent forest (obligation of forest management)		Non-permanent forest (no obligation of management, except community forests)		
State forests	Council forests	Communal forests	Community forests	Private forests

The permanent forest estate (PFE) is the private estate of the state (or local council) and is designated to remain forested in the long term. Shifting cultivation is strictly forbidden in permanent forests and local use of forest resources is restricted. These forests must first be reserved and then managed under the supervision of the Ministry of Forestry (hereafter referred to as the ministry). The forest law (Article 24) proposes several kinds of PFE, the main types being production forest, which is dedicated to timber exploitation, and protection forest, which conserves the natural environment.

The non-permanent forest estate (NPFE) consists of forestland that may be allocated to uses other than forestry. It includes communal forests, community forests and forests belonging to private individuals. Communal forests, which represent almost all of the NPFE, are neither gazetted nor subject to specific management plans. Most of the time, they are managed according to local 'traditional' rules. On the contrary, community forests must be used according to simple management plans. It is in the NPFE that most of the villagers' uses of natural resources must take place, especially all agricultural activities.

So far, the division between PFE and NPFE has been proposed only for southern Cameroon. In 1995, the state approved a zoning plan. Of the 14 million hectares planned, about 9 million hectares are proposed to constitute the PFE, of which production forests account for 6 million hectares. Permanent forests will be created in those forest areas considered free of any human occupation. The main element in delineating these 'free' forest zones is the anticipated extension of cultivated areas in view of demographic extrapolations. To this end, each village has been given a buffer zone sized according to the number of its inhabitants, which will serve as an agricultural land reserve for future fields and plantations. A minimal buffer strip has also been established along main roads. All other forest spaces are proposed to be included in the PFE. Hunting, Non-timber forest products (NTFP) gathering, fishing and ancient village sites have been completely overlooked in this technocratic approach, and, according to the zoning plan, most of these activities and sites are now located within the PFE.

How the classification procedure is applied in the field

The classification of the permanent forest has a direct and substantial impact on rural livelihoods since it restricts or prohibits local use of large forest areas.

It therefore meets several difficulties when applied at village level.

An approach that contradicts the local representation of forest space

The zoning plan reflects a particular definition of how forest should be managed: the sustainable use of resources requires a strict spatial delineation of the permanent forest. This concept is prevalent in sub-Saharan Africa, at least in forest administrations, and is at variance with local populations' perception of their forest areas. This considerable divergence regarding forest representations and associated modes of exploitation is one major difficulty in achieving sustainable forest management in Central Africa (Diaw and Oyono, 1998; Karsenty and Marie, 1998). This results in the underassessment of local uses in official management plans for permanent forests.

What makes a land claim legitimate?

According to the ministry staff, the classification procedure provides for compensation for the loss of customary property rights held by the local population in only three cases: operating plantations, holy sites and areas covered by landownership deeds. In these three cases, locals are entitled to apply for financial compensation (rarely granted) or may claim for a boundary shift. No other claims (food crops, hunting campsites, community forest, etc.) warrant changes to the zoning plan proposal. Consequently, in large parts of the country, local populations find their forest resources absorbed within the PFE, which implies that their rights will be curtailed or even withdrawn.

An effective change in the permanent forest's boundaries

In spite of its bias against rural populations' concerns and interests, the classification process may lead to substantial changes in the outlines of the permanent forest. Villagers' claims are never entirely accepted; but authorities do not generally dismiss them all out of hand. In practice, rural populations generally ask for an extension of the agro-forestry area that takes the form of a 5km wide buffer zone on each side of the road (Lescuyer and Emerit, 2005). This request, which would excessively reduce the area of permanent forest, is not accepted as is by authorities; but it serves as an argument to make marginal changes to the boundaries on the basis of the most urgent and legitimate claims. On average, the outcome is a 15 per cent decrease in the area of the permanent forest. Apart from this extension, the main benefit of classification for local people is that they are officially granted an agro-forestry area that can be more or less freely used (at least for their own consumption).

Impacts on rural livelihoods

The positive and negative impacts of forest law enforcement (or non-enforcement) on rural livelihoods are summarized in Table 8.4. Forest classification remains a hot potato for the political authorities in Cameroon. The stakes are high since it may disturb rural populations when they discover that agricultural practices are highly restricted in what they still consider their

Table 8.4 *Classification procedure: Contribution to rural livelihoods*

Legal instrument		Contribution to rural livelihoods	
		Positive	Negative
Classification procedure	Implemented	• Agro-forestry area secured for resource use by villagers • Marginal shift of permanent forest estate (PFE) boundaries	• Ban on all agricultural practice in the PFE • Restriction of forest resource extraction in the PFE • Neglect of villagers' representation of forest space and of associated modes of resource exploitation • Reduced area for community forest
	Non-implemented		• Under-information of local people

forests. One measure may be proposed to balance the current negative impact of permanent forest classification on rural livelihoods: starting to think about integrated management of the PFE and NPFE, in which rural populations would play an active and effective role (Lescuyer et al, 2001).

Hunting and gathering of forest products

The letter of the law

Apart from restricting future access to land for agricultural purposes in large parts of the country, one major social consequence of the forest law is to limit hunting and NTFP gathering by the local population. These practices are regulated in three main respects:

1. Hunting and NTFP gathering is allowed throughout the national territory (except for protection forests) as long as it remains a traditional activity

intended only for self-consumption. Commercialization of forest produce is forbidden without a permit delivered by the ministry.
2 Protected plant and animal species must not be extracted. Similarly, many endangered species can be collected only under strict conditions.
3 Locally commonly used hunting techniques, such as trap hunting with steel-wire cables and hunting with the use of guns, are forbidden (Article 80). Forest law indicates that weapons and traps may only be made out of plant materials. Night hunting is also forbidden.

To counterbalance this restrictive regulation, three possibilities are open to rural populations for drawing economic benefits from hunting:

1 Rural communities may ask the ministry to establish a community hunting territory. Hunting territories are established and managed on the same basis as the community forest model. Such areas are taken out of the NPFE and may cover up to 5000ha.
2 Especially in northern Cameroon, the ministry may lease some areas in the PFE to professional hunting guides.
3 Discussions among the Ministry of Forestry, World Wide Fund for Nature (WWF) and the Gesellschaft für Technische Zusammenarbeit (GTZ) have recently led to the proposal of a new hunting concept (not yet considered in the forest law): the community management zone. Its objective is to allocate large hunting areas to local populations so that they directly manage and benefit from game meat. Consideration is being given to establishing such hunting areas on both the PFE and NPFE (Roulet, 2004).

Two outlawed activities: Hunting and gathering

It is proving difficult in Cameroon to abolish the illegal activities of hunting and NTFP gathering. There are three reasons for this. First, the legal framework is non-enforceable. Prohibiting bush-meat sales is, quite simply, unrealistic because it is an age-old and customary practice for rural people and the ministry has no means of controlling these commercial networks. Furthermore, prohibiting the use of steel-wire traps and guns is pointless since they are used throughout the national territory. Ministry staff do not even attempt to apply this measure. Second, the intended measures for establishing and managing both hunting territories and hunting areas have not been published (van der Wal and Djoh, 2001). Third, some poachers practise commercial hunting with impunity since they have the support of powerful political or economic players.

Impacts on rural livelihoods

The overall impact of the forest law on hunting and NTFP gathering activities is unclear (see Table 8.5), but it appears that its strict enforcement would be vigorously rejected by the rural population. Given the discrepancies between the law and actual practices, several legal considerations may be contemplated:

- Ordinary hunting equipment should be recognized.
- Considering the lack of resources at the ministry, any game management must integrate and, indeed, be built on local knowledge and capacities.
- Some tentative combinations of timber management plans (under the responsibility of the logging company) and game management plans (under the responsibility of communities) may produce results that will be of use in moving towards integrated management of tropical forest resources.
- The Convention on Biological Diversity (Article 8j) confers intellectual property rights on local populations who, through their practices, generate valuable knowledge or resources. This right is not yet clearly integrated within national law.

Table 8.5 *Assessment of impact of hunting and NTFP gathering regulations on livelihoods*

	Legal instrument	**Contribution to rural livelihoods**	
		Positive	**Negative**
Regulation of hunting and non-timber forest product (NTFP) gathering	Implemented	• Hunting territories and hunting areas	• Banning or restriction of resource extraction in and around protected areas
	Not implemented	• Income from commercial hunting and NTFP gathering • Free use of any hunting techniques or materials	• Non-native poachers • Impossibility of defining and ratifying community hunting territories

Commercial logging

Commercial logging of timber is characterized by the exploitation of large forest areas, the use of heavy equipment and, generally, timber processing to make high-quality products (sawn wood and plywood), often oriented towards the export market. In Cameroon, the exploitation concession is the main type of permit that allows this kind of activity.[1] Its main features are depicted in Table 8.6.

The application of the exploitation concession is under the control of the Ministry of Forestry (with the support of Global Witness as independent observer during the period of 2001 to 2005). Their purpose is to verify the compatibility between forest management official documents and field reality. This situation has produced three important outputs:

Table 8.6 *Main features of the exploitation concession*

Area	<200,000ha in the permanent forest estate (PFE), under the form of a forest management unit (FMU)
Duration	15 years, renewable once
Allocation	Through public bids in which logging companies must submit technical and financial offers
Legal requirements	Forest management plan Specifications (*cahier des charges*) established by the authorities on the basis of both the quality of the forest and inhabitants' requests Processing of at least 70% of the logged timber by local industry
Taxes and fees	Felling taxes based on the logged timber volume Annual area fee based on the FMU area and the financial proposal by the logging company (minimum of 1000 Communauté Financière Africaine francs (CFAF) per hectare)

1. increasing transparency inside the Cameroonian commercial logging system;
2. reduction of the ecological impacts of logging with the application of forest management plan guidelines; and
3. better assessment of timber tax amounts, especially felling taxes.

The Ministry of Forestry–Global Witness initiatives, however, have little impact on rural communities since most of the commercial logging repercussions on the rural environment are not investigated by independent observers. These impacts include:

- a direct impact through the involvement of local populations in establishing and implementing the forest management plan, and the specifications that constitute the framework for commercial logging operations;
- an indirect impact through a decentralized forestry taxation system, especially the annual area fee, 40 per cent of which is paid to the rural councils in the forest zones being logged and 10 per cent to the neighbouring village communities.

First direct contribution: Involvement of local communities in the management process

A forest management plan must be drawn up for all forests located in the PFE. It must precisely define how the forest is to be used in order, on the one hand, to optimize logging (in the case of a production forest) and, on the other hand, to ensure long-term maintenance of forest resources. Local populations must be a full member of the structure in charge of the forest management plan and must especially be involved in three important stages:

1 Socio-economic surveys must be carried out to take into account the traditional rights and practices of resident populations in relation to forest resources.
2 The forest management plan must indicate what rural populations may expect from the logging company in terms of social infrastructures.
3 The forest management plan must specify how participatory management of the forest will be effective in the field.

Apart from the obligation to draw up a forest management plan, the logging company also has a legal obligation to set up a processing factory in its logging area. Since the exploitation concession will last at least 15 years, this installation is supposed to provide employment for rural populations. Another benefit is to provide local populations with processed wood scraps, either free of charge or for a nominal price, which may be used for personal or commercial purposes.

As of March 2005, 46 forest management plans were agreed upon by the relevant authorities. Two impacts of forest management plans on local livelihoods are clear:

1 The obligation to set up local processing factories and employ local residents is not really applied. The necessity to maximize financial margins prompts producers to establish themselves close to urban markets and to recruit trained workers.
2 The scope of the socio-economic surveys is still marginal. They are still considered a constraint by the logger, rather than a means of benefiting from local knowledge and practices (Lescuyer and Essiane Mendoula, 2001). The point is that local populations have little say in the establishment of forest management plans and cannot exert any real influence on the practical consequences.

Second direct contribution: Specifications for logging operations

All commercial logging operations are framed by specifications that are fixed by the authorities with a view, amongst other things, to answering the local population's requests. In practice, the content of the section of the specifications regarding socio-economic infrastructures varies substantially. Its application also varies, especially since, for many logging companies, these direct social contributions to local populations are now redundant given the decentralized taxation measures.

As a result, all socio-economic commitments by the logging company are to be negotiated with the local populations. These negotiations generally take place during village meetings (*tenue de palabres*) that gather most villagers together with the logging company and the administrative authorities. From both a legal and practical point of view, however, these discussions about social infrastructures come late in the forest management process: whatever the local population may say, the forest has already been allocated to loggers. Local populations no longer have the power to deny them access to the timber in the forests. The point of such meetings is merely to reach an agreement on the 'gifts' to be provided by the company. Moreover, these 'negotiations' do not take place on an equal footing as the logging company has an advantage since:

- Most of the local people do not know anything of the timber's commercial value.
- There often is a prior informal agreement between the logging company and some of the administrative authorities, so the latter prevent the local population from 'speaking loudly'.
- The document that officially ratifies the negotiation agreement is almost always lost in the administration archives.

The annual area fee

Under Article 68 of the law, the annual area fee (AAF) is the central tenet of the decentralized taxation system. All forest management units are liable for payment of the annual area fee, based on the financial proposal made in the bid. Three factors explain why the AAF has skyrocketed (see Figure 8.1): many FMUs were allocated in 2000 and 2001; the logging companies have tended to increase their AAF rate proposals; and the tax services have been made more efficient in order to collect forestry taxes properly.

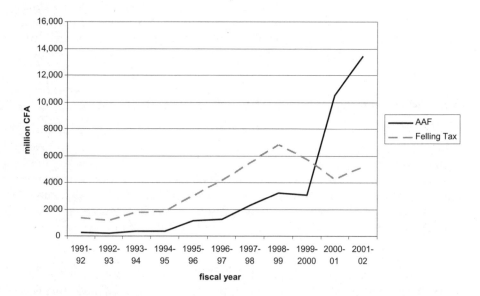

Figure 8.1 *Evolution of the annual area fee over the last decade*

Source: author

The impact of the annual area fee on the rural population comes from its distribution of 50 per cent for the treasury, 40 per cent for the local council where the logging takes place, and 10 per cent for the population in and around the logging area. Management arrangements at local level for these funds are established by Joint Order No 122/MINEF/MINAT of 29 April 1998:

- The 10 per cent share is allocated to nothing but local socio-economic projects in such a way as to promote community development.
- Revenues collected through the annual area fee remain public funds subject to the control of the tax office.
- An area fee management committee is to be established in every beneficiary community to manage the 10 per cent destined for the communities. The mayor is the chairman of the committee and the person authorized to pass accounts. Article 5(1) provides for village communities to be represented on the committee by six people chosen by the villagers themselves. In addition, the committee generally includes civil servants and local elected representatives. It is under the supervision of the administrative authorities (often the subdivision officer).
- A public call for bids must be issued for all projects in order to select the most competitive contractor.

The management of the AAF has been under close scrutiny for several years. Studies have generally concluded that the impact of the decentralized taxation system in terms of social infrastructures is very low as the revenues have been misused (Milol and Pierre, 2000; Bigombé Logo, 2004). Some recent events temper this rather gloomy analysis. It seems, for instance, that a few mayors who were criticized for their biased management of the annual area fee were replaced in the 2002 council elections by new candidates who committed themselves to more transparent and efficient use of forest revenues (Poissonnet, 2005). The flipside of the coin is the mayors' excessive commitment to managing the AAF at the expense of the other members of the rural council or of the forestry fee management committee (FFMC) (Nzoyem et al, 2003; Ngoumou Mbarga, 2005). While fund misappropriation is apparently decreasing, community representatives on the area fee management committee still have a rather figurative role. Consequently, the mayors tend to use the annual area fee as a personal political support and favour a sprinkling of micro-projects over the council territory, rather than triggering real development dynamics among rural communities.

Impacts on rural livelihoods

A summary assessment of the enforcement of legal mechanisms concerning commercial logging is given in Table 8.7, combining direct and indirect contributions. The drafting of forest management plans in Cameroon is recent and does not provide a precise idea of how they will affect the livelihoods of rural populations. However, several factors can be pinpointed:

Table 8.7 *Contributions of commercial timber exploitation to livelihoods*

Legal instrument		Contribution to rural livelihoods	
		Positive	Negative
Direct contributions: • involvement in the management process • specifications	Implemented	• Employment (direct or indirect) • Completion of socio-economic infrastructures • Regular payment of the annual area fee	• Projects funded by the annual area fee do not generate development dynamics
	Not implemented		• No real integration of customary rights within the management plan • Specifications are incomplete or largely not respected • Predominant role of mayor in decentralized forestry tax management

- Socio-economic surveys may be used for leverage to really take into account communities' concerns and to reduce local conflicts.
- Specifications remain an important issue since they constitute the main doorway to direct discussions between logging companies and local populations. However, they are mainly defined by the central services at the Ministry of Forestry, and their application (or not) often depends upon the administrative authorities – hence the recurrent discontent among local populations. To introduce more transparency into this process, it may be suggested that the document (*procès-verbal*) that ratifies the village negotiation agreement is collectively signed and then distributed to the various actors.

Similarly, the use of the 10 per cent annual area fee share could be improved in several ways, by:

- increasing involvement of community representatives in the area fee management committee – their role is to be expanded and specified;

- issuing a regular publication that summarizes and follows up on the projects funded; and
- investing in enhancing local capacities rather than in numerous disconnected micro-projects.

Beyond the law: Small-scale logging

The diminishing legal scope for small-scale logging

Commercial logging is only part of the timber utilization story in south Cameroon. The forest law is also open to much more reduced timber production activities carried out by individuals in order to satisfy their basic personal needs or to develop a (relatively small) business centring on wood products. Such logging operations are conducted with light equipment, such as chainsaws or mobile saws. Five small-scale logging activities are defined in the forest law:

1. Traditional right of use: resident populations can log as many trees as needed for their domestic use.
2. Direct exploitation of the community forest by the community (*exploitation en régie*) is permitted when using only light equipment.
3. Timber recovery special authorization (suspended): this permit allows any Cameroonian to recover abandoned and anonymous timber found along the roadside.
4. Individual felling authorizations (suspended) are issued to Cameroonians to allow them to cut a maximum of 30 cubic metres of wood in the NPFE, for non-commercial uses only.
5. Exploitation permits (suspended) are authorizations to exploit a maximum volume of 500 cubic metres located in a particular area of the NPFE.

As a consequence of substantial abuse, the Ministry of Forestry suspended the latter three permit types in July 1999. It is now aiming to focus logging operations on more controlled permits, such as exploitation concessions, while encouraging small-scale loggers to become involved in community forests. This measure has unfortunately forced all of the small-scale activities into the informal (i.e. illegal) sector for two reasons:

1. The domestic timber and wood products markets are growing and require low-priced commodities. For instance, yearly consumption of small sawn-wood pieces amounts to 80,000 cubic metres in Yaoundé and 100,000 cubic metres in Douala (Plouvier et al, 2002).
2. The commercial timber sector's requirements clearly exceed what sustainable management of the Cameroonian forests can provide. Even considering the legal logging operations allowed in the NPFE, processing capacity exceeds the current timber harvest by a minimum of 350,000 cubic metres per year (Abt et al, 2002). The illegal exploitation of the NPFE is thus a short-term solution to supplying industrial sawmills.

As a result, total annual production of illegal sawn-wood pieces is estimated at about 500,000 cubic metres (CIRAD and Institutions et Développement, 2000), which corresponds to one fifth of commercial timber production. The turnover amounts to 20,000 million Communauté Financière Africaine francs (CFAF) (US$30 million) per year.

A substantial contribution to rural livelihoods

The small-scale timber sector involves sawyers, handlers, edgers, truck drivers and a well-organized network of sellers and wholesalers. Excluding sales operations, this sector provides 4000 direct full-time jobs, most of them in rural areas (Plouvier et al, 2002). In almost every case, the timber for these operations is logged in the NPFE, mainly in the agro-forestry area set aside for local residents.

The cost structure and revenues accruing from all small-scale timber production activities have been assessed at several sites.[2] Manual workers (loggers, handlers and drivers) are paid according to the workload; but the average wage does not fall below a threshold of 30,000 CFAF per month. For most of them, the objective is to save enough money to be able, in a few years, to start up a similar operation on their own. As illustrated in Table 8.8, small-scale timber production appears to be a profitable business. At Ngambe Tikar, the delivery of one 20-tonne truck (containing around 27 cubic metres) generates a 400,000 CFAF profit for the producer. The margin is probably smaller at Lomié because of higher production costs (Plouvier et al, 2002).

The profile of small-scale timber producers is relatively homogeneous. Most of them are young (20–35 years old), relatively educated, natives or long-term residents of the area that they exploit, and quite dynamic from an economic point of view. It is the rule rather than the exception that some of

Table 8.8 *Costs and revenues from small-scale timber production*

Cost categories	Exploitation at Lomié, sale at Yaoundé (CFAF per m³)	Exploitation at Ngambe, sale at Bafoussam (CFAF per m³)
Tree purchase	6000	2500
Labour: logging, sawing and handling	22,300	15,000
Equipment and depreciation	20,000	3700
Transport	13,000	9000
Total cost	61,300	30,200
Sale price (within the district)		44,500
Producer margin		14,300

their profits are reinvested in semi-collective projects at village level. One man, for instance, has managed to provide the village with a corn mill, a bar and a mobile disco, which all give more or less regular employment to three villagers. Such semi-collective investments help to maintain good relationships with the people to whom he will have to apply to obtain more timber.

To conclude, Table 8.9 presents an overview of the impact of small-scale timber production on rural livelihoods. According to our analysis, this activity, which has been banned, has essentially positive impacts on rural living conditions. The recent restrictive regulations concerning small-scale timber production should at least be adapted. Rather than simply banning this activity, measures could be proposed to progressively encourage individual loggers to practise a legal activity (Plouvier et al, 2002).

Ambiguous development of community forests

Community forest: A legal revolution

Under forestry law (Article 37), a community forest (CF) is defined as 'a forest forming part of the non-permanent forest estate, which is covered by a management agreement between a village community and the forestry administration'. It can extend to 5000ha. Forest products of all kinds resulting from community forest management belong solely to the village communities concerned. In contrast to a landownership title, the management agreement between the state and the community does not confer any forest ownership rights on the latter. The community enjoys the use of the forest according to a simple management plan that stipulates:

- the beneficiary community;
- the CF boundaries and its main uses;
- a description of the forest;
- the operating programme; and
- forest and wildlife management instructions.

Table 8.9 *Contribution of small-scale timber exploitation to rural livelihoods*

Legal instrument		Contribution to rural livelihoods	
		Positive	Negative
Small-scale timber exploitation	Implemented	• Extraction of timber resources for domestic use	
	Not implemented	• Substantial and direct revenue	
		• Partial investment of this revenue in collective infrastructures	

The management plan lays the foundations for the management agreement between the community and the ministry: it is usually planned for 25 years and must be revised every five years.

In spite of the new possibility for local populations of obtaining their own forests, this legal instrument took time to bear fruit. In 1998, only two CFs were allocated. Three linked events at the end of the 1990s helped to put the CF concept into practice. First, the publication of the 1998 *Manual of the Procedures for the Attribution and Norms for the Management of Community Forests* offered a practical translation of the legal mechanisms. Second, the ministry created a community forest unit that was in charge of both popularizing the CF concept and supervising practical initiatives. Lastly, several NGOs decided to commit themselves to the implementation of CFs in the field. As a result, in 2003, 63 CFs were running, 56 reserved and 120 requested (MINEF, 2004).

The sudden increase in the number of CF requests must not be allowed to hide the high technical, financial and labour costs of such projects. The establishment of a CF involves complex administrative procedures and represents a substantial investment for communities in terms of time and money (Etoungou, 2003; Oyono, 2004). Without the help of an influential NGO, it is almost impossible for a community to invest the necessary sum into a process whose positive outcome is not guaranteed. To overcome this financial hurdle, many communities have contacted logging companies for support (Karsenty, 1999). The logging company would finance the launch and follow-up of the CF request, and would be repaid with timber logged in the CF. The ministry considered these partnership schemes as a flawed use of the CF concept and spirit: rather than fighting rural poverty, the CF became a means of providing commercial logging companies with timber from the NPFE. The response of the ministry with a view to halting this trend was the suspension of the industrial exploitation of CFs and the promotion of CF exploitation by communities through small-scale production techniques in 2001.

The same year, communities were given the right of pre-emption over proposed sales of standing volume areas in order to enable them to set up a CF instead. Communities seem to highly appreciate this new right, as seen during the following public bid, when out of 143 sales of standing volume proposals, 107 communities used their pre-emption rights to reserve these areas for future CFs.

A troubled success

The current success of community forests in Cameroon has raised many questions, notably from officials of the Ministry of Forestry.

First, the ongoing CF projects tend to have little impact on the community as a whole. One reason often quoted is the absence of any real community on a village scale since the notion of 'community' remains a contested concept (Brown, 1999; Lescuyer, 2000), with an uncertain capacity to mobilize a category of people capable of exercising a common claim to 'ownership' of the resource. As described in Klein et al (2001), the first benefits accruing from a CF are generally used for individual purposes, whereas the CF collective institutions, such as common initiative groups and associations, appear to be hypothetical rather than real. In this context, common initiative groups and associations are often a new means for the elite of benefiting from CFs, while masquerading as community empowerment.

Second, to compensate for their low technical capacities to deal with CFs, communities are often involved in quite perplexing partnerships with NGOs and forestry companies. Some NGOs play an ambiguous role in community forestry (Etoungou, 2003). For some of them, community forests comprise a new activity whose main advantage is that it is generously funded by bilateral or multilateral cooperation agencies. The main issue, however, is the leading behaviour of many NGOs in the CF process. Rather than providing temporary assistance, NGOs often induce villagers to become involved in CFs, provide them with technical and financial assistance, and take care of follow-up until the CF is reserved. In relation to forestry companies, informal alliances between Cameroonian logging operators and village elites are still common. The latter are interested in logging CFs as quickly as possible, and primarily for their own benefit.

Third, as a result of an inconsistent legal framework and consequent abuses, there has been a multiplication of administrative mechanisms for CF operations, leading to potential legal contradictions. The multiplication of legal mechanisms regarding CFs has a tendency to strengthen the discretionary power of ministry staff. Given the administrative complexity and variability, it is not unusual for certain civil servants to be convinced by more subjective arguments, of which money is far from the least significant. One precise example illustrates how high the unofficial costs of obtaining a CF without external assistance can be (see Table 8.10).

Impacts on rural livelihoods

Community forests are presented in the law as the main tool to alleviate rural poverty in southern Cameroon. In practice, the actual impact is more limited since CFs are somewhere between legality and illegality. Moreover, the legal mechanisms are changing rapidly and partly hinder the secure and sound management of CFs. The main impacts of the CF regulations are summarized in Table 8.11.

Given the numerous types of drift in the application of the CF process, many people are now considering its temporary suspension. If applied, this legal break will provide an opportunity to reformulate certain procedures, as well as redefine the roles of the players concerned. In August 2003, the Ministry of Forestry put new restrictions on the issuing of transport permits for timber. They can be issued only if CF leaders demonstrate that the timber volume to be transported is compatible with the exploitable volume stated in the management plan.

Table 8.10 *The hidden costs of community forests*

Community forest (CF) step	Means of 'facilitating' the procedure
Financial cost of the consultation meeting (excluding contributions in kind)	150,000 CFAF
Follow-up of the CF request at the Ministry of Forestry decentralized services	900,000 CFAF
Follow-up of the CF request at the ministry central services	250,000 CFAF, plus personal contacts
Follow-up of the management plan at the ministry decentralized services	1 million CFAF
Approval of the management plan by the ministry central services	Part of what is paid to the planning office (15 million CFAF) is used to 'motivate' certain people
Signature of the management plan by the ministry decentralized services	250,000 CFAF
Signature of the management plan by the senior divisional officer	2.7 million CFAF, plus political contacts

Table 8.11 *Contribution of community forests to livelihoods*

Legal instrument		Contribution of community forests (CFs) to rural livelihoods	
		Positive	Negative
Community forest	Implemented	• Revenue for collective goods and services (when supported by NGO) • Safe access to forestland • Substantial and rapid revenue when in partnership with professional loggers	• Exorbitant initial investment
	Not implemented	• Substantial and rapid revenue when in partnership with professional loggers • Most CF procedure costs are borne by NGOs	• Risk of benefits accruing to elites • Administrative and legal haze generates informal expenses for all official procedures

Conclusion

Research issues

The contribution of forest law enforcement to rural livelihoods is a new and politically sensitive issue. Many aspects of this problem have not been thoroughly investigated and several items of data are confidential. Besides the current gaps in the available information, four issues are to be researched in relation to the role of the forest law in alleviating rural poverty.

First, in this chapter, we directed our attention to the short-term socio-economic impacts of forest law enforcement. Our approach was to disregard all ecological consequences of practices observed in the field. These environmental effects are significant variables, however, with regard to the quantity and quality of forest resources in the long run. A helpful complement to this socio-economic study would be to consider the relationships between law enforcement and forest health. This analysis would provide an opportunity for questioning the relevance of the law for achieving sustainable forest management, as well as for considering the performance of the current illegal practices in this respect.

Second, more flexible means to enforce the forest law may be investigated. In Cameroon, following the French legal tradition, the legal system is a series of prescriptive norms that aims to comprehensively regulate human activities. But paradoxically, the shortage of enforcement means and the inappropriateness of several legal norms are important factors in generating illegal forest activities. Instead, what would be needed is law with sufficient flexibility of implementation in order to work in differing contexts. This approach would require the central administration to establish basic forest management norms and guidelines, which could be marginally adapted to local circumstances under supervision of decentralized authorities (Karsenty, 1998). Such a flexible process to enforce forest law is already working for forest management plans as logging companies modify several parameters (timber species, diameter, etc.) to fit the field ecosystem reality while abiding by administrative minimum requirements. This kind of adaptable instrument may be usefully extended to other (currently illegal) forest activities.

Third, forest law enforcement provides local players with considerable amounts of money. Most of these funds, however, have been invested in largely unproductive projects and have failed to trigger any real development in rural areas. Nor have they enabled improved management of forest resources. One reason for this failure may be the allocation of these funds to rural people without any commitment from them. The process is not subject to any real negotiations in which stakeholders might agree on how to share both the benefits and the costs. Instead of a periodic distribution of funds among the villages concerned, negotiated and contractual processes could be promoted in order to really involve the community in sustainable management of the forest. Rural stakeholders may also carry out such negotiations with external private players, such as logging companies interested in ensuring timber supplies to their processing factories (Mayers and Vermeulen, 2002) or environmental organizations willing to pay to compensate for the public benefits provided by tropical forests (Karsenty, 2004; Lescuyer, 2005).

Lastly, the consequences of devolving forest management to local stakeholders warrant further research. In the case of Cameroon, it may be asked whether the recent trend towards decentralization provides rural players with powerful means of defining and implementing convenient management of the forest or, on the contrary, constitutes only a new opportunity of procuring part of the income from logging. Close monitoring of the community forest initiatives will give some insights into this fundamental issue.

Policy recommendations

The compilation of the heterogeneous impacts of the forest law on rural livelihoods presented in this chapter does not produce unambiguous results. The contribution of law enforcement varies from relatively positive (e.g. the decentralized taxation system) to quite negative (e.g. restrictions on hunting activities); but the overall impact remains mitigated. In light of this examination, one may wonder whether strict implementation of the forest law and regulations would have a real net positive impact on rural livelihoods. In the field, the common practice is to proceed with the 'adaptive' application of the law, which always leaves some room for arrangements between players (Brown, 1999; Lescuyer, 2002). It seems to us that such flexible enforcement of the forest law stems from two major facts.

First, strict application of the law would prevent many people from carrying out their traditional economic activities (hunting, small-scale forestry, etc.) and would significantly and directly impair rural livelihoods. Important or critical revenues are at stake. In consequence, many harsh regulatory measures are hardly acceptable to rural residents who sometimes do not hesitate to intimidate local Ministry of Forestry officials.

Second, the administrative procedures are undoubtedly too complex in relation to the financial, educational and organizational means of the stakeholders concerned. In addition, the numerous regulatory documents do not constitute a consistent package since certain texts may contradict others. As a result, while everyone recognizes the homogeneity of the forest law, the texts covering its application are sometimes confusing, notably for the decentralized ministry officials.

These two fundamental issues require changes to and adaptation of the forest law. There is no doubt that several mechanisms are clearly unreasonable since they are too disconnected from reality – the best example is the prohibition of hunting with rifles. Several marginal aspects of the law should be modified.

Notes

1 Another permit is the sale by standing volume that extends over 2500ha. It lasts one year and can be renewed twice. It is not subject to a forest management plan.

2   The initial investment consists of a second-hand chainsaw, an oil barrel, cash to pay the first salaries and professional contacts. The total expense amounts to less than 1 million CFAF.

References

Abt, V., Carret, J. C., Eba'a Atyi, R. and Mengin-Lecreulx, P. (2002) *Etude en vue de la définition d'une politique sectorielle de transformation et de valorisation du bois. Première partie: Analyse de l'adéquation entre la production de la forêt camerounaise et la capacité des usines de transformation*, Paris, Ecole Nationale Supérieure des Mines de Paris

Bigombé Logo, P. (2004) *Le retournement de l'Etat forestier. L'endroit et l'envers des processus de gestion forestière au Cameroun*, Yaoundé, Presses de l'UCAC

Brown, D. (1999) *Principles and Practice of Forest Co-Management: Evidence from West/Central Africa*, London, European Union Tropical Forestry Paper 2

CIRAD (Centre de Coopération Internationale en Recherche Agronomique pour le Développement) and Institutions et Développement (2000) *Audit économique et financier du secteur forestier au Cameroun*, Yaoundé, MINEFI

Diaw, M. C. and Oyono, P. R. (1998) 'Dynamiques et représentations des espaces forestiers au sud-Cameroun', *Arbres, Forêts et Communautés Rurales*, vols 15 and 16, pp36–43

Ekoko, F. (2000) 'Balancing politics, economics and conservation: The case of the Cameroon forestry law reform', *Development and Change*, vol 31, no 1, pp131–154

Etoungou, P. (2003) 'Decentralization viewed from inside: The implementation of community forests in east Cameroon' in Ribot, J. and Larson, A. (eds) *Environmental Governance in Africa*, Washington, DC, WRI, Working Paper No 12

Karsenty, A. (1998) 'Entrer par l'outil, la loi, ou les consensus locaux?' in Lavigne-Delville, P. (ed) *Quelles politiques foncières pour l'Afrique rurale ? Réconcilier pratiques, légitimité et légalité*, Paris, Karthala

Karsenty, A. (1999) 'Vers la fin de l'Etat forestier? Appropriation des espaces et partage de la rente forestière au Cameroun', *Politique Africaine*, vol 75, pp147–161

Karsenty A. (2004) 'Des rentes contre le développement? Les nouveaux instruments d'acquisition mondiale de la biodiversité et l'utilisation des terres dans les pays tropicaux', *Mondes en Développement*, vol 127, no 3, pp59–72

Karsenty, A. and Marie, J. (1998) 'Les tentatives de mise en ordre de l'espace forestier en Afrique centrale' in Rossi, G., Lavigne-Delville, P. and Narbeburu, D. (eds) *Sociétés rurales et environnement: Gestion des ressources et dynamiques locales au Sud*, Paris, Karthala

Klein, M., Salla, B. and Kolk, J. (2001) *Forêts communautaires: Les efforts de mise en œuvre à Lomié*, London, DFID, FRR, ODI, Rural Development Forestry Network Paper 25f(ii)

Lescuyer, G. (2000) *Evaluation économique et gestion viable de la forêt tropicale: Réflexion sur un mode de coordination des usages d'une forêt de l'est-Cameroun*, Paris, Ecole des Hautes Etudes en Sciences Sociales

Lescuyer, G. (2002) '"Tropenbos" experience with adaptive management in Cameroon' in Oglethorpe, J. (ed) *Adaptive Management: From Theory to Practice*, Cambridge, IUCN, SUI Technical Series Volume 3

Lescuyer, G. (2005) 'La biodiversité, un nouveau gombo?', *Natures Sciences Sociétés*, vol 13, pp311–315

Lescuyer, G. and Emerit, A. (2005) 'Utilisation de l'outil cartographique par les acteurs locaux pour la gestion concertée d'une forêt au sud du Cameroun', *Cahiers Agriculture*, vol 14, no 2, pp225–232

Lescuyer, G., Emerit, A., Essiane Mendoula, E. and Seh, J. J. (2001) *Community Involvement in Forest Management: A Full-Scale Experiment in the South Cameroon Forest*, London, DFID, FRR, ODI, Rural Development Forestry Network Paper 25C

Lescuyer, G. and Essiane Mendoula, E. (2001) 'La variable humaine dans la gestion de la forêt tropicale: Intérêt et résultat des enquêtes socio-économiques pour l'aménagement de la zone de recherche Tropenbos', in Faohom, B., Jonkers, W. B. J., Nkwi, P. N., Schmidt, P. and Tchatat, M. (eds) *Sustainable Management of African Rain Forest*, Wageningen, Tropenbos Foundation

Mayers, J. and Vermeulen, S. (2002) *Company–Community Forestry Partnerships: From Raw Deals to Mutual Gains?*, London, International Institute for Environment and Development, Instruments for Sustainable Private Sector Forestry Series

Milol, A. and Pierre, J. M. (2000) *Impact de la fiscalité décentralisée sur le développement local et les pratiques d'utilisation des ressources forestières au Cameroun: Volet additionnel de l'audit économique et financier du secteur forestier*, Yaoundé, MINEFI

MINEF (Ministry for the Environment and Forests) (2004) *État des lieux de la foresterie communautaire au Cameroun*, Yaoundé, MINEF, Direction des Forêts–Cellule de Foresterie Communautaire

Ngoumou Mbarga, H. (2005) *Etude empirique de la fiscalité forestière décentralisée au Cameroun : Un outil de la gestion durable de la forêt?*, Montpellier, ENGREF et CIRAD

Nzoyem, N., Sambo, M. and Majerowicz, C. H. (2003) *Audit de la fiscalité décentralisée du secteur forestier camerounais*, Châtenay-Malabry, Institutions et Développement

Oyono, P. R. (2004) *Institutional Deficit, Representation, and Decentralized Forest Management in Cameroon: Elements of Natural Resource Sociology for Social Theory and Public Policy*, Washington, DC, WRI and CIFOR, Environmental Governance in Africa Working Paper 15

Plouvier, D., Eba'a Atyi, R., Fouda, T., Oyono, P. R. and Djeukam, R. (2002) *Etude du sous-secteur sciage artisanal au Cameroun*, Yaoundé, MINEF-PSFE

Poissonnet, M. (2005) *Mise en œuvre de la gestion forestière décentralisée au Cameroun: Impacts politiques, socio-économiques et environnementaux d'un processus en apprentissage*, Montpellier, ENGREF et CIRAD

Republic of Cameroon (1994) *Law No 94/01 of 20 January 1994 to Lay Down Forestry, Wildlife and Fisheries Regulations*, Yaoundé, Republic of Cameroon

Republic of Cameroon (1995) *Decree of Implementation No 95/531 of 23 August 1995 to Lay Down the Application Clauses for the Forest Regime*, Yaoundé, Republic of Cameroon

Republic of Cameroon (1996) *Master Law 96/12 of 5 August 1996 Relating to Environmental Management*, Yaoundé, Republic of Cameroon

Republic of Cameroon (1998a) *Joint Order No 122/MINEF/MINAT of 29 April 1998 to Lay Down the Utilization Clauses of Forest Exploitation Revenues by the Resident Village Communities*, Yaoundé, Republic of Cameroon

Republic of Cameroon (1998b) *Manual of the Procedures for the Attribution and Norms for the Management of Community Forests*, Yaoundé, Editions CLE

Roulet, P. A. (2004) 'Chasseur blanc, coeur noir'? *La chasse sportive en Afrique centrale*, Orléans, Université d'Orléans

van der Wal, M. and Djoh, E. (2001) *Territoires de chasse communautaire: Vers la décentralisation de la gestion cynégétique. Observations relatives au village de Djaposten*, London, DFID, FRR, ODI, Rural Development Forestry Network Paper 25e(iv)

9

Forest Law Enforcement and Rural Livelihoods in Bolivia

Marco Boscolo and Maria Teresa Vargas Rios

Introduction

Bolivia is one of the poorest countries in Latin America with over 80 per cent of its rural population living below the poverty line (Pacheco, 2001). At the same time, Bolivia's forests cover over 53 million hectares. From these two realities, the concept that sustainable forest management and community forestry could become key instruments for sustainable development and poverty alleviation has gained considerable attention in recent years (ARD, 2002).

The link between forest use and rural livelihoods is, in part, mediated by the laws, regulations and norms that dictate how, when, where and by whom forest resources can be used. This chapter explores the positive and negative relationships between forest laws and rural livelihoods. Livelihoods can be affected by the way in which forest policies are designed, violated and enforced. Therefore, we explore the ways in which forest legislation, illegal forest activities and forest law enforcement affect livelihoods. These relationships are complex because of the diversity of social groups that rely on forest use, the diversity of livelihood strategies that they employ and the dynamic nature of the regulatory framework that affects forest use.

Several livelihood dimensions need to be considered (see Chapter 6). Because of limited time and data availability, this chapter focuses, in particular, on the following impacts of forest law enforcement on livelihoods: legal access to forests, sustainable management, participation in political processes and forest products income. Other indicators, such as wage income, agricultural income and consumption, government services, physical security, vulnerability, cultural heritage and economic growth, are mentioned only for those specific circumstances where information was available. The analysis presented in this chapter is based on a review of relevant documents, the direct experience of the authors and a series of interviews carried out in July 2003.

The chapter is structured as follows. First, we provide a brief overview of the key policy elements that affect forest use. Second, we briefly describe the main social groups that depend upon forest use for their livelihoods, together with a typology of such use. In the next three sections, we describe how regulatory design, illegal forest activities and forest law enforcement affect the various dimensions of livelihoods mentioned above. We conclude by presenting some general policy recommendations. However, we note that there are information gaps that will need to be filled before more detailed policy options can be meaningfully suggested and evaluated.

Elements of the Bolivian forest regulatory framework

The Bolivian forest regulatory framework is largely based on a set of laws and regulations passed since the mid 1990s. Key among them is the forest law (No 1700/96), passed in 1996, which describes the principles by which forests may be accessed, managed and taxed, and which establishes the institutions responsible for forest law enforcement, including their financing and the roles to be played by local governments.

Forest policy reform aimed at improving key aspects of sector development includes the following goals, to: promote sustainable forest management; introduce a less corruptible system of forest fees; reduce illegality; stimulate community forest management; increase land security for a variety of stakeholders; legalize traditional and legitimate forest uses by indigenous peoples and small-forest users; and create a stronger institutional framework (Pavez and Bojanic, 1998; Contreras and Vargas Rios, 2002).

Implementation of the forest law has been inextricably linked to the effective implementation of other laws, particularly the agrarian law (also known as Instituto Nacional de Reforma Agraria, or INRA, law; see Box 9.1), the environmental law (No 1333/1992), the protected areas law, and the popular participation and decentralization law.

In the following sections, we summarize how the forest and the agrarian laws affect key aspects of forest use in Bolivia: land tenure and ownership; forest access; forest management; taxation; regulation and enforcement; institutions and governance.

Land tenure and ownership

In order to engage in legal forest use, Bolivian operators must demonstrate legal ownership or tenure (e.g. having a concession contract) of the land. Documentation of ownership has long been a major problem in Bolivia since multiple claims and titles frequently existed for the same land parcel and a national-level property register was only beginning to be developed. The INRA law, also passed in 1996, attempted to correct this problem (see Box 9.1).

> **Box 9.1 The National Institute for Agrarian Reform (INRA) law**
>
> Before passage of the INRA law, most of Bolivian forestland was under unclear land tenure and ownership, promoting insecurity, conflict and mining of forest resources. The INRA law:
>
> - refined the process for title regularization to be integrated with a property register;
> - acknowledged the exclusive right of indigenous communities to claim communal ownership of land and removed the possibility of overlapping use claims;
> - created a new institutional apparatus composed of INRA, the Agrarian Superintendence (in charge of classifying lands according to their potential use and ensuring that lands are used according to such classification) and the Agrarian Judicature;
> - identified alternative mechanisms to clarify and title landownership; and
> - recognized preferential access to public lands for small farmers and communal groups.

Social groups such as indigenous peoples benefited most from these reforms since they were given priority in the clarification of territorial claims. This process turned out to be longer and more expensive than anticipated. By the year 2002, out of 64.5 million hectares in need of *saneamiento*, only 14 million hectares had been titled (see the INRA website, www.inra.gov.bo). Remaining confusion regarding land tenure and ownership continues to be a source of conflict and insecurity, as well as an important barrier for many actors 'entering' the formal forest regime. Furthermore, even lands with clear titles continue to suffer invasions from landless peasants.

Forest access

Along with the new process for clarifying landownership, the new forest regime also changed the way in which forests could be accessed. First, it extended access rights to previously excluded groups. Second, it introduced a more transparent process for awarding new use rights (e.g. concessions). Third, reforms eliminated the possibility of overlapping use rights. For example, before 1996, concessions for timber extraction were frequently awarded on territories claimed by indigenous groups.

These changes made commercial forest use accessible to previously disenfranchised groups such as indigenous communities, small timber extractors and private landowners. Under the previous regime, these groups could not participate in forestry except as illegal operators. By 2002, groups

Table 9.1 *Forest access by right (hectares managed according to authorized plans)*

Years	Industrial* concessions	Local community associations	Long-term* concessions	Indigenous territories	Private properties	Total
1997	5,498,017	0	361,721	0	0	5,859,738
1998	5,516,615	0	339,000	121,609	93,443	6,070,667
1999	5,330,853	0	294,022	141,150	199,791	5,965,816
2000	5,302,520	0	294,022	238,259	239,670	6,074,471
2001	4,972,447	407,721	112,000	444,406	351,344	6,287,918
2002	4,443,012	423,203	112,000	555,681	561,911	6,095,807

Source: SF (1997–2002)

other than concessionaires were actively managing about 1.6 million hectares of forest (see Table 9.1). A description of forest access in Bolivia is provided in Box 9.2. Forest access is now available in five categories: industrial concessions; long-term concessions; local community associations; indigenous territories; and private properties. The following provides a brief description of what these are (see also the section on 'Forest users').

Industrial and long-term concessions. The forest law allowed holders of concession contracts signed before 1996 to either maintain the old regulations (e.g. fees paid on a per volume basis, concession contract valid for 20 years) or to adopt the new regulations (e.g. fees paid on an area basis, concession contract valid for 40 years). Those that chose to be treated according to the new regime are here described as 'industrial concessions'. Those that chose to be regulated according to the old regime are here called 'long-term concessions'.

Local community associations, or Asociaciones Sociales del Lugar (ASLs), are groups of traditional forest users (e.g. ex-*motosierristas*),[1] peasant communities and indigenous populations legitimized by the forest law to legalize their commercial activities. Organized in ASLs, these operators could now have legal access to municipal forest reserves.

Indigenous territories, or *Territorios Comunitarios de Origen* (TCO), are legal denominations of indigenous groups with the right to commercially use the forest resources on their own land.

Institutions and governance

The new forest law replaced the *Centro de Desarrollo Forestal,* the corrupt and inefficient regulatory agency under the old regime, with the regulatory agency, Superintendencia Forestal (SF). The SF has the mandate to regulate and control commercial forest activities[2] and to protect public and private lands (the key functions of the SF are described in Box 9.3). The SF was designed to ensure its independence from political influence. Its direct financing from

Box 9.2 Forest access in Bolivia

Private firms may access public forests via a transparent international auction system. Concession contracts are awarded for 40 years, and they are transferable and renewable. To operate in public forests, private firms need to:

- be legally constituted and registered;
- win a forest concession at public auction (the only exception being pre-1996 contracts adopted into the current regime);
- annually pay the area tax (*patente forestal*) – the area tax is US$1 per hectare per year for concessions adopted from the previous regime; for new concessions awarded through open bidding, the area tax amount offered is variable and constitutes an important factor in winning the public auction; and
- develop a forest management plan and have annual operating plans approved by the regulatory agency, Superintendencia Forestal (SF).

One should note, however, that not a single auction has taken place in Bolivia since the passing of the forest law. All of the existing industrial concessions have been adopted from contracts awarded administratively before 1996.

Public forests can also be accessible to local community associations. To gain legal access, a group must undertake a rather lengthy process involving the legal creation of an Asociación Social del Lugar (ASL), the legal identification and delimitation of municipal forest reserves, and the development and approval of a management plan. Each ASL must have at least 20 members with a history of forest use of at least five years.

On private forestlands,[3] commercial use requires:

- ownership title (*título de propiedad ejecutorial*) or, for small properties, a certificate of the municipality, a document issued by INRA that the owner has made a claim of ownership (*proceso en trámite*) and a certificate issued by the municipal forest unit (with little verifying capacity) that the parcel does not overlap with other rights;
- registration of the property;
- a map showing the location and coordinates of the property;
- legalized copies of all documents;
- support and counter-signature of family members;
- verification of the registration of the professional who signs all the documentation presented;
- a forest management plan approved by the SF;
- an annual forest operational plan approved by the SF; and
- payment of the area tax.

Indigenous territories are treated legally like private properties.

forest taxes also linked the sustainability of the SF to that of the sector. It is led and run by officers widely perceived to be technically competent and honest (Boscolo and Vargas Rios, 2002).

> **Box 9.3 Key functions of the Superintendencia Forestal**
>
> - To award concession rights on private lands to companies and the local community association (ASL).
> - Evaluate and approve forest management plans and authorize harvesting on public and private lands.
> - Authorize clearing permits on land classified for conversion.
> - Issue regulations and norms to ensure sustainable use of forest resources.
> - Collect area taxes, fines and revenues from the auction of seized products, and distribute them to prefectures, municipalities and a national forestry development fund, *Fonabosque*, according to pre-established percentages.
> - Issue permits (*certificados forestales de origen*) for the transport of forest products.
> - Control and enforce (inspections and audits) regulations related to the harvest and transport of forest products.
> - Resolve conflicts and controversies over forest rights.
>
> The SF is also mandated to make information about the functioning of the public administration public and accessible through the means of public hearings and by providing free access to information.
>
> *Source:* SF (1997–2002)

A new national forestry development fund, *Fonabosque*, was created to help finance the transition from the old to the new regime. It was expected that many technical and financial resources would have been needed to finance forest projects, educate forest users about the principles and practices of forest management, and fund the development of management and utilization plans.

The law envisioned increased roles for local governments, including in enforcement activities (see the following section). Prefectures, operating at the departmental level, are responsible for public forest education, for forest research and for intervening in areas where illegal logging or clearing is being carried out. Municipalities are mandated to select public areas to be allocated to the ASL. Municipal forest units were also to be established to assist in the management of municipal forest reserves, and to complement the SF in implementing the forest law (particularly in assistance to the ASL) and in promoting the adoption of better forest practices.

The public can also become directly involved in forest law enforcement. For example, private citizens may use a special authorization or warrant granted by the SF to inspect field operations (Contreras and Vargas Rios, 2002).

Forest management regulations and enforcement

The new forest law also introduced a new set of standards for the sustainable utilization and protection of forests. For private landowners and forest concessionaires with areas greater than 200ha, the law introduced the mandatory requirement of a forest management plan (FMP), annual harvesting plans (*planes operativos de aprovechamiento forestal*, or POAF), a 20-year minimum felling cycle, and a comprehensive set of prescriptions for how to conduct most forest planning and harvesting operations. Under no circumstance does the law allow the use of chainsaws to process logs (e.g. to square them in *cuartones*) in the forest. These regulations were intended to ensure the continued utilization and protection of forest resources over time. To a large extent, the regulations were effective. Boscolo et al (2003) have shown that industrial concessions currently adopt several good forestry practices. By the end of 2002, about 4.5 million hectares under industrial concessions were managed according to forest management plans.

The SF can use a variety of instruments to enforce forest regulations (Cordero, 2003), applicable at various stages of forest planning, harvesting, transport and processing. In conjunction with the SF, enforcement responsibilities are shared with prefectures, municipalities, forestry professionals and civil society. For example, forestry professionals are personally responsible before the law for the proper preparation and execution of management and harvesting plans.

Taxation

The new forest law also replaced a tax system based on harvested volume with a regime based on the area under concession or management (*patente*).[4] This choice was motivated by the desire to curb corruption, reduce speculative holdings of public land and encourage more intensive harvesting. Concessions adopted from the previous regime were to be taxed at the rate of US$1 per hectare on the total area of the concession (which, considering the minimum 20-year felling cycle, translated into US$20 per harvested hectare). Forest use on private land (including indigenous territories), on the other hand, was to be taxed at US$1 per harvested hectare.

Small-scale timber harvesting operations working within municipal forest reserves were to be taxed at a rate corrected for forest richness, accessibility and size of the concession – around US$2.5 to $8.5 per harvested hectare (STCP Engenharia de Projetos Ltda, 2000). Areas to be converted to agricultural uses were to be taxed at the rate of US$15 per hectare plus 15 per cent of the value of standing timber. An additional 15 per cent was then to be paid by the eventual buyer of timber, if timber was commercialized. The SF carries out tax collection and then distributes the proceeds to local governments according to a sharing arrangement specified in the forest law.

This reallocation of fiscal revenues was intended to finance a larger process of decentralization, as well as to provide local governments with financial incentives to promote sustainable forest management. Pacheco (2000) notes, however, that these transfers have not provided enough incentives since they are only loosely linked to performance.

The differentiated fiscal treatment of uses on private and public lands has resulted in strong (but unsuccessful) lobbying efforts by industrial operators to privatize forest concessions. Claiming unfair treatment and economic difficulties, in 2003 the industry did succeed in obtaining the passing of a supreme decree that modified the way in which forest use is taxed in industrial concessions. The new decree establishes that industrial concessions are to be taxed at US$1 per harvested hectare, plus a regulation fee (*tarifa de regulación*) varying between US$3.6 and US$4.6 per harvested hectare that accrues entirely to the SF.

In sum, the set of reforms initiated in the mid 1990s produced significant positive outcomes:

- *Improved accountability and transparency.* A survey conducted among industrial concessionaires by Boscolo and Vargas Rios (2002) concluded that the SF is now generally perceived as honest and technically competent. The SF attempts to respond to the needs of forest users, particularly the most vulnerable ones.
- *Reduced conflicts over forest resources.* Overlapping claims have been reduced.
- *Improved sustainable forest management.* The new regime promoted a more sustainable way of using forest resources (Fredericksen et al, 2003). About 6 million hectares are currently managed according to forest management plans. About 1 million hectares are independently certified.
- *Reduced illegality.* The SF is increasingly taking legal action, and many traditional and legitimate forest uses have been legalized.
- *Stimulated community forestry.* The forest law introduced a new legal framework for the development of community forestry in Bolivia. Following a slow start, local communities and indigenous groups currently manage about 1 million hectares of forest.

Forest users

Bolivian forests cover 53.4 million hectares, about 48.6 per cent of the country's total area. Three main natural regions characterize the country: the *altiplano*, the valleys and the lowlands. About 80 per cent of Bolivian forests are located in the lowlands, the remaining 20 per cent in the valleys (Pacheco, 2001). Forests are primarily situated in the departments of Santa Cruz, Beni, Pando and the tropical areas of La Paz and Cochabamba.

About 35 per cent of Bolivia's total population of 8.1 million live in the lowlands and valleys. Of the population of this area, about 51 per cent live in urban centres. The remaining 49 per cent, or between 1.4 and 1.6 million people, live in rural areas and depend to various degrees upon forest use for their survival (INE, 2001).

To our knowledge, there exists no study that has attempted to provide an overall assessment of the relationship between forest use and rural livelihoods in Bolivia. A limited number of studies have assessed the contribution of forest use to income and consumption (e.g. Stoian, 2000; Godoy et al, 2002); but these studies have focused on specific locations or on a specific social group. In the following section, we describe the attributes of the main social groups that depend upon forests for their livelihoods and briefly summarize their dependence upon forests. Partly based on a classification by Pacheco (2001), forest-dependent social groups can be disaggregated into six categories. Table 9.2 summarizes the main attributes of these groups.

Indigenous peoples

Indigenous peoples constitute an extremely poor group (Pacheco, 2001; Godoy et al, 2002). According to the 2001 national census, indigenous peoples comprise a total population of about 450,000, of which about 180,000 live in rural areas (INE, 2001). They are a quite heterogeneous group of people in terms of language, culture and forest dependency.

It is difficult to assess the forest contribution to the livelihood of this group. Indigenous groups in various locations have access to forests that differ in richness in terms of both timber and non-timber forest products (NTFPs), infrastructure and access to markets. They also adopt a variety of livelihood strategies that vary with seasons, and they combine forest use (e.g. NTFP extraction), small-scale cultivation, wage labour and, in some areas, the occasional selling of timber. Quantitative studies are scarce. An exception is a recent study by Godoy et al (2002) that estimates the contribution of rain forest to household consumption and earnings of two villages belonging to the Tsimané Amerindians of the department of Beni. The two villages chosen were at different distances from the closest market town. Forest goods accounted for about 53 per cent of the total value of household consumption. Forest goods included fish and game (about 36 per cent) and plants (about 17 per cent). The remaining part came from farm products (36 to 44 per cent) and purchases (4 to 11 per cent). The mean annual value of household earnings was the equivalent of US$286–385. The total forest contribution to household earnings (measured as sale of forest goods plus wage labour in forest activities) was 21 per cent and 45 per cent in the two villages, respectively. These numbers suggest a strong dependency upon forest resources. Yet, other groups closer to urban centres or with access to poorer forests rely more heavily upon agricultural income and wage labour.

NTFP extractors

Another group that is heavily dependent upon forest use for its livelihood consists of forest dwellers and peri-urban communities in northern Bolivia.

Table 9.2 *Attributes of forest-dependent social groups and contribution of the forest to livelihood*

Social groups	Location	Forest-dependent population	Area (thousand hectares)	Forest use	Importance
Indigenous groups	Dispersed throughout Beni, Pando, Santa Cruz and parts of La Paz and Cochabamba	180,000	22,483	Small-scale slash-and-burn agriculture, extensive use of non-timber forest products (e.g. Brazil nuts and palm hearts)	Highly dependent upon forest resources, mostly for subsistence
Small farmers involved in non-timber forest product extraction	Beni, Pando and northern La Paz	25,000–30,000	292 (?)	Slash-and-burn agriculture in combination with seasonal extractive activities	Highly dependent upon forest resources, mostly for income
Small-scale timber producers	Beni, Santa Cruz and northern La Paz	500*	800 (?)	Small-scale logging of marketable species	Highly dependent upon forest resources, mostly for income
Small farmers in valleys	Chuquisaca, Cochabamba, Santa Cruz and Tarija	700,000–800,000	5790	Fuelwood collection, where available	Highly dependent upon fuelwood
Colonos (settlers)	Chapare, Alto Beni, northern Santa Cruz and Yungas	500,000–600,000	3192	Clearing for conversion	Use forests for fuelwood and fodder
Wage workers	Forest areas in the lowlands	5600	N/A	Various extractive activities	Highly seasonal dependence upon forest resources for income

Note: * See the section on 'Small-scale timber producers'.

Source: adapted from Pacheco (2001)

This group relies on the extraction of NTFPs, such as Brazil nut and heart of palm, as the main source of cash income. It also depends upon subsistence-level agriculture. A small proportion of these people collect NTFPs within their parcels, while a larger proportion is hired to collect NTFPs within large Brazil nut establishments known as *barracas*.

A ballpark estimate of the size of this group is about 25,000–30,000 people (Pacheco, 2001). Bojanic (2001) provides slightly more conservative figures, while others (e.g. Prisma Institute, 2000) put this number at 40,000.[5] This group operates in an area of about 10 million hectares within the department of Pando, the Vaca Diez Province in Beni and northern La Paz. In this area, sometimes referred to as northern Bolivia, extraction, processing and export of NTFPs is highly important to the local economy. In 1999, two-thirds of national forest exports came from this region (US$48 million out of US$73 million), primarily from Brazil nuts (US$31 million) and palm hearts (US$2 million). By comparison, timber exports from this region amounted to US$15 million in 1999.

According to Stoian (2000), about one quarter of the income of NTFP extractors in northern Bolivia comes from rural activities (primarily NTFP extraction, but also agriculture production and agriculture wage labour). Rural income from these activities varies between US$1300 (about 11 per cent of total household income) for extra-regional migrants and US$600 (about 20 per cent of total household income) for town residents in Riberalta. The poorest and most forest-dependent group are ex–forest dwellers with a rural income of US$770 (about 32 per cent of total household income).

Small-scale timber producers

An unknown number of small-scale timber producers are distributed in a variety of semi-urban centres across the lowlands. Before the forest law was passed in 1996, they used to obtain a significant part of their incomes through informal logging operations conducted in public forestlands, concessions, indigenous lands, private forestlands and even protected areas. A large part of this group relies on a broad number of complementary economic activities (Pacheco, 1998, 1999, 2000). The number of members organized in associations is about 500; but this number is likely to be a gross underestimate (Pacheco, 2001). In some municipalities, they are now organized in ASLs, while in others, an ASL could not be formed due to lack of people or resources, or the organizational difficulties that creating an ASL comports.

Small farmers

A relatively large group consists of small farmers, or *campesinos*. It is estimated that this population amounts to 500,000–600,000 people. Many of them are Andean descendants who came to the lowlands driven by poverty. The livelihood of this group depends primarily upon agricultural activities carried out on lots smaller than 50ha,[6] but usually less significant than this.[7] While a small percentage of this group have legal title to the land, the great majority do not. Their dependency upon forest resources is limited to fuelwood and fodder collection. They do, however, substantially contribute to deforestation.

Industrial concession holders and larger landowners

Finally, two important groups of forest users are industrial concession holders and private landowners of areas larger than 200ha. Before 1996, a small number of families held 86 concessions covering an area of about 22 million hectares. Currently, the number of concessions held by these families has been reduced to 75, covering an area of 4.4 million hectares (see Table 9.1). This sharp reduction in concession area was allegedly a consequence of the change in taxation regime. The shift from a volume-based tax to an area tax discouraged the speculative holding of large forest areas. The remaining forestland reverted back to state control. Compared to the other groups, this one is relatively better off, although some variability exists within it.

We do not have estimates of the number of households who control areas (including forests) larger than 200ha or the extent of the area that they control. Their main activities are agriculture and cattle ranching. Steininger et al (2001) suggest that they are the main driver of deforestation in Bolivia, at least in the department of Santa Cruz.

Forest laws and livelihoods

The groups affected by the forest law

Indigenous peoples. With the agrarian law of 1996, the government of Bolivia recognized the exclusive right of indigenous peoples to own and control their ancestral lands, legally referred to as TCO. Before 1996, indigenous peoples had *de facto* control of their land for hunting, fishing and collection of NTFPs; but much of the land that indigenous groups would later claim as TCO was considered public land. This situation exacerbated conflicts with cattle ranchers, farmers and timber extractors, both small-scale and industrial operators. For example, many indigenous territories overlapped with forest concessions, national parks and claims of private ownership, creating tensions that have remained unresolved (e.g. CFB, 2000). Situations have, at times, escalated to the point where indigenous peoples and their leaders have been kidnapped, tortured, threatened and even killed. Such abuses have greatly diminished since, with the new regime, conflicting claims became more limited. Nevertheless, there are still isolated reports of indigenous peoples suffering physical abuse when they try to protect their timber resources from theft (Anonymous, 2003).

Indigenous peoples have, to date, demanded ownership of about 22.3 million hectares. The titling process, however, has been extremely slow and, by May 2003, had been completed for about 3.3 million hectares (CIDOB,

2003, unpublished data).[8] Within TCO lands, more than 5 million hectares are affected by settlements of small farmers not recognized as TCO members and without land titles (CIDOB, 2003, unpublished data).

As mentioned above, the commercial use of forests within TCO requires compliance with the same forest management regulations that apply to forest concessions. They include the development and approval of forest management plans and annual harvesting plans before harvesting can take place. Commercial forest use is also subject to the tax of US$1 per harvested hectare. Traditional and non-commercial forest uses, however, do not require government authorization and are not taxed (Lobo and Duchén, 1999).

NTFP extractors. The new forest regime treats the extraction of NTFPs in ways similar to the extraction of timber products. Extraction of NTFPs requires a management plan and payment of an area fee.[9] Most NTFP extraction is in the hands of families with past use of the land but no legal title. The law allows the awarding of concessions for the extraction of NTFPs. To date, however, only one concession has been awarded for the extraction of palm hearts in the Bajo Paragua region. NTFP concessions were 'grandfathered' (adopted, as was the case for timber concessions). They could not form ASLs.[10]

With the new law, families involved in NTFP extraction ended up being neither concessionaires nor owners. In the opinion of some respondents, the new forest regime increased the level of conflict between this and other groups (e.g. indigenous groups and farmers) that also claimed access to the same resources. The situation is particularly severe in the north where land tenure is especially confused and where overlapping claims are widespread (e.g. Henkemans, 2001).

Small-scale timber producers. Under the previous regime, small-scale timber producers operated largely illegally. In an attempt to 'formalize' this group of forest users within the forest regime and to legalize their access to public forest resources, the new forest law introduced provisions for the establishment of ASLs. Once formed, such local community groups could gain access to municipal forest reserves defined by municipal administrations, together with INRA and the Ministry for Sustainable Development and Planning (MSDP). Each municipality could allocate up to 20 per cent of its forest area as municipal reserve for such use. Following the law, norms were issued both to regulate the creation of ASLs and to establish municipal forest reserves. By the end of 2002, 53 ASLs had been created and 16 concessions awarded to ASLs in municipal forest reserves, with authorizations for harvesting covering an area of 420,000ha (SF, 2003).

Small farmers. Article 23 of the forest law gives landowners the right to clear forests (*chaqueo y desmonte*) for agricultural uses and to sell the timber cut in the process.[11] Policy developers expected these activities to be secondary and allowed them to avoid the waste of timber already cut. Alternatively, to cut trees, landowners could develop a forest management plan. Because the latter requires a land title, it is technically complex and costly to prepare, and it is subject to a lengthy approval process. Thus, many landowners turned to deforestation permits even when their main motive was to harvest timber

(Cronkleton and Albornoz, 2004). By the late 1990s, it became clear that these small landowners had neither the capacity nor the resources (and probably not the interest either) to develop forest management plans. After 1996, harvesting on one's own land continued, first by using the temporary measures instituted to ease the entry into the new forest regime, then by using deforestation permits and later by taking advantage of other mechanisms made available by the SF, such as 3ha authorizations. Intermediaries and larger forest operators often coordinated these harvesting activities.

Deforestation permits and 3ha authorizations allow small landowners (facilitated by intermediaries) to sell their timber while avoiding cumbersome and lengthy bureaucratic procedures. These permits, however, are not cheap. When the timber is sold, a deforestation permit costs farmers US$15 per hectare, plus 30 per cent of the value of the wood. For these reasons, many operators continue to harvest and transport timber without authorization in areas such as northern La Paz. In these areas, clashes with the SF have, at times, turned violent.

Strengths of the current system

As mentioned above, forest policy reform in Bolivia was guided by the vision to introduce better resource management and conservation, improve governance, recognize the rights of vulnerable groups such as indigenous peoples and provide a more equitable access to forest resources. It could be said that improving livelihoods for the rural poor was an implicit objective of these reforms. From a livelihood perspective, our opinion is that some strengths characterize the current system. They include the following.

Attentiveness to disenfranchised groups, including indigenous peoples. As noted earlier, under the previous regime several groups (e.g. indigenous communities and small timber extractors) could not participate in forestry activities except as illegal operators. The current system has changed this and enabled these groups to legally access, manage and market forest resources.

Attention devoted to ensure a strong, independent and capable central regulatory agency. It is the impression of many observers that the SF has played a critical role in ensuring an honest effort to balance the above objectives. Key elements of the SF way of operation include the following.

Commitment to honesty, transparency, accountability and sustainable forest management. For example, a survey conducted in 2001 has shown that, although various stakeholders resent the rigidity of many forestry regulations, the personnel of the SF is perceived as honest, committed and qualified.

Responsiveness to users' needs. Over the past seven years, the MSDP and the SF have struggled to retain a balance among the various, often conflicting, objectives of sustainable forest management, local development and industry competitiveness. At times, regulations were passed to improve management. At other times, certain regulations were relaxed to respond to the needs of certain groups. For example, the SF allowed various actors (e.g. ASLs, the TCO and companies) to conduct forest operations on their own land without a management plan, but simply with the commitment to develop one (*corta a cuenta*). In other circumstances, the SF chose not to apply the law – for example, when failing to seize trucks transporting illegal timber or to cancel concessions for which area taxes went unpaid.

Openness to self-criticism. There is an increasing acknowledgement within the SF that the forest law, although superb in many respects, was aimed at regulating the behaviour of certain actors (particularly small landowners and small timber extractors operating at the municipal level) without a good understanding of their existing organization, motivations and constraints. In the case of the ASL, for example, the law imposed an organizational structure based on shared rights and responsibilities that is quite different from the way in which people used to operate. The SF appears quite open to discussing these difficulties with the interested stakeholders and to finding innovative solutions.

Weaknesses of the current system

Although introduced with the implicit intention of improving livelihood for the most vulnerable groups, the current forest policy framework has yet to deliver on its promises. In our view, the problematic areas include the following.

High direct costs of compliance. The direct costs to prepare and have approved a forest management plan are sizeable. For example, Cabrera (1999) estimates that the elaboration of the plan for the area of Santa Maria (2433ha) was about US$20,000. To this cost, one must add annual costs (from the second year onward) of US$8000 to carry out the census and prepare the operational plan. Because all areas above 200ha are subject to similar regulation, these costs are independent of the intensity of harvesting and the capacity of forest users to develop and follow them.

Excessive bureaucratic requirements. Operators perceive the process for plan approval as complex, lengthy and bureaucratic. For example, complying with regulations requires a clarified status of landownership provided by ministries or the INRA, institutions that tend to operate slowly. Furthermore, in many cases, such certificates cannot be obtained because conflicts over landownership have not yet been resolved.

Unrealistic prohibitions. A problem that keeps being discussed is the legal prohibition to process timber in the forest using chainsaws (Article 75). Under the previous regime, a large amount of timber was extracted and commercialized by *cuartoneros* – small-scale timber extractors without large harvesting and transport equipment. The prohibition on the use of chainsaws in timber extraction was passed on environmental grounds since it was believed that chainsaw processing was wasteful. Unpublished research showed, however, that while chainsaw processing did result in larger amounts of wood waste, it also allowed for better utilization of the tree trunk and branches, with an overall positive net yield. Stocks (1999) concluded that the prohibition of chainsaw use for primary processing was also an obstacle to developing viable community forestry among indigenous groups.

Institutional failures. Complying with regulations also requires extensive technical expertise and financial resources not readily available to most TCO, ASLs and small landowners. In its design, the new forest regime considered various institutional mechanisms to face these challenges. For example, *Fonabosque* was to make available financial resources to enable the development of these plans. In practice, *Fonabosque* was never operationalized. At the same time, prefectures and municipalities, which were mandated to provide technical assistance for plan development and implementation, fell way short of meeting their mandate and expectations.

This state of affairs has left most operators with neither the knowledge nor the resources to do change things. This institutional failure, in time, translated into a form of 'dependency' upon NGOs for technical and financial resources. Because different groups have unequal access to outside support, compliance with the law is more difficult for some groups than others.

Ad hoc taxation. The new taxation regime differentiated between industrial concession holders, small timber extractors, private owners and extractors of NTFPs. This regime aimed at facilitating the entry of private owners (including indigenous groups) and small timber producers into the formal forest sector. Yet, the type and level of taxation have rarely been supported by rigorous analyses. As a result, debates continue about the unequal treatment of various operators. Furthermore, no one can tell to which degree taxation is related to the value of the resource or to what degree the level of taxation is linked to illegality. Such ad hoc taxation is, in part, related to a government failure: between 1996 and the present no single hectare of public land has been auctioned. An international and transparent auction would have provided some indication of the market value of Bolivian forests.

Ad hoc regulatory adjustments. Earlier, we praised the SF for not applying rigidly the letter of the law and for being responsive to users' needs. We balance this view with the impression that many of the 'exceptions' made by the SF have been unsupported by any study or analysis. For example, it appears that the 3ha authorizations were not subjected to consultations with important stakeholders or to analytical scrutiny.

Is there a good balance between sustainability and competitiveness? Finally, one cannot ignore the fact that the improvements in forest management, institutions, governance and basic rights for poor groups have yet to translate into long-term economic gains. The right balance between good environmental management, improved livelihood and long-term competitiveness has, in Bolivia, yet to be found.

We have attempted to qualitatively illustrate the main impacts of forest laws on the livelihoods of the four groups in Table 9.3. Given the absence of precise definitions, the qualitative descriptors used should be considered estimates, or working hypotheses.

Table 9.3 *Impact of the new forest regime on the livelihood of four forest user groups*

Livelihood criteria	Indigenous peoples	Small-scale timber users	Non-timber forest product (NTFP) extractors	Small farmers
Forest products income	+/−	+/−	+/−	+
Wage income	+	+	=	=
Agricultural income and consumption from clearings	=	n.a.	n.a.	−
Government services	+	+/−	+	n.a.
Physical security	+	+/−	+	=
Legal access to forests	+	+	+/=	+
Less vulnerability	+	+/−	+/=	n.a.
Sustainable management	+	+	=	n.a.
Participation in political process	+	+	+/=	n.a.

Notes: + means that the impact, overall, was positive;
− means that the impact, overall, was negative;
+/− means that the impact was positive for some members of this group but not for others;
= means that the impact was insignificant or it could not be assessed;
n.a. = not applicable.

The impact of forest legislation on livelihoods

It is difficult to draw conclusions on an issue characterized by so many dimensions and players. It can be deduced, however, that forest regulations affected various groups in different ways.

For operators who had been involved in forest activities before 1996, the new regulations introduced new requirements on forest access and use, thereby reducing their capacity to earn cash income. Concessionaires, small-scale timber producers and the chiefs (*dirigentes*) of certain indigenous groups, such as the Multiethnic Indigenous Territory and the Chiman Indigenous Territory, were negatively affected by the new management regulations. These groups had, in the past, benefited from the selective (and unregulated) harvesting and marketing of high-value species such as mahogany, cedar and oak.

The new regime extended the possibility of engaging in forestry to previously disenfranchised groups. For private landowners, small farmers and most indigenous groups, the new regime meant the possibility of legally harvesting the timber on their lands.

The forest law introduced greater transparency, accountability and provisions for a more equitable distribution of economic benefits. Therefore, even if in absolute terms the income accruing to a given group may have declined, it is now being distributed more equally. Difficulties in accessing detailed economic data have made it impossible in this project to assess the net economic impact of the law on the income of these groups.

Among other livelihood dimensions, the current forest policy has improved things for most social groups. These dimensions include the following (see also Nittler, 1999):

- a more transparent and accountable system of governance (clearer tracking of government revenues and their use, and clearer process of resolving land tenure conflicts and providing access to forest resources, etc.);
- a process of policy-making that is responsive to the needs of various stakeholders (although many claim that the government acts under pressure, citing, for example, the supreme decree that allowed the sale of illegally cut wood in the TCO Multiethnic Indigenous Territory and Chiman Indigenous Territory, the supreme decree to abolish the need for a deforestation plan on areas smaller than 5ha, and the supreme decree that allows timber harvesting on properties of less than 200ha without ownership title, etc.);
- legal access to forest resources by indigenous communities and private landholders;
- increased physical security, particularly for indigenous peoples and their land.

Illegal forest activities and livelihoods

Because of the points mentioned above, many groups have little choice but to engage in illegal forest activities. These activities are simpler, within the reach of forest users' capabilities, cheaper and socially acceptable. However, illegal forest activities may also affect rural livelihoods in negative ways (see Chapter 6). We consider here the two main types of illegal forest activities:

1. illegal activities related to timber harvest, transport and commercialization; and
2. illegal clearings (usually with fire).

Illegal logging and transport

Conflicting views exist on the extent and importance of illegal logging and transport in Bolivia. On the one hand, the recorded harvest in recent years has

Table 9.4 *Authorized versus actual harvesting in Bolivia (cubic metres)*

	Authorized	Actual	Ratio (percentage)
1998	1,379,324	797,220	58
1999	1,586,659	502,427	32
2000	830,717	495,835	60
2001	965,000	559,159	58
2002	1,636,440	581,782	36

only been between 36 per cent and 60 per cent of the authorized volume (see Table 9.4). In 2001, for example, the SF authorized the harvesting of 965,000 cubic metres; but official harvesting was only 559,159 cubic metres. Using data from the SF, the Corporación Andina de Fomento (CAF, 2003) estimated that this roundwood produced 256,000 cubic metres of sawn wood (about 45 per cent yield), out of which about 34,000 cubic metres were exported, while 222,000 cubic metres were consumed domestically. If one adds to these official figures, the estimates by the International Tropical Timber Organization (ITTO, 1996) and the Engenharia de Projetos Ltda (2000) that about 50 per cent of the wood harvested, transported and commercialized in the country has no authorization, the total amount of sawn wood that is commercialized in Bolivia is about 450,000 cubic metres. Assuming that domestic- and export-quality sawn wood have the same recovery rates (45 per cent), one arrives at a total harvest of about 1 million cubic metres in 2001 – a figure quite similar to the amount authorized. Looking at these figures, the volume of illegality in Bolivia does not appear severe.

Others, however, consider illegal logging an important problem for various reasons. First, it ignores planning requirements and fails to respect vulnerable areas and species. Therefore, it threatens the sustainability of resource use and the jobs that depend upon it, particularly in some frontier areas. Second, illegal actors compete with operators who attempt to manage forests sustainably without having to incur all of the management costs. Third, illegal activities result in lost government revenues that could be directed towards better forest law enforcement, environmental education and extension services.

Unfortunately, the interest in illegal logging has not yet translated into a comprehensive assessment of the problem and its causes in Bolivia. It remains a theme that is not well studied or understood. Existing studies tend to be qualitative in nature (an exception is Cordero, 2003) and to focus on a specific site (e.g. Roper, 2000; Uberhuaga, 2001). Stocks (1999) estimates that the wood confiscated by the SF for lack of authorization amounts to less than 10 per cent of the volume cut illegally.

More recently, using a survey of the *Unidad Operativa Forestal* of the SF, Cordero (2003) compared legal and illegal extractions during 2001 in indigenous territories in the area of San Ignacio de Velasco in the department of Santa Cruz and concluded that 88–98 per cent of the wood harvested in that area was taken illegally. Cronkleton and Albornoz (2004) detected severe abuses of the 3ha authorization in Guarayos Province.

We are unaware of any quantitative study that looks at the impact of illegal logging on livelihoods in Bolivia. From the interviews we conducted and the documents we reviewed, we found no evidence clearly suggesting the way in which illegal logging impacts upon livelihoods. Our opinion is that these impacts are circumstantial. There are cases where illegal logging amounts to blatant theft from the rightful owners of the land. There are other cases where illegal activities are carried out in accord with, and with the help of, landowners. When illegal harvest occurs on indigenous lands, we suspect that the deleterious effect on indigenous institutions documented by Tecklin (1997) and Roper (2000) continues. Finally, there are cases where illegal harvest takes place in areas of 'open access' or abandoned concessions, with harm being inflicted primarily on the state and the resource. When logging occurs in abandoned concessions, the operators who most benefit from these activities are intermediaries and companies, not the rural poor. Therefore, we suspect that their positive impact on livelihoods is rather limited.

Illegal deforestation

While estimated illegal logging is lower than authorized logging, illegal deforestation is much higher than legal deforestation. During the past three years, the SF has authorized annual clearings of 13,000–34,000ha. During the same years, the SF has detected violations and initiated administrative processes against illegal clearings in areas of 6500–37,000ha (SF, 2002). These annual detection data grossly underestimate the extent of unauthorized deforestation that takes place. A study funded by Bolfor[12] concluded that annual deforestation between 1993 and 2000 was about 200,000ha in Santa Cruz alone (Camacho et al, 2001). Santa Cruz is the department where deforestation is most severe and where over 60 per cent of national deforestation takes place (Steininger et al, 2001).[13]

Much of this deforestation has occurred in areas classified for conversion. For example, of the 1.4 million hectares deforested in Santa Cruz during the past seven years, about 58 per cent were situated in areas classified for agropastoral use (Camacho et al, 2001). Yet, significant deforestation has occurred within the permanent forest estate and protected areas. The reasons behind these illegal activities are complex and have to do with institutional failures (e.g. in the clarification of land rights); lengthy, cumbersome and expensive bureaucratic requirements (including the costs to prepare a land-use plan, a deforestation plan and the payment of taxes); pursuit of private gains without fear of penalties; and even some perverse incentives that still exist in the system.[14]

In these areas, the environmental damage caused by clearing may have been justified on economic grounds and sanctioned by the government. The economic damage, on the other hand, may have been more serious as perpetrators may well have failed to pay the applicable taxes and fees for harvest and commercialization. For example, if deforestation taxes were paid on 200,000ha instead of 34,000ha (as happened in 2002), revenues from this activity would increase sixfold from 10 million bolivianos to 56 million bolivianos, or about US$1.5 million to US$8 million, adopting a conversion rate of 1 boliviano = US$0.145. Yet, one cannot assume that increased government revenues would automatically translate into improved livelihoods. In fact, tax collection from deforestation activities accrues to *Fonabosque*, prefectures and municipalities, none of which have a track record of functioning properly.

It is difficult to say who benefited from this illegal deforestation and by how much. Steininger et al (2001) suggest that the majority of the deforestation that took place in Santa Cruz occurred to the benefit of large landowners and the agro-industry. Small landowners (e.g. indigenous migrants from the Andes), however, have also been a significant actor in illegal deforestation.

Some amount of illegal deforestation has also taken place in areas destined for long-term forest production. Camacho et al (2001) estimate that approximately 185,000ha have been cleared on such fragile land in the department of Santa Cruz between 1993 and 2000.

To date, the Camacho et al (2001) study is the only one that has estimated deforestation at the national level. Not much can be inferred about 'illegal' deforestation from the study, however, since it relies on data from 1993, three years before the new regime came into being. The SF is in the process of developing deforestation maps for 1996. These maps will determine the extent of deforestation after 1996 and will depict who and what areas suffered the most from these activities.

Conclusions

Policy options

During the late 1990s and the beginning of the 2000s, the Bolivian government started to think more in terms of incentives than in terms of rigid regulations and began a re-evaluation of how forest law enforcement was carried out. As an example of the former, in 2001 the MSDP passed a supreme degree specifying that wood for public projects (e.g. school buildings) must be purchased only from regulated sources. As an example of the latter, the SF has recently begun a process of institutional restructuring that aims to evaluate what did and did not work well over the past seven years. Key elements of this process of institutional restructuring include the following.

Clarify agency mission, goal and strategy. Within this process, there is increased recognition that the subjects of regulation should play a more important role in defining how to achieve the goals of ecological, economic and social sustainability. For example, the process that led to the development of the forest law excluded some important forest user groups (e.g. organizations of *motosierristas*), while others were considered but not sufficiently understood (e.g. small timber extractors).

Improve consistency between central and local offices. Officers operating at the local level are often unaware of new directives issued by the central office or apply them without full knowledge of the regulations. Even within the SF in Santa Cruz, the various *intendencias* are, at times, operating without a coherent vision.

Build internal capacity to simplify bureaucratic procedures. Reduced bureaucracy is increasingly seen as necessary not only to make forest management easier for forest users, but also to reduce regulatory costs and other barriers so that informal operators may enter the national forest regime. As part of this process, some revision of the current control mechanisms (largely based on the use and control of permits for the transport of forest products) will take place.

Improve the existing information system. Over the past seven years, the SF has collected a large amount of information. Yet, because critical information available in Santa Cruz is unavailable in local offices or at road posts, opportunities exist for fraudulent use of many regulatory instruments. For example, quarterly reports (*informes trimestrales*) containing information on sources and destination of wood are prepared and filed periodically with the SF. Yet, nobody is processing them or using them to improve forest management and timber processing control.

Promote a programme of awareness-building. Heightened awareness is needed at the national level among operators and professionals (who are often the ones who suggest doing things illegally), and even within the SF. The latter needs more professionals in communication, sociology, dispute resolution and negotiation. It needs to work with people in order to look for alternatives. It also needs to change its attitude from that of a police officer to that of a problem solver. This process has already begun, but must be improved and institutionalized. In a country in which the capacity to control is vastly inferior to the claims that the state makes, it is essential to establish better relationships between regulators and forest users.

Create and strengthen alliances with local NGOs. Forest law increased the possibility for NGOs and the public to participate in various facets of forest policy implementation. Yet, this opportunity has not been fully utilized by either the NGO community or the SF. Attempts at correcting this situation are under way.

While the above initiatives are being implemented, the following steps would also improve the relationship between forest law enforcement and livelihoods.

Extend coordination and clarification of goals and actions among various government agencies. While the SF is currently evaluating its goals and effectiveness, it is essential that other institutions (prefectures, municipalities and the MSDP) all align behind a congruent and coherent goal for forest-sector development and livelihood improvements. More coordination and streamlined bureaucratic requirements are needed not only at the central level, but also at local levels.

Reduce open access areas by transferring rights more efficiently. There are still several million hectares of forest that afford essentially open access.

Increasing the area under clear allocation of rights would reduce the extent of open access forests. The process of *saneamiento* has been slow. Parallel approaches to resolving existing land-use conflicts could be considered. For example, besides technical and administrative procedures, consensus-building or dispute resolution approaches could be used to settle disputes in some areas. Better coordination between INRA and the SF is also called for.

Reconsider the purpose of management plans. If forest management is envisioned as enhancing the livelihood of the rural poor, livelihood indicators need to be inserted in management plans. To do so, forest management plans should be simplified and modified in order to incorporate goals such as income generation and job creation, besides forest management. The Bolivian forest model is based on the assumption that technical norms will be sufficient to ensure ecological and economic sustainability. The validity of this assumption is questionable. There is an increasing perception that those who complied with the law did so at a cost. Was this necessary? Was it the best way of achieving this goal?

Develop indicators for monitoring and evaluating forest law enforcement impacts on livelihoods. By developing and monitoring livelihood indicators, it will become possible to learn about the impact of specific regulations on various social groups. The monitoring of a carefully chosen set of indicators will also enable quantification of the trade-offs among different livelihood dimensions or among different social groups. Knowledge of where we stand, where we are going and of these trade-offs will be an important input for future forest policy decisions.

Gaps and open questions

Two basic pieces of information are missing in order to better understand the implications of forest law enforcement and illegal activities for livelihoods in Bolivia. First, the scarcity of baseline information makes it difficult to determine in which direction livelihood dimensions – particularly the ones related to income and consumption – are moving. Second, few quantitative data exist on many livelihood indicators. Without these pieces of information, it remains difficult to assess the net impact of alternative regulations on a given group. It is also difficult to assess the existence of trade-offs among different livelihood dimensions or among various social groups. This chapter has highlighted several questions that could be the subject of future research.

How can institutional performance be linked to livelihood protection? This study illustrates that, in many respects, the way in which forest law enforcement affects livelihoods depends upon forest policy design and upon how the enforcing agency uses its discretionary powers. The SF has maintained high standards of credibility and accountability even while operating in a complex and difficult situation. What are the conditions that enabled the SF to operate as it did? To what degree can the effectiveness of the SF be attributed to inspired leadership, independence from political influence or support from the donor community?

What conditions needed to be in place to turn Bolivia away from poor governance, bad resource policies and disrespect for the rural poor onto a more virtuous course? Bolivia's progress in improving forest governance and recognizing the rights of indigenous groups appears to have been impressive. The catalysts for change, however, and the costs associated with the reforms are less well understood. What conditions needed to be in place? What coalitions needed to be built and how? What steps were taken and in what sequence? And how long will it take before such changes in basic rights translate into quantifiable changes in income and consumption? In short, how did Bolivia break the cycle of bad governance and disregard for the most vulnerable parts of its population?

Are livelihood improvements for selected groups and economic development conflicting with or reinforcing objectives? The reforms undertaken in Bolivia have produced significant changes. A considerable area is currently managed according to forest management plans, and – with about 1 million hectares of natural forest under independent certification – Bolivia is the tropical country with the largest percentage of such forest. Exports of high value-added products have also increased significantly and, in 2000, represented 70 per cent of total timber products exports. Legal access to forest resources has been opened to previously disenfranchised groups, and the fiscal benefits from forest use now reach local municipalities and communities. Corruption has greatly diminished and the regulatory agency, the SF, is generally viewed as honest and competent.

On the other hand, the sector currently faces serious difficulties. During recent years, total forest products exports have fallen and foreign investment has been minimal. Some forest concessions have been returned to the state and many industrial concessionaires have been unable and/or unwilling to pay their forest fees. Furthermore, poverty continues to afflict the majority of the rural population, particularly among indigenous groups. These conflicting outcomes raise important questions as to whether (and how) it is possible to balance the objectives of better management and improved livelihoods with long-term competitiveness and economic health.

What promise does sustainable forest management hold as a development strategy for rural communities? The chapter has shown that the Bolivian forest policy framework attempted to pursue a range of complex goals, from sustainable forest management to improved transparency and accountability in policy-making and use of the fiscal revenues, to promotion of community forestry and strengthening of the institutional framework. Implicitly, but just as importantly, it sought to make forest management competitive with alternative land uses. Are all of these objectives reconcilable? What lessons can be learned from this experience? The question of whether sustainable forest management can respond to the development needs of rural communities in Bolivia is now critical.

Notes

1. Operators who have only a chainsaw to carry out harvesting operations.
2. Forest activities for private consumption are exempt from regulations.
3. Under the previous regime, timber harvesting could not take place on private lands unless the owner had constituted and registered a company to legally harvest and commercialize timber.
4. The rationale, advantages and pitfalls of the change from a volume-based tax to an area tax have been described by many, including Contreras and Vargas Rios (2002).
5. This number, however, includes processors, retailers and intermediaries.
6. We picked 50 because a number of government programmes redistributed land to landless households. The allocation of land ranged from 20ha to 50ha.
7. There are other farmers who own or work land up to 200ha. We group them together with small farmers since the law treats them equally.
8. In July 2003, about 800,000ha have also been titled, bringing the total area fully titled to over 4 million hectares.
9. The proposed area fee of US$0.30 per hectare was never implemented because it was considered too high a tax. NTFP extractors currently pay nothing, while processing companies pay a per volume or per unit tax.
10. Contracts for the extraction of NTFPs can also be made with concession holders or private owners. Nevertheless, we are unaware of a case where such agreements have been made.
11. This legitimate activity was not 'legal' under the previous regime since only registered companies could legally harvest and commercialize timber.
12. Bolfor is a sustainable forest management project that started in 1992 with the support of the government of Bolivia and the US Agency for International Development.
13. High illegal deforestation is made possible by the low capacity of the SF to enforce clearing regulations. Since there is no organized system that contains information on who owns what, perpetrators of illegal clearing cannot usually be found. The situation is slowly improving now as INRA advances with the *saneamiento* process and as databases are being created. With the development of 1996 forest maps, the SF has begun to identify the owners of about 400,000ha who have illegally cleared their land.
14. For example, lands that are claimed as private but are not 'sanitized' (the titling process is incomplete, or a title exists but is insufficient) could revert to the state if the owner cannot demonstrate that the current land use fulfils a socio-economic role (*función economica y social* or FES). While there are many ways to demonstrate FES (including sustainable forest management, conservation, tourism and NTFP collection) the most traditional way is with agricultural and pastoral use. Because most INRA staff have an agricultural background and are accustomed to establishing the FES according to agricultural criteria, their judgement is biased in favour of agricultural use.

References

Anonymous (2003) 'Golpean a indigenas Bolivianos por impedir tala de bosque', *Quechua Network*, 8 April

ARD (2002) *USAID/Bolivia's Country Strategy 2004–2009: Bolivia Country Analysis of Tropical Forests and Biological Resources*, Report submitted to the United States Agency for International Development, Burlington, VT, ARD

Bojanic, A. (2001) *Balance Is Beautiful: Assessing Sustainable Development in the Rainforests of the Bolivian Amazon*, Bolivia, PROMAB Scientific Series 3

Boscolo M., Snook, L. and Quevedo, L. (2003) *Adoption of Sound Forestry Practices by Concession Holders in Bolivia: Practices' Attributes and Adopters' Characteristics*, Draft report, CIFOR

Boscolo M. and Vargas Rios, M. T. (2002) *Improving the Competitiveness of Sustainable Forest Management in Bolivia: What Instruments Do We Really Have?* (in Spanish), Bogor, Indonesia, CIFOR, BOLFOR and USAID

Cabrera, W. (1999) 'Plano general de manejo forestal zona agraria Santa Maria (Canton Yotau, Municipio El Puente, Provincia Guarayos, area 2433 ha)', Santa Cruz, Bolivia

CAF (Corporación Andina de Fomento) (2003) *Estudio de identificación, mapeo y análisis competitivo de la cadena de maderas y manufacturas en Bolivia. Programa Andino de Competitividad*, Caracas, Venezuela, Corporación Andina de Fomento

Camacho, O. M., Cordero, W. Q., Martinez, I. T. and Rojas, D. M. (2001) *Tasas de deforestación en el departamento de Santa Cruz*, Santa Cruz, Bolivia, BOLFOR

CFB (Cámara Forestal de Bolivia) (2000) 'Organizaciones campesinas imponen al gobierno revisión de la ley forestal', *Bolivia Forestal*, vol 2, no 14

Contreras, A. and Vargas Rios, M. T. (2002) *Social, Environmental and Economic Impacts of Forest Policy Reforms in Bolivia*, Washington, DC, CIFOR and Forest Trends

Cordero, W. (2003) *Control de operaciones forestales con enfasis en la actividad illegal*, Documento Tecnico no 120, Santa Cruz, Bolivia, BOLFOR

Cronkleton, P. and Albornoz, M. A. (2004) 'Uso y abuso del aprovechamiento forestal en pequeña escala – Provincia Guarayos', Informe presentado a la Superintendencia Forestal de la Nación, Santa Cruz, Bolivia, CIFOR

Duchén, R. (1998) *Seguridad jurídica para el aprovechamiento del bosque*, Boletin no 13, March, Santa Cruz, Bolivia, BOLFOR

Fredericksen, T., Putz, F. E., Pattie, P., Periona, W. and Pena-Claros, M. (2003) 'Forestry in Bolivia: From planned logging towards sustainable forest management', *Journal of Forestry*, vol 101, no 2, pp37–40

Godoy, R., Overman, H., Demmer, J., Apaza, L., Byron, E., Huanca, T., Leonard, W., Pérez, E., Reyez-García, V., Vadez, V., Wilkie, D., Cubas, A., McSweeney, K. and Brokaw, N. (2002) 'Local financial benefits of rainforests: Comparative evidence from Amerindian societies in Bolivia and Honduras', *Ecological Economics*, vol 40, pp397–409

Henkemans, A. B. (2001) *Tranquilidad and Hardship in the Forest: Livelihoods and Perceptions of Camba Forest Dwellers in the Northern Bolivian Amazon*, Bolivia, PROMAB Scientific Series 5

INE (Instituto Nacional de Estadistica) (2001) *National Census*, La Paz, Bolivia, INE

INRA (Instituto Nacional de Reforma Agraria), www.inra.gov.bo, accessed in 2003

ITTO (International Timber Trade Organization) (1996) *Promoción del desarrollo forestal sostenible en Bolivia*, Yokohama, Japan, ITTO

Lobo, R. and Duchén, R. (1999) 'El manejo forestal en las TCOs', *Boletin Forestal* 16, Santa Cruz, Bolivia, BOLFOR

Nittler, J. (1999) *Encuesta sobre la ley forestal y su implementación*, Documento Técnico 76A/1999, Santa Cruz, Bolivia, BOLFOR

Pacheco, P. (1998) *Estilos de desarrollo y su impacto en la deforestación y degradación de los bosques de las tierras bajas de Bolivia*, Serie Bosques y Sociedad 2, La Paz, Bolivia, CIFOR, Centro de Estudios para el Desarrollo Laboral y Agrario and Taller de Iniciativas en Estudios Rurales y Reforma Agraria

Pacheco, P. (1999) 'Influencia de las políticas públicas sobre los bosques', Paper presented to the seminar on La Política de Bosques en los Países de la Comunidad Andina, March, Lima, Peru

Pacheco, P. (2000) *Avances y desafíos de la decentralización de la gestión de recursos forestales en Bolivia*, La Paz, Bolivia, CIFOR and BOLFOR

Pacheco, P. (2001) *The Role of Forestry in Poverty Alleviation*, Rome, Forestry Department, FAO

Pavez, I. and Bojanic, A. (1998) *El proceso social de formulación de la ley forestal de Bolivia*, La Paz, Bolivia, CIFOR, CEDLA, TIERRA and PROMAB

Prisma Institute (2000) *Informe sobre la situación del sector forestal*, La Paz, Bolivia, Instituto Prisma

Roper, J. M. (2000) 'Whose territory is it? Resource contestation and organizational chaos in Bolivia's multiethnic indigenous territory', Paper prepared for the 2000 Meeting of the Latin American Studies Association, Miami, 1–18 March

SF (Superintendencia Forestal) (1997–2002) *Informe anuales*, Santa Cruz, Bolivia, SF

SF (2003) *Memoria 6 años de gestión Superintendencia Forestal de Bolivia*, Santa Cruz, Bolivia, SF

STCP Engenharia de Projetos Ltda (2000) *Plan estratégico para el desarrollo forestal de Bolivia*, Curitiba, Brazil, Informe para la Cámara Forestal de Bolivia

Steininger, M. K., Tucker, J. T., Townshend, J. R. G., Killeen, T. J., Desch, A., Bell, V. and Ersts, P. (2001) 'Tropical deforestation in the Bolivian Amazon', *Environmental Conservation*, vol 28, no 2, pp127–134

Stocks, A. (1999) *Iniciativas forestales indígenas en el trópico Boliviano: Realidades y opciones*, Documento Técnico 76/1999, Santa Cruz, Bolivia, Proyecto BOLFOR

Stoian, D. (2000) *Variations and Dynamics of Extractive Economies: The Rural–Urban Nexus of Non-Timber Forest Use in the Bolivian Amazon*, PhD thesis, Freiburg, Germany, University of Albert-Ludwigs

Tecklin, D. R. (1997) *The Mahogany Frontier: Timber Extraction and Regulatory Project in the Bolivian Amazon*, MA thesis, Berkeley, CA, University of California

Uberhuaga, P. (2001) *Legal or Illegal? Networks and Forest Regulations in Timber Commercialization: An Analysis of Social Capital and Interfaces among Ethnic Groups in the Tropic of Cochabamba, Bolivia*, MSc thesis, Wageningen, Germany, Wageningen University

10
Sustainable Forest Management and Law Enforcement: A Comparison between Brazil and Finland

Sofia R. Hirakuri

Introduction

The tropical region has the largest area of forest decline in the world. The causes of tropical deforestation are complex and are directly connected to the economic, social and political problems of developing countries (Leonard, 1985; Rietbergen, 1993; Bromley, 1999; Palo, 1999). These causes vary from one region to another, including clearing of the forest for agriculture or shift cultivation, cattle ranching, demand for fuelwood and commercial logging (FAO, 1999).[1] In many cases, the main causes of deforestation in Latin America have been policy choices by governments. For example, governments have often favoured the conversion of forests to agriculture, cattle ranching and other land uses (WRI, 1985; Repetto, 1990).

The deforestation rate in Brazil is not as high as in South-East Asia. Brazil has experienced uncontrolled timber exploitation, however, which began to accelerate during the early 1980s. In order to avoid further unplanned development, it is critical at this point to discover how to stop the trend of unsustainable timber exploitation.

The Brazilian government has made efforts to regulate forest management since the 17th century (Wainer, 1991). The federal government is responsible for the stewardship of a healthy environment[2] and the protection of flora[3] by making laws and applying them. Nevertheless, illegal predatory logging predominates in the Amazon, as documented by three major government reports[4] and many other studies. Consequently, national and international demand for better and more effective logging control has increased, and there is a requirement for more transparency in the process of restructuring public policies related to forest and non-forest policy (Ecoporé, 1992; GTA/Amigos da Terra, 1997). Recognizing that it is necessary to change the legislation and the mechanisms of application for forestry legislation, the Brazilian government has begun various initiatives to intensify control and to revise forest policies. Among the initiatives is the revision of forest management law to promote the adoption of sustainable management of private forested areas (SAE, 1997; MMA, 1998).[5] However, no data exist on the actual state of logging control to guide this revision process. The challenge is to ensure that forest management is conducted properly to guarantee the sustainability of forest resources. This will require appropriate enforcement actions by the government.

The Brazilian government has made efforts to enforce the forestry laws in many ways, but they have still ended in failure. In the Brazilian case, logging practices are highly regulated; the system itself is contradictory, however, because compliance with the laws to protect the forest is actually more difficult than non-compliance. The reasons for this failure are problems related to legal and administrative procedures, human resources and financial resources. These include:

- a complicated administrative procedure to procure logging permits;
- a deficient process of forest control;
- a low rate of compliance with forest laws;
- a deficient legal system to impose effective penalties;
- institutional problems within the enforcement division;
- the scanty budget allotted to field enforcement at the Brazilian Institute of the Environment and Renewable Natural Resources (IBAMA); and
- economic incentives unfavourable for sustainable forest management.

In order to deal with these problems, a model of successful experience is needed. This model should be adaptable and applicable in Brazil. Among various countries with a long tradition in forestry, Finland is an exemplary case of a successful model. Finland has high compliance: 96 per cent of forest owners complied with the forestry laws in 1997. In the past, for instance, from 1935 to the 1960s, the non-compliance level was relatively high and stable. Since 1965, however, Finland has experienced a substantial decrease in the number of violations and achieved the present state. Today's high compliance is a result of the emphasis on forest extension, economic incentives and many years of committed enforcement.

The following two sections review the forest administration system in Finland and Brazil. Each section considers the country's forest resources, the legal framework and the enforcement system. The third section carries out a comparative analysis of the two countries. This section considers lessons from the Finnish experience that could be adapted to Brazil and factors that are more difficult, if not impossible, to adapt.

Forest administration in Finland

Finland's forest resources

Finland has vast forest resources and is the most densely forested country in Europe, followed by Sweden, Austria, Germany and France (FAO, 1996). The forests are mostly coniferous and cover 71 per cent of the total land area.[6] Forest management is based on long rotation periods, generally 70 to 100 years in southern Finland, and 100 to 140 years in the northern area, depending upon the species (Hakkila, 1995; MAF, 1999b).

Finland was the first country to conduct a systematic national forest inventory, carried out between 1920 and 1923.[7] Since then, inventories have been carried out at ten-year intervals and show that Finland's forest resources have not diminished (Parviainen, 1992). The country's current growing stock volume is the largest recorded since 1917. Annual felling is smaller than net increment even in forests used for wood production (METLA, 1998a). Efficient forest management, rehabilitation of poorly growing forests, drainage, increasing density of forests and fertilization largely explain the increase in the forest stock growth (FFA, 1998).

In the past, however, Finland's forests were not so secure. The government became concerned about timber shortages and founded the National Board of Forestry in 1851 to monitor the situation (FFA, 1998; Reunala, 1999). A study conducted by a forestry expert found that the only pristine forests left were in Lapland and in the eastern region of Finland. Slash-and-burn cultivation had been the major cause of deforestation. Extensive tar production and cutting timber for shipbuilding, charcoal and fuelwood were other causes. Since the state of the forests was poor, forest management became an important issue. As a result, the government imposed restrictions on the utilization of forest resources, and field-oriented forestry education was recommended (Parviainen, 1992; FFA, 1998; Shepherd et al, 1998; Reunala, 1999). Sustainable production of wood has now replaced exploitation (Hakkila, 1995).

The forest industry developed considerably during the early 20th century. As a result, timber supply became an important national interest and the government made many efforts to promote forest management. It established the Forest Management Association, Tapio, in 1907 and the district forestry boards (currently, forestry centres) in 1928. In 1950, it passed the Forest Management Association Act, which created an association in every local government to provide professional services to forest owners (Reunala, 1999). The 1950s also witnessed massive investment in the forest products industry, which increased the demand for wood. This made it crucial to intensify wood production. To that end, the government provided public funding for forest improvement in the 1960s (Niskanen and Pirkola, 1997). This policy was so successful that 'during the latter half of the 20th century, forest-destroying cuttings and the threat of overexploitation of the forest resources [were] eliminated' (Hannelius and Kuusela, 1995, p134). Finland has not experienced any deforestation in the past 50 years (METLA, 1998a).

The forest sector accounts for one third of Finland's gross export income. The forest sector's share of gross national product has been gradually decreasing.[8] Forestry is still important, however, particularly with regard to exports. In 1998, the country exported US$38 billion total goods, of which forest products accounted for 29 per cent.[9]

Legal framework of forestry regulation in Finland

Constitutional framework
The Finnish Constitution establishes the principle that 'everyone is responsible for preserving biodiversity' and gives the public the right to freely access any forests without permission from the forest landowner. It permits all recreational activities provided that they do not harm property or nature, except in some restricted areas.[10]

Major laws governing forestry
Three major laws govern forestry in Finland:

1 the Forest Act of 12 December 1996
2 the Act on the Financing of Sustainable Private Forestry of 12 December 1996; and
3 the Nature Conservation Act of 12 December 1996.

The most important is the Forest Act, which guides and regulates silviculture in forests belonging to all kinds of forest owners and which has been in force since 1997. The Act on the Financing of Sustainable Private Forestry provides financial support to private forest owners. The Nature Conservation Act aims to maintain biological diversity, to conserve natural beauty, and to promote the sustainable use of natural resources and the natural environment, awareness of general interest in nature and scientific research.[11]

Evolution of forest management law
The first Forest Act was enacted in 1886. It prohibited the destruction of forests. The Private Forest Act enacted in 1928 followed the same principle. It prohibited forest owners from actions that would decrease the forest resource. The law also introduced many other innovative provisions. First, it required forest owners to reforest after clear-cutting (Greeley, 1953). Second, it prohibited companies from buying forestland so that forest ownership would not become concentrated. This provision had a fundamental impact on how private forest ownership evolved in Finland (MAF, 1999b). Third, the law introduced forestry organizations to supervise compliance with the act and to give assistance to private forest owners (FFA, 1998). The provision of the 1928 Forest Act that prohibits forest landowners from destroying the forest remains in effect in the current Forest Act of 1996 (hereafter the Forest Act).[12]

The goal of the Forest Act is 'to promote economically, ecologically and socially sustainable management and utilization of the forests in such a way that the forests provide a sustainable satisfactory yield while their biological diversity is being maintained'.[13] Until recently, the concept of sustainability was limited to wood production and the main goal of forest management was to increase volume growth. But the scope of the law has changed significantly since 1992. Following international trends, it now focuses more on social and ecological aspects, in addition to sustainable timber production.[14] The first law obliging forest owners to consider biological diversity was passed in January 1994. Since then, efforts to find a balance between the productive use of forests and protection of biodiversity have become crucial to forest management planning in Finland (Hakkila, 1995). Today, the main objective of sustainable forest management is to produce a stand of good quality and commercial value, while taking into account the biodiversity of forest ecosystems.

Enforcement system

Although the government enacted its first Forest Act in 1886, initially the law was not enforced effectively because the task was left exclusively to the government, as opposed to cooperative efforts between the government and private organizations. With limited personnel, the government could hardly supervise 20 million hectares of forests appropriately. The results did not meet the government's expectations (Portin, 1998). Consequently, the government revised the Forest Law and enacted the Private Forest Act in 1928 to improve the quality of forestry and the enforcement system. The 1928 act established district forestry boards (currently, forestry centres) to enforce the law (MAF, 1999b). Significantly, enforcement authorities started to emphasize forest extension services over punishment.

Administrative organization of forest control

The highest authority in forest policy is the Ministry of Agriculture and Forestry. The Ministry's Department of Forestry has four divisions. The first two, the Forestry Development Centre Tapio and the forestry centres, are important for forest management control.[15] The other two are the Forest and Park Service and the Finnish Forest Research Institute.[16]

The Forestry Development Centre Tapio supervises the enforcement of laws relating to forests by the forestry centres. It also develops Finland's forest policy to promote sustainable forest management and other forestry-related activities, including wood production, the use of forest resources and the protection of the forest's diversity. In addition, it guides and supervises state-financed forest management and improvement projects, and forest planning and extension (MAF, 1999a). The forestry centres are the authorities actually responsible for checking compliance with, and enforcement of, the Forest Act. They are responsible for implementing and monitoring forest legislation; drawing up forest management plans; giving advice on forestry regulations; promoting sustainable forest management by offering consulting, development, information, publication and training services; and providing forest tree seed procurement.[17] An independent group of decision-makers called the 'disassociated authority' oversees each forestry centre's decisions.[18] Among the responsibilities of the forestry centres described above, forest extension service and inspection have been the major means of ensuring that forests are managed properly and that the forest management requirements of the Forest Act gain compliance.

Forest extension approach
This law enforcement approach emphasizes forest extension more than punishment. The forest extension service provides technical support services to help forest owners carry out forest management practices, including silviculture and logging practices and identification of important habitats for the protection of biodiversity (Hänninen, 1999).

In cooperation with the forest management associations, forestry centres transfer technical knowledge to forest owners, forest workers and other entities involved in forestry through instruction and practical field demonstrations (Brevig, 1997). In Finland, forestry centres distinguish between two different target audiences: the forest owners and the forest workers. To do this, they use mass media,[19] arrange regional or local meetings and advise individual forest owners and forest workers (Brevig, 1997; Hänninen, 1999). Providing extension services by giving individualized instruction and information has proven most efficient. Through personal contacts, Finnish extension organizations were able to contact 82 per cent of the private forest owners over a period of five years during the late 1980s (Nikunen and Ranta, 1991; Hänninen, 1999).

Forest extension gives continuing training and education that helps forest owners to adapt to new circumstances or demands, as required by legislation or changes in society. Finland has been committed to promoting sustainable development since the early 1990s, when international pressure began to demand more careful forest ecosystem and nature management. Since then, Finland has included forest ecosystem and nature management in ongoing forestry practices, and the forestry centres have provided training on forest ecosystem management and identification of key biotopes.[20] In the fall of 1997, almost 6600 officials, machine contractors and forest workers took part in field training to identify the habitats that the Forest Act defines as valuable (FDC Tapio, 1997c).

Forest field inspection
The forestry centres conduct inspections of forest management operations. In the past, they used to conduct field inspections of 60 per cent of the total forest area. Today they inspect only a sample of 3 to 5 per cent of forestlands, which corresponds to about 8000 to 10,000 forest areas. They also inspect when forest-use declarations seem suspicious. The Ministry of Agriculture and Forestry defines the target of 3 to 5 per cent per year (FDC Tapio, 1999a). This goal of 5 per cent inspection has been a subject of debate in recent years. Some researchers consider 5 per cent insufficient to maintain a good quality of forest management.[21] The quality of forest management, especially forest regeneration, also concerns the government (FDC Tapio, 1997b).

The Forestry Development Centre Tapio developed the system for inspecting and monitoring sustainable forest management. As discussed above, however, the local forestry centres conduct the field inspections. To ensure the consistency of control measurements, all of the forestry centres apply an equivalent system to supervise the implementation of the Forest Act.[22] Finland's forest field inspection system is a tool to promote better forestry, rather than a means of punishing forest owners.[23]

Forest administration in Brazil

Brazil's forest resources

Brazil is a country with vast forest resources that occupy 551 million hectares, or 65 per cent of its total area (FAO, 1999). The Brazilian rain forest comprises 30 per cent of the world's forested areas (Mahar, 1989). Almost all of Brazil's rain forest area is in the Amazon basin.[24] Approximately 64 per cent of the Amazon is covered by rain forests; other types of vegetation such as savannas (*cerrado*) and natural prairies cover the remaining 36 per cent (Pandolfo, 1994). The forest is rich in biodiversity.[25] In addition to timber, the forest is rich in natural resources, such as aluminium, copper, tin, nickel, iron, gold, manganese and natural gas (Pandolfo, 1994).

The Amazon rain forest was almost intact until the 1960s. The first cause of deforestation in the coastal area, where temperate forests predominate, was the introduction of sugarcane plantations (Magalhães, 1998). Later on, the exhaustion of forests in the south-east region was due to massive plantations of cotton, coffee and other crops (Wainer, 1991).

The Brazilian forest sector has been important for the country's economy, although the contribution to the gross national product has fluctuated during the past 30 years.[26] Brazil is the largest exporter of forest products in Latin America, and Brazil's share in the world trade of forest products is gradually increasing.[27] According to FAO data, the export of forest products in Brazil has been increasing over the past 15 years.[28] The increase in exports of forest products in 1995 coincided with the year in which Brazil recorded its highest deforestation rate.

The amount of deforestation in the Amazon has been significant. According to data from the National Institute of Spatial Research (INPE), the total deforested area in the Amazon is equivalent to 14 per cent of its original forest cover.[29] INPE's data also show that the deforestation rate has been fluctuating since 1990 (see Figure 10.1). The peak was registered in 1995, when the deforested area totalled 29,050 square kilometres; since then, the deforestation rate has been gradually increasing, reaching 26,000 square kilometres in 2004. Experts predict that the forest will be reduced to one quarter in 50 years.[30]

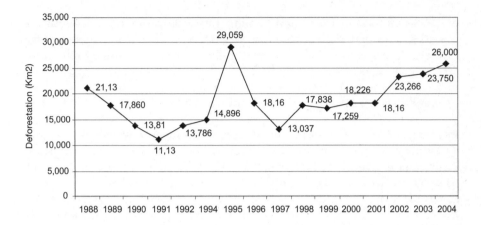

Figure 10.1 *Deforestation rate in the Amazon from 1988 to 2004*

Note: The Instituto Nacional de Pesquisas Espaciais only reports the average deforestation rate between 1993 and 1994.

Source: Instituto Nacional de Pesquisas Espaciais

Legal framework of forestry regulation in Brazil

Forest legislation

Legal instruments are differentiated according to their legal power. Laws and law decrees have the highest authority. Second highest authority is granted to the executive decree, which has the same normative value as any law but can be invalidated judicially. Least powerful are administrative acts (Freitas, 1995; Meirelles, 1995). The implementation of environmental legislation resides with the federal executive agency IBAMA. The problem inherent in this system is that the majority of forest-related regulations are based on administrative acts, and these lack stability because they can easily be altered without the approval of congress or be summarily invalidated judicially.

Major laws governing forestry

Three major laws govern forestry in Brazil. First, the Forestry Code, Law No 4.771, is the principal legal document of general forest policy and was established in 1965. Second, the National Environmental Policy, Law No 6.938 of 1981, establishes the basic principles of governmental action to maintain ecological balance, planning and law enforcement for the rational use of natural resources. The overall objective of this law is to preserve, improve and restore environmental quality appropriate for living standards that will ensure socio-economic development.[31] Third, the Environmental Crime Law of 1998 regulates criminal and administrative penalties for behaviour and activities harmful to the environment.[32] This law imposes strict sanctions upon forest-related violations and sets fines ranging from a minimum amount of 50 Brazilian real (US$31) to a maximum of 50 million Brazilian real (US$31 million).[33] Until the advent of this law just a few years ago, all fines were based on administrative acts that were easily overturned in the courts.

Evolution of forestry legislation and forest management law

The bulk of current forestry legislation is based on the Brazilian Forestry Code (Law 4.771/65) of 1965, which established that native forests could only be exploited with technical plans for forest management.[34] But the Brazilian government has responded slowly to the cause of forest management, taking 29 years to activate this legal provision. Meanwhile, forest management issues were handled through administrative acts without the compelling power of administrative decrees and normative legislation.[35]

Until the mid 1980s, little legislation regulated alternative land use and forest management. Between 1965 and 1985, only an average of 3.1 legal instruments per year were established, most of them related to economic development from forest-based industries. As a result of unplanned economic development during the 1970s, pressure on tropical forests in the Amazon started to accelerate. This pressure increased dramatically in the 1980s, when the depletion of forests became almost total in the south and south-east regions of Brazil – that is, the states of Paraná, São Paulo and Minas Gerais. During the second half of the 1980s, demand for better land-use regulation increased and included both alternative land-use and forest management.

In response, the number of legal instruments also increased, reaching an average of ten per year and peaking in 1989, when 15 legal instruments were established. The federal government established more rigid regulations for forestland use. It eliminated fiscal incentive mechanisms for the establishment of forest-based industries in 1988.[36] In 1989, it created requirements for the registration of legal reserve areas by notary public[37] and temporarily suspended fiscal incentives and credit lines for agriculture and cattle ranching.[38] The first executive decree to regulate the Forestry Code provision concerning native forest management was established in 1994.[39] In 1996, the government instituted a major change in one of the most visible forest protection laws. Until that time, the Forestry Code permitted landowners to deforest 50 per cent of their land; they were required to leave the other half in forest.[40] But the 1996 Provisional Measure reduced allowable deforestation to just 20 per cent in the hope of limiting further forest loss.[41]

The enforcement system

Forest control administration

In 1981, the National Environmental Policy Law provided for the establishment of the National Environmental Management System (SISNAMA) to implement environmental policies. SISNAMA maintains agencies at federal, state and municipal levels.[42] At the federal level, the implementing unit is IBAMA.[43] The Ministry of the Environment (Ministério do Meio Ambiente, or MMA), established in 1992, is responsible for the overall planning, coordination and implementation of national environmental policies, including the definition of general forest policy.[44] Created in 1989, IBAMA is the major institution directly responsible for forest administration, implementation and enforcement of forest laws.[45] IBAMA is an amalgamation of the major federal executive agencies that deal with natural resources.[46] According to Decree No 97.946/89, Article 1, IBAMA's role related to forestry includes 'registration, licensing, enforcing and regulating the sectors that utilize raw materials coming from natural resources (XI), as well as regulating the utilization of these resources and products or sub-products resulting from the exploitation (XIII)'. Although forest policy implementation falls primarily under the federal government's responsibility, state governmental agencies may also become involved in control of forest management plans as a consequence of new constitutional provisions (Hummel, 1995).

Decentralization of administrative organization

Until the advent of the new Brazilian Constitution in 1988, overall environmental policy, including forest policy, was administered exclusively by the federal government. However, the new constitution delegated some powers to state and local governments in an attempt to decentralize administrative responsibilities.[47] Almost immediately, states and municipalities responded to this initiative by creating environmental councils based on respective state constitutions and municipal laws (Ribas, 1991).

Although IBAMA was formed in 1989 – following the new constitution with its emphasis on decentralization – it adopted a centralized administration model like the federal government before it. At first, IBAMA's priority was to plan policies and handle governmental subsidies, functions which require a strong centralized structure (IBAMA, 1997). Since these two jobs are no longer the main responsibilities of IBAMA, the agency has strived to decentralize its administrative organization. But despite ten years of restructuring to achieve decentralization at the state and municipal levels, IBAMA has not yet finished instituting a more efficient administration of natural resources (SBS, 1995; IBAMA, 1997).

At the same time, forest control still falls under the centralized responsibility of the federal government. Organizations seeking permits to log or clear-cut in the Amazon must obtain them from the federal forestry control agencies. But some states[48] have initiated processes to control logging within their forests as a result of the new constitutional provisions (Hummel, 1995). Yet, municipalities have taken little or no responsibility for the implementation of forest policy, including enforcement (Ribas, 1991).

A comparative analysis of Finland and Brazil

Forest management law implementation and enforcement are very complex and differ from one country to another. This section examines whether the forest management enforcement system that has been successful in Finland can be adapted to the situation in Brazil. The enforcement of forest policy and the control of forest management plans are handled very differently in these two countries. Nevertheless, some of the approaches used in Finland could be applied to Brazil and other tropical forest countries. These can be divided into three broad categories: the regulatory approach, the market-oriented approach and the consensus-oriented/social-control approach. Besides these approaches, there are also those factors culturally specific to Finland that cannot be applied to Brazil.

Brazilian forest management is chaotic at present. Non-compliance with forestry laws is the main cause of this crisis. Finland, in contrast, has high compliance with forestry laws because of its emphasis on forest extension to promote good management of forest tracts. Compliance is also a result of many years of enforcement. The compliance rate among Finnish timber companies in 1997 was 96 per cent. Three per cent of the companies were given comments on their plans, providing them with recommendations to improve their forest management performance. Only 1 per cent of Finnish plans were judged to be irregular. In contrast, government audits in Brazil found the compliance rate with forest management laws to be just 30 per cent in 1996. Causes of non-compliance were many and varied, including corruption within IBAMA, deficient legal structure, the lack of any forest extension service and various institutional problems within IBAMA. And even among the 30 per cent of projects that appeared to fully comply with forest management laws, the management plans were not designed as a tool to produce sustainable timber, but rather as a means to satisfy legal requirements to procure logging permits (Silva, 1997).

Table 10.1 summarizes the key features of the Finnish system and the current problematic Brazilian system. These can be divided into three categories:

1 practices that can be easily implemented;
2 practices that are somewhat difficult and may take time to implement; and
3 practices that are more difficult to implement and may take a long time because they require changes in attitude towards the forest resources or changes in perceptions about the legitimacy of the governmental agencies involved in forest management.

Some policy-makers may argue that these culturally entrenched factors are impossible to change. However, it is always possible for any problematic behaviours or attitudes to change because the culture is changing as well. These reforms will require a series of sequential steps. Once one change begins to take place, it is easier to move on to the next one; thus, attempts at reform must start with the easiest steps.

Factors that can be changed

Regulatory approach

Forest management plans
Generally speaking, for the Finns, a forest management plan is a tool for forest management. Forest plans detail the measures that the industry takes during and after harvesting and describe the effects of those steps in promoting a healthy and sustainable forest estate. For many timber companies in Latin American countries, however, a forest management plan is simply a hoop that the company must jump through in order to obtain a permit to harvest timber.

The Forest Act in Finland does not require forest owners to have forest management plans; but there are two advantages to having one. First, forest owners can project economic profitability through a forest management plan (FDC Tapio, 1997a). Having a management plan also enables forest owners to obtain access to government financing to promote sustainable forestry (FDC Tapio, 1999b). A crucial feature of the forest management plan in Finland is that it is prepared by the government or by the forest management associations. That is, an agency which does not have a direct interest in the outcome of the forestry operations prepares the forest management plan. Since agents in Finland are independent, the plans that they recommend are more likely to be ecologically sound and environmentally sustainable. This should then result in good forest management if the owner implements the plan with care.

By contrast, the Forestry Code in Brazil requires forest owners to have forest management plans (Portaria No 486/86, Article 2). Ideally, a forest management plan is a tool to guide foresters or forest managers in conducting good forestry practices. Forest management plans in Brazil, however, are typically not used as careful guidelines for sustainable timber production, but only to meet legal requirements to obtain logging permits (Silva, 1997). In Brazil, a private forest engineer hired by the forest owner prepares the forest management plan. As an employee of the owner, the engineer frequently drafts a forest management plan for the benefit of the company without regard to the sustainability of the forests. Illegal logging practices exist because professionals, such as forest engineers, prepare, submit and implement fraudulent forest management plans. So, the punishment of forest engineers for fabricating or falsifying plans would be a valuable deterrent. A forest engineer can draft and follow up, effectively, three or four forest management plans. In Brazil, however, it is common that forest engineers submit and claim to follow up dozens of plans. Forest management plans are approved by unprepared officers, whose supervision is under the management of political appointees. In general, the head of the administration is nominated by a political party whose aim it is to protect the economic interests of certain classes.

Table 10.1 Comparison of the main features of enforcement approaches between Finland and Brazil

Enforcement approaches and their instruments	Finland	Brazil	Recommendations for Brazil	Degree of difficulty to change*
Compliance rate	96%	30%		
Regulatory approach				
Forest management plan	Not required by law; but 60% of forest owners have a plan	Required by law, but just a formality	Establish punishment for making fraudulent forest management plans	1
			Establish punishment for approving fraudulent forest management plans	1
			Establish an internal audit system	2
			Establish an external audit system	2
Forest-use declaration	14 days in advance	Not required		
Logging permit	Not required	Required, but just a formality	Notification of the start of logging	1
Field inspection	3%–5% of forest areas	Required for 100% of logging projects, but not conducted professionally	Random inspection	1
			Conduct inspection during logging	1
			Check the timber transportation system at strategic points	1
Legal sanction	Rarely applied	Applied, but low penalties	Activate sanctions established in the 1998 Environmental Crime Law	2
			Apply alternative penalties – make IBAMA a law and practice controller	2
			Improve the performance of public prosecutors	2
Institutional coordination	Clearly defined	Unclear	Clarify the separate state and federal responsibilities	1
			Establish an interagency coordination system	2
			Establish an independent audit system	2
Market-oriented approach				
Economic incentives such as:		No economic incentives for forest management	Establish public funding for forest improvement	2
Direct subsidies	Building forestry-related		Establish tax subsidy for forest management	2

Low-interest loans	Subsidies for forest rehabilitation projects and silviculture treatments	Establish economic incentives for forest management projects	2
	Road construction, fertilization, forest regeneration, prescribed burning, harvesting of energy wood, etc.		
Forest tax subsidy	Capital income taxes, yield taxes and property taxes	Establish a lower tax for timber from managed forests	2
		Place surcharge on timber from deforestation areas	2
Forest certification	Active	Incipient	Promote forest certification widely	2
Social control/ consensus-oriented			
Forest extension service	Widely used	None	Set up a forest extension service unit within the government	2
Advice	Private forest owners association (since 1920?) (Federal, regional and local levels)	None	Establish forest management association in every local government	2
Negotiation	Pivotal to resolve forest violations	None	Difficult to carry out	3
Other factors and background info			
Forest culture	Strong value of forest	No forest culture in general	Increase awareness about forest's value	3
Land tenure	Secure (90% by inheritance)	Insecure	Clarify the land tenure system	3
Attitude towards government authority	High legitimacy	Low legitimacy	Expand participation of society in decision-making process	3
Corruption index[†]	10	3.9	Establish punishment for public authorities involved in corruption	3
Forest management practices	70 years	25 years[‡]		
Land-use alternatives	Mainly forestry	Forestry, agriculture, cattle ranching and mining		

Notes: State government must be involved in forest administration, including permit approval, inspection, export control and control of international obligations.

* Degree of difficulty to change: 1 = easy to change; 2 = somewhat difficult to change; 3 = difficult to change.
† According to Transparency International, corruption index score ranges from 10 (least corrupt) to 0 (most corrupt).
‡ Forest management plans started to be required in 1977.

Logging permits

Whereas the Forest Act in Finland does not require a logging permit, the Forestry Code in Brazil calls for a logging permit to use forest resources. The Forest Act in Finland requires forest owners to present a 'forest-use declaration', a document equivalent to a logging permit, which contains all necessary information on logging operations, the regeneration method, planting and the preservation of forest biodiversity. The 'forest-use declaration' is an important tool for ensuring compliance with the law.[49] This declaration must be presented 14 days before any logging operations begin. This requirement makes the process of inspection effective and efficient. As a result, forest management control in Finland has been successful.

In contrast, in Brazil, the procedures required to secure a logging permit have been problematic. Applicants must work their way through a complicated administrative procedure to procure logging permits (see Figure 10.2). The bureaucracy is convoluted and slow, and legal loopholes make it easier for loggers to push through poorly designed plans. Logging permits can be either permits to harvest logs through forest management of natural forests or permits for clear-cutting (see Table 10.2).[50] The main problem is that it is easier to be granted a clear-cutting permit than it is to obtain a permit for logging under forest management. To follow the guidelines for forest management is considerably more costly and time consuming than simply clear-cutting; there are extra costs required to prepare the forest management plans, to pay field inspection fees and to prepare other paperwork. In addition, the current system requires a pre-logging field inspection of all medium- and large-scale properties for forest management. A lack of inspectors to conduct the field inspection results in long delays which, in turn, encourage the use of bribes and fraud to speed up the process. However, if someone seeks only to clear-cut a tract of land, no field inspection of any kind is required, either pre- or post-logging.[51] As a result, an entrepreneur spends an average of five months to obtain a logging permit under sustainable forest management, while permits for clear-cutting can be obtained in a matter of days (Barreto and Hirakuri, 1999). Furthermore, logging permits issued for small-scale clear-cutting of particular areas are often used illegally to log in areas other than those specified. It is vital to eliminate these barriers to forest management in order to protect the natural forest resources.

Forest field inspection

Finland also differs from Brazil with respect to its forest inspection system. In Finland, forest field inspections are conducted randomly rather than being made on all properties; in contrast, forest inspections in Brazil are supposed to be done on all properties.

Rather than using inspections to catch and punish forest owners for violations, forest field inspections in Finland are used to promote better forestry. The forestry centres inspect only a sample of 3 to 5 per cent of forest lands after logging. They also inspect when forest-use declarations seem suspicious. There are concerns among scholars that the goal of a 5 per cent field inspection is perhaps insufficient to maintain good forest management; but the overall results have been satisfactory, as shown by the high compliance rate. The Forestry Development Centre Tapio developed a uniform forest field inspection system to ensure the consistency of control measurements.

In comparison, the current Brazilian system is supposed to inspect all properties. However, factors such as deficient legal structure, lack of human resources and lack of financial resources cause enforcement problems for the sustainable management of forests. All of these factors in Brazil make it unrealistic to expect foolproof inspections. As noted earlier, logging through sustainable forest management is subject to a prolonged process of control before and after logging. The enforcement of forest management projects occurs in three stages after a logging permit is issued:

1 on the transportation of logs;
2 on timber industries; and
3 on the transportation of processed timber (Barreto and Hirakuri, 1999).

At none of these stages does the inspector visually check the area where the timber was supposedly extracted. It is crucial in any inspection and enforcement system to check the forest itself, observing and monitoring the logging activities as they occur. This would include careful pre-logging inspection to verify that the forest composition is accurately described in the proposal, as well as later inspection of the forest during and after logging. The results of audits carried out by the Brazilian Agricultural Research Corporation (EMBRAPA) showed that none of these inspections were conducted properly (Silva, 1997). Another problem is a lack of field inspection of clear-cut and illegal logging areas. Field inspections are carried out principally on properties that are in the process of authorization or already authorized. A significant part of logging, however, takes place in unauthorized forest areas using the documentation of authorized areas (Barreto and Hirakuri, 1999). This fraud is apparently difficult to detect in the field because, as mentioned earlier, the inspectors of timber transport do not check the origin of the timber.

Legal sanctions
Forest sanctions in Finland are not strict. They have rarely been applied because most conflicts can be solved through negotiation. The Finnish Forest Act does establish penalties such as fines and prison sentences; however, fines are seldom levied. When they are applied, most take the form of day fines, in which the fine for each day corresponds to one third of the net income of the violator.[52] The most commonly applied penalties are restraining orders or orders to restore the degraded forests.

By contrast, in Brazil fines are the most common penalty for infractions of forest law. Nevertheless, the legal system in Brazil is deficient in imposing effective penalties. Fines levied by IBAMA are not an adequate instrument to promote compliance by logging industries with their forest management plans. The comparison of levied and collected fines shows that violators pay

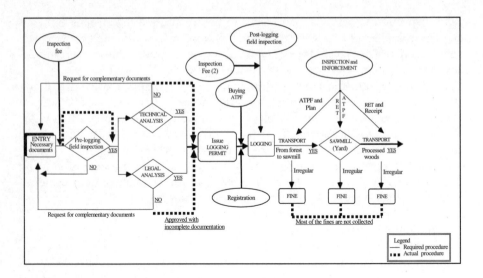

Figure 10.2 *Required and actual logging permit procurement process, inspection and enforcement actions in Brazil*

Notes: Inspection is not conducted during harvesting.
ATPF: Forest Product Transport Authorization
RET: Special Transport Regime

very low fines for illegal behaviour. For instance, illegal logging practices have been punished with low fines (the average fine being equivalent to US$243; the maximum, US$4000) due to a weak legal basis (Hirakuri, 2003). Thus, the fine system does not constitute a deterrent to violators. The actual collection of fines is only 30 per cent of the total amount levied.[53] The amount of collected fines and the amount spent by IBAMA's enforcement units were compared. Taking all states of the Amazon region together, enforcement activities in forest management represent a loss for the federal government. The amount of fines collected in 1997 covered half of the enforcement costs in only two states of the Legal Amazon. Thus, the establishment of an effective system for collecting fines and for imposing fines sufficiently punitive to deter the illegal behaviour of forestry industries is urgently required to improve Brazilian forest management.

Institutional problems
Generally speaking in Brazil, institutional problems such as a lack of financial resources, lack of institutional coordination, lack of personnel and training, and lack of sustainability of governmental institutions have led to poor enforcement. These four problems often are so intrinsically linked that it is difficult to analyse each issue separately. This section discusses the lack of

Table 10.2 Legal requirements and limitations of logging permits in the Amazon

Legal requirements	Forest management			Clear-cutting permit			
	Large scale	Simplified	Simplified		Property		Collectively owned
	Any size of property	Community area < 500ha per year	Private area < 500ha	< 100ha	> 100ha ~ < 200ha	> 200ha	Extractive area > 100ha
Technical project	Complete management (a)	Simplified management	Simplified management	Simple requirement	Simple requirement	Logging operation plan	Simple requirement
Annual operation plan	X	X	X				
Harvest limitation	Local productivity rate	Extract five trees per hectare	Extract five trees per hectare	Maximum 3ha per year	20% of total area		Maximum 5ha per year
Environmental impact assessment (EIA)	(b)					Property > 1000ha (c)	
Legal reserve	Logging allowed	Logging allowed	Logging allowed	20%–50%	20%–50%	20%–50% (d)	20%–50% (e)
Keep a permanent preservation area	X	X	X	X	X	X	X
Inventory of 100% of tree species to be harvested	X	X	X				
Prior-logging field inspection	X	X	X			X	
Follow-up field inspection	X						
Procurement of log transport permit (ATPF) (f)	X	X	X	X	X	X	X

Notes: (a) A complete forest management project includes the establishment of cutting cycle and measurements of forest regeneration after logging.
(b) An EIA is not required for forest management as defined by the National Council for the Environment (CONAMA). It is necessary, however, to describe any negative impacts that logging may cause and remedial measures, which are included in the forest management plan.
(c) State environment agencies may require EIAs for activities in smaller areas with high potential impacts – for example, intensive use of toxic substances.
(d) In savannah areas, the legal reserve area should be only 20 per cent of the property (Medida Provisoria 1.885-38, of 29 June 1999).
(e) It refers to 50 per cent of the property. However, the clear-cutting area must not be over 20 per cent of the property (Medida Provisoria 1.885-38, of 29 June 1999).
(f) The Forest Product Transport Authorization (ATPF) is used by IBAMA to control the origin of timber and harvested volumes.

Source: based on Decree 1.282/94; Portaria 48/95; Decree 2.788/98; Instrução Normativa No 5/98, No 6/98 and No 7/99.

financial resources and institutional coordination since these two problems are common to Finland and Brazil.

Lack of adequate financial resources. Lack of adequate financial resources is considered one of the major constraints on forest law enforcement. Although both Finland and Brazil lack adequate financial resources, there is a major difference between the Finnish and Brazilian budget allocations.

Finland experienced a major decrease in general public expenditures during the early 1990s. This resulted in cuts in the government's inspection budget and forestry extension budget (Hänninen, 1999). The budget cutbacks started to affect the quality of forest management. Because of the problems detected in the quality of forest management and changes in the Forest Act requiring more forest extension assistance due to new requirements regarding environmental issues, the government reassessed the budget for forest enforcement. The budget trend of the forestry centres during the years 1999 and 2000 shows that the government emphasized forest management, as well as forestry information, extension and training, increasing the budget allocation by 60 and 80 per cent, respectively (Hirakuri, 2003). It should be noted that the trend in budget allocation has been considerably higher for forest extension than for inspection itself.

Like Finland, low budget allocation for enforcement of forest laws has been one of the causes of poor enforcement in Brazil. Over the past several years, the budget allocation for enforcement has decreased relative to the number of approved forest management projects.[54] The financial resources available to regional offices are insufficient to support the actions that they are legally required to implement. The costs of implementing the present enforcement system, including pre-logging and post-logging field inspection and timber industry inspection, are not covered by the financial resources allocated for this purpose. The most costly stage of the forest management project control is the pre-logging field inspection (Hirakuri, 2003). Eliminating this preliminary field inspection stage would cut operational costs by almost half. In contrast to Finland, where the government emphasizes forest extension and forest management, in Brazil the enforcement budget designated for forest management is small and is non-existent for forest extension.

Institutional coordination. Institutional coordination, such as communication between various agencies and internally among divisions, plays an important role in effective implementation of forest control because it prevents overlapping responsibilities. Whereas the Finnish government has clearly defined institutional coordination, the Brazilian government presents a confusing and uncoordinated institutional structure.

In Finland, the responsibilities of each governmental agency in charge of forest law enforcement are clearly defined. The Forestry Development Centre Tapio, among other responsibilities, supervises the enforcement of forest laws by the forestry centres and develops Finland's forest policy. The forestry centres are the authorities actually responsible for checking compliance with and enforcement of the Forest Act. By contrast, institutional coordination in Brazil has floundered. Brazil has a number of governmental agencies involved

– none with complete authority to issue an infraction report – to inspect forest activities or to confiscate illegal forest products. The infraction report illustrates a typical example of this problem. IBAMA has exclusive authority to issue infraction reports.[55] So, when Fundação Nacional do Índio (FUNAI) agents encounter illegal logging practices, which often occurs within indigenous reserves, or when the federal or civil police encounter an illegal forest-related activity, FUNAI agents or police officers have limited powers to charge the violators since IBAMA's agents must issue the infraction reports.[56]

In Finland, the authority responsible for forest control works closely with the local forest management organizations and this unique institutional structure has played a major role in forest control. This development of cooperation skills, 'linking up the enforcing agencies with established organizations, such as forest management associations, has made technical services and skills available to every forest owner coming under the law' (Greeley, 1953, p106). In addition, forestry centres, which are responsible for implementing and monitoring forest legislation and drawing up forest management plans, are subject to an audit carried out by an independent group of decision-makers, the 'disassociated authority', which oversees each forestry centre's decisions. By contrast, no systematic external audit system exists in Brazil. As a result, problems such as technical gaps in the logging permit issuance process are never resolved. The only audit ever carried out was a field survey by EMBRAPA in 1996. This study revealed that forest management projects in the region of Paragominas (Pará State) did not follow silvicultural treatments, logged trees that had diameters smaller than legally authorized, overestimated the volume of commercial timber, and did not present prospecting inventories of timber species to be harvested. In short, the industries failed in a variety of ways to implement the planned forest management techniques (Silva, 1997). This audit suggests that forest management plans present widespread problems. This kind of survey was unprecedented and it was not repeated because there is no systematic, independent auditing system of forest control in Brazil.

Market-oriented approach
Important in the analysis of forest resources management are fiscal and financing instruments, and the establishment of forest certification as a market instrument. This section discusses those instruments as used in Brazil and Finland to promote forest management.

Taxes, subsidies and low-interest loans
Government policies, such as taxes and subsidies, have a strong impact on forest management. In Finland, governmental policy has had positive impacts in promoting sustainable forestry; on the other hand, the policies in Brazil have had a negative impact on the forest, causing deforestation and degradation.

In Finland, economic incentives for forestry supplement a regulatory approach. They have been essential supports for forest owners in carrying out the forest improvement work that is necessary for timber production. Finland started to use this instrument during the 1950s and it reduced illegal forest activities. The 1997 Act on Financing Sustainable Forestry broadens the objectives not only to ensure the sustainability of timber production, but also to maintain biological diversity and to undertake forest ecosystem management.[57] One of the main requirements for the eligibility of financial support is to have a forest management plan. So, the Financing Act indirectly obliges the forest owner to have a management plan, which helps in national forest planning. By contrast, the current system in Brazil provides no economic incentives to promote sustainable forest management. Rather, the government established fiscal and financial incentives to encourage economic development in the Amazon, favouring the establishment of agriculture and cattle ranching. In terms of forestry, during the 1960s, forest-based industry development was encouraged by forest-sector tax incentives. These incentives were provided to promote investment in large-scale industrial forest plantations rather than forest management of natural forests.[58] This initiative resulted in the establishment of about 6 million hectares of forest plantations, mainly concentrated in the central-southern region of the country. In the 1970s and 1980s, other programmes for the expansion of pulp and paper production, and charcoal production for the iron and steel industry, were introduced.[59] These fiscal incentives were gradually phased out; but they were replaced with other economic incentives designated for agriculture and cattle ranching[60] that do not favour forest management of natural forests.

Nevertheless, this discouraging situation seems to be improving. One of the major financial institutions in Brazil, the National Bank of Economic and Social Development, was planning to provide the first low-interest loan for a forest management project in the Amazon.[61] As we have seen, the public policies of fiscal incentives and bank credit lines favoured deforestation with no counterpart for positive results (Hummel and Minette, 1990). The lack of financial mechanisms to promote forest management is jeopardizing the forests. According to public policy specialists, the subsidies given in Latin American countries have proven not only to be uneconomical and environmentally unsustainable, but have also created opportunities for corruption (Hafner, 2000).

An economic incentive approach benefits the forest landowner and the industry. It does not burden forest landowners with heavy management costs. Finland put this approach into practice during the 1960s. Brazil, however, is just beginning to introduce proper economic incentives for promoting sustainable forestry. The evidence is strong that economic-oriented approaches result in greater 'environmental' gains than most direct on-the-ground regulations, and they are less costly (Hyde, 1998).

Forest certification
The forest certification system in Finland reinforces sustainable forest management. The system recognizes the special characteristics of forest ownership in Finland – namely, that most forest holdings are small and privately owned. Its main feature is that certification is carried out in the form of regional group certification, which differs from other countries. Forest owners in Finland have

widely accepted the forest certification system, as shown by the large number of hectares certified since its establishment in late 1999.[62] On the other hand, certification has not led to any major changes in forest management practices. It appears that forest certification in Finland has largely been carried out to gain greater access to timber markets for export and has increased the total cost of the system with few immediate improvements in forest management. It is a good sign that a forest certification process is gradually developing in Brazil. The objective of forest certification in Brazil is to have wide access to new markets, rather than to promote forest management. One particular initiative indirectly linked with the forest certification scheme is the establishment of the Group of Buyers (Grupo de Compradores) of Certified Timber.[63] Forest certification will provide opportunities for timber companies to have access to markets for sustainable products in Brazil (Amigos da Terra, 1997).

Although the forest certification system seems very promising for the sustainability of forests, there are many legal and institutional problems that must be overcome, such as legal deficiencies in the enforcement procedures, lack of financial resources and lack of institutional coordination.

Social control/consensus-oriented approach

Forest extension service

The forest extension service, as discussed earlier, helps forest owners to carry out forest management practices, including silviculture and logging practices, and to identify important habitats for the protection of biodiversity. Therefore, the forest extension service is one of the essential factors in achieving sustainable forestry. Forest extension has been the backbone for sustainable forestry in Finland; in contrast, forest extension services on a large scale do not exist in Brazil.

The forest extension service offers advice to forest owners on good forest management practices and practical forestry field training. In Finland, the constant training of forest owners has been at the core of forest extension. The emphasis on the forest extension service makes a difference with regard to compliance with the law. Forest owners comply with legal requirements because they know what they are supposed to do – for example, in terms of forest management and the protection of biodiversity. The establishment of forest management associations in every local government has been vital in carrying out forest extension. By contrast, in countries such as Brazil, Bolivia and Costa Rica, forest owners and forestry-related workers have not received enough – if any – training or information on forest management. Particularly in Brazil, a lack of forest-sector organization that offers forest extension to conduct forest management has resulted in low compliance with forest laws because forest workers do not know what they are supposed to do. This lack of a forest extension service has jeopardized the sustainability of forests. Forest extension service programmes are just beginning on a pilot project basis through private initiatives in the Amazon region;[64] but this will not be enough to serve extensive territories such as Brazil or other large tropical forest countries.

It should be noted that the need for forest extension services is greater in Brazil, where the workers (loggers) themselves are not the forest owners; the loggers are mostly local people hired seasonally and have no background in forestry. There is a risk that a forest owner may train a group of workers and that these workers will be gone the next season. Since a high turnover of workers is expected, there is an urgent need to disseminate forest extension services as widely as possible so that the owners' risk will diminish. As discussed above, one way of expanding forest extension is to ensure that local organizations promote forest management.

Negotiation

A consensus-oriented approach consists of negotiation with the forest landowners to encourage them to comply with forestry laws. Whereas in Finland negotiation with the owner has been pivotal in resolving most forest violations, Brazil has relied heavily on the regulatory approach.

In Finland, the consensus-oriented approach in gaining the cooperation of forest owners has contributed to sustainable forestry. This approach has been emphasized since 1928, when the first Forest Act was established. Negotiations are possible in Finland because forest owners are aware of the legal requirements as a result of the government's education efforts. Therefore, the forest owners know what is at stake and what they need to do. The outcomes of negotiations have ranged from planting at the owner's cost to prohibiting further cutting until past infractions were resolved.[65] In contrast, there is no negotiation system in Brazil. Brazil has relied mostly on regulations such as the permit and fine system, which have failed to promote forest management. Negotiation in Brazil would be difficult to implement because it requires a high degree of compliance as a foundation, which is not common in Brazil. In addition, there are problems of widespread corruption. Realistically, the government is more a reflection of economic power than a neutral entity with the ability to act in the public interest. In the case of Brazil, emphasis on negotiation does not make sense because its governmental officials are especially swayed by political power, which serves economic power more than the needs of the general public.

Factors that are more difficult to change

The different approaches to law enforcement, discussed in the previous sections, affect the degree of compliance. Besides those approaches, which can be applied to other countries, there are also factors that are culturally specific to Finland which cannot be applied to Brazil. These are forest culture, forest ownership and government legitimacy.

Forest culture

The forest is strongly connected with the Finnish way of life. Forests have played an important role in shaping Finland's economy and culture. Besides economic benefits, Finns have enjoyed and respected the non-material values that forests offer. These values include beliefs, myths and customs related to the forest. In fact, the forest is reflected in every aspect of Finnish culture, including literature, music and paintings.[66] Moreover, the recreational use of forests has been a part of Finland's traditions. By contrast, 'forest culture' is almost non-existent in Brazil. The forest in Brazil has not been considered an important asset. The lack of a forest culture resulted in the conversion of forestlands to other land uses because forests were not valued. The forest has been considered unfriendly and synonymous with disease and terror, and the clearing of forests has been regarded as a sign of progress (Nascimento, 1988; Nascimento and Kengen, 1988; Pandolfo, 1994). But this view, of course, disregards the perspectives of indigenous peoples and probably also the viewpoints of *ribeirinho* (riparian communities).[67]

Forest land tenure
Land tenure has been a concern of sustainable forest management because property rights are a prerequisite for the application of economic instruments, such as bank credit subsidies or governmental direct subsidies. Finland and Brazil differ in forest ownership structure. Forest ownership in Finland is based on private smallholdings and secure forestland tenure. Quite the opposite, Brazilian forest ownership is based on private large holdings and insecure forestland tenure.

In Finland, forestry based on small-scale and private forestry has favoured sustainable forest management. Private ownership resulted from a particular historical situation involving a long tradition of an independent peasantry (Ripatti, 1996). Thus, Finnish forestry is largely family forestry with secure land tenure. Private forestry has played an important role in the forest industry's timber supply, accounting for 75 per cent of the wood used by the industry (METLA, 1998a; MAF, 1999b). Therefore, forest policies have been implemented to support private small-forest owners. Finland has succeeded in establishing a system of forest law enforcement and economic instruments that do not discriminate against small-forest landowners, as occurs in many tropical forest countries.

By contrast, insecure land tenure is a barrier to investment in a long-term project such as forest management in Brazil. Unlike Finland, where small holdings averaging 36ha in size predominate, in Brazil the size of landholdings is large, averaging 1000ha.[68] Most of these lands are held as a real estate investment. As a result, the clearing of large properties is often motivated by land speculation, rather than promoting forest management. The acquisition of forestlands in the Amazon region is spurred by the weak land tenure system that favours deforestation. The requirement for sustainable forest management is a long-term commitment to maintaining forests. In Brazil, investment in forestland is very risky due to frequent forest land tenure disputes.

Legitimacy of the government
The political and economic system in Finland can be characterized as a developed democracy. The legitimacy of public authority is high because of the high level of education and the tradition of democratic political culture. As

a result, Finnish people's attitude towards the government and its legislation is positive – that is, they comply with laws and administrative guidelines. Forest owners follow guidelines because they trust the government. In Finland, it is culturally traditional to obey laws. Another factor accounting for the high legitimacy is the legal structure, which provides for the oversight of administrative organizations at the federal, regional and local levels.[69]

By contrast, in Brazil the level of legitimacy of public authority is very low. Corruption has been one of the major reasons for difficulties in enforcing forest laws.[70] Some of the conditions that encourage corruption are intricate bureaucracy, which results in delays in obtaining documents and services; loopholes that allow evasion of both forest inspections and timber transport inspections; and the low salaries of public servants (Júnior, 1997). Obviously, legitimacy of government cannot simply be transplanted to Brazil because its political history is uniquely different from that of Finland and other developed countries.

Notes

1   Commercial logging primarily for internationally traded species of tropical hardwoods is a cause of deforestation in most tropical forest countries, although it varies from country to country (Leonard, 1985).
2   Article 225, *caput.*, Constituição da República Federativa do Brasil.
3   Article 225, §1, VII, Constituição da República Federativa do Brasil.
4   First, evidence accumulated by the Brazilian Institute of Agriculture Research (EMBRAPA) shows that the federal government has failed to enforce forestry regulations in relation to forest management, indicating widespread failure to implement forestry laws (Silva, 1997). Second, the technical screening of forest management plans, as conducted by the Brazilian Institute of the Environment and Renewable Natural Resources (Instituto Brasileiro do Meio Ambiente e Recursos Naturais Renováveis, or IBAMA), shows that 75 per cent of the plans presented irregularities (IBAMA/MMA, 1997). Third, the report of the Federal Strategic Affairs Secretariat shows that 80 per cent of timber production in Brazil comes from illegal sources, indicating the widespread illegality (Secretaria de Assuntos Estratégicos, 1997).
5   The first step to revise the law was promulgation of Decree 2.788 on 29 September 1998, which altered Articles 1, 2, 3, 5 and 6 of Decree 1.282/94 related to forest management.
6   Water covered 9.9 per cent; agricultural land, 8.1 per cent; built area, 2.8 per cent; and other areas, 11.2 per cent (MAF, 1999a).
7   In many other countries, national forest inventories started to be carried out during the 1950s.
8   Thirteen per cent in 1970; 11.5 per cent in 1980; 7.4 per cent in 1990; 9.1 per cent in 1995; 7.4 per cent in 1996; and 7.9 per cent in 1997 (METLA, 1998a).
9   The metal industry accounted for 53 per cent; the chemical industry, 9 per cent; and other industries, 9 per cent (MAF, 1999a).
10 The restrictions apply to cultivated fields and plantations, around people's home and to protected areas such as nature reserves (Ministry of the Environment, 1999).

11 The Nature Conservation Act 1096/1996.
12 The Forest Act 1093 of 12 December 1996 (Metsälaki 1093/1996) (hereinafter Forest Act 1093/1996).
13 Forest Act 1093/1996.
14 This was a prominent feature of the United Nations Conference on Environment and Development held in Rio de Janeiro in 1992 and the Ministerial Conference on the Protection of Forests in Europe held in Helsinki in 1993.
15 There are 13 regional forestry centres:

   1 coast;
   2 south-west Finland;
   3 Häme-Uusimaa;
   4 Kymi;
   5 Pirkanmaa;
   6 south Savo;
   7 south Ostrobothnia;
   8 central Finland;
   9 north Savo;
   10 north Karelia;
   11 Kainuu;
   12 northern Ostrobothnia; and
   13 Lapland.

16 The Forest and Park Service is in charge of managing state-owned forestlands and protected areas such as nature reserve, wilderness and conservation areas. It also operates recreation services such as information centres and offers services such as accommodation, hiking and fishing licences (MAF, 1999b). The Finnish Forest Research Institute (METLA) was founded in 1917 with its headquarters in Helsinki. It is divided into eight regional research stations. Its objective is to solve forest-related problems through scientific research (METLA, 1998b). Detailed information about METLA is available at www.metla.fi.
17 The other responsibilities are to draw up multiple-use, landscape and environmental plans; develop economic activities based on forestry; provide advice on forestry planning and extension for forest owners; advise on guidelines on governmental support programmes for private forestry; build forest roads and renovate ditching; and maintain biological diversity of forests (MAF, 1999a).
18 The independent group intervenes in decisions such as giving financial support for certain types of work or owners (Forestry Centre of North Karelia, 1999).
19 The mass media include radio, television, newspapers, the internet, periodicals, pamphlets, etc.
20 Klaus Yrjönen, biodiversity project chief at the Forestry Development Centre Tapio, interview on 8 October 1999, Helsinki.
21 Harri Hänninen, senior researcher at the Finnish Forest Research Institute, interview on 21 September 1999, Helsinki.
22 The uniform report consists of information on felling, establishing a new tree stand, tending of young forest, forest road construction, and forest ditching maintenance (FDC Tapio, 1999c).
23 Ari Niiranen, head of the North Karelia Forestry Centre, interview on 30 September 1999, Joensuu, Finland.
24 The Amazon Basin, or Continental Amazon, comprises six states (Acre, Amapá, Amazonas, Pará, Rondônia and Roraima), with a total area of approximately

3.6 million square kilometres. For regional planning and economic development policy purposes, however, the government incorporated four more states (Mato Grosso, Tocantins, parts of Maranhão and Goiás) within Continental Amazon, establishing the so-called Legal Amazon, with a total area of about 5.2 million square kilometres (Pandolfo, 1994, p34). In this chapter, the term 'the Amazon' refers to the 'Legal Amazon', and these terms are used interchangeably.

25 For instance, 1ha in the Amazon contains up to 230 tree species, compared with 10 to 15 species normally found in a hectare of temperate forests (WRI, 1985). The overall estimate is that the Amazon rain forest contains about 30,000 plant species (Viana, 1997), and hosts two-thirds of the world's flora and fauna (Mahar, 1989).

26 The contribution of the forest sector to the gross national product decreased from 6.3 per cent in 1970 and 5.8 per cent in 1974 to 4 per cent in the early 1990s. This decrease is a consequence of the growth of other industrial and service sectors (Kengen, 1999). In recent years, however, the forest sector, composed of segments such as processed silviculture and extraction from natural forests, wood and furniture, iron and steel, and pulp and paper, was responsible for an estimated annual revenue of US$53 billion in 1993–1995, equivalent to 6.9 per cent of the gross national product (Lele et al, 2000).

27 The total world forest product market represented approximately US$130 billion in 1998; forest products from Brazil represented 1.7 per cent, which corresponds to US$2.3 billion. Brazil's share has been increasing, reaching 2 per cent in 2000 (World total US$145 billion; Brazil US$3 billion) and 2.6 per cent in 2004 (World total US$179 billion; Brazil US$4.7 billion) (FAO, 2005).

28 In 1990, the total export value was US$1.4 billion; it gradually increased, reaching US$2.1 billion in 1993, and by 1995 the export value had almost doubled, reaching US$3.2 billion. Since then, it has been growing steadily, reaching US$3 billion in 2000; and, the export value further increased to US$4.7 billion in 2004 (FAO, 2005).

29 The 14 per cent was calculated based on the total deforested area by 1988 presented by Dennis Mahar, plus the total deforested area from 1989 to 1997 presented by INPE (Mahar, 1989; INPE/IBAMA/MMA, 1998; Pandolfo, 1994).

30 *Veja*, 7 April 1999, São Paulo, Brazil, p112.

31 Law No 6.938/81, Article 2.

32 Law No 9.605 of February 1998 (hereinafter Law 9.605/98).

33 Law 9.605/98, Article 75. In March 1999, US$1 corresponded to 1.6 Brazilian real. Other administrative penalties according to Article 72 of Law 9.605/98 include:

- warning;
- single fine;
- daily fine;
- seizure of animals, products and by-products of fauna and flora, instruments, supplies, equipment or vehicles of any nature used in violation;
- destruction or immobilization of the product;
- suspension from selling or manufacturing the product;
- embargo on works or activity;
- demolition of the works;
- partial or total suspension of activities; and
- restrictions on rights.

The restrictions on rights include:

- suspension of registration, licence or authorization;
- cancellation of registration, licence or authorization;
- loss or restriction of tax incentives and benefits;
- loss or suspension of participation in lines of credit at official credit establishments; and
- prohibition on contracting with the government for a period of up to three years (Law 9.605/98, Article 72, para 8).

34 Article 15 of the Forestry Code establishes that 'Exploitation of the primitive forests of the Amazon Basin is forbidden without the necessary safeguards; they may only be exploited in accordance with technical plans and proper management as established by law, *to be enacted after the period of one year*' (emphasis added).
35 Administrative Decree No 10 of 1975 was the first legal instrument to use the term 'sustainable forest management'; other major administrative acts that dealt with forest management were Normative Instruction No 001 of 1980, Administrative Decree No 302-P of 1984 and Normative Instruction No 80 of 1991.
36 Provisional Measure No 21 of 6 December 1988.
37 Law No 7.803 of 18 July 1989.
38 Decree No 97.628 of 10 April 1989.
39 Decree No 1.282 of 19 October 1994.
40 Forestry Code, Article 44.
41 Provisional Measure No 1.511 of 25 July 1996.
42 The National Environmental Management System (SISNAMA) comprises the National Council for the Environment (Conselho Nacional do Meio Ambiente, or CONAMA) formed by representatives of the states, unions, NGOs and experts; the Ministry of the Environment and Legal Amazonia (MMA) and an executive agency (IBAMA) linked to MMA; other public agencies concerned with environmental matters; and state and municipal environmental agencies (Huber et al, 1998).
43 Law No 7.735 of 22 February 1989 (hereinafter Law 7.735/89).
44 Provisional Measure No 09 of 16 October 1992.
45 Law No 7.735/89.
46 The federal executive agencies are the Special Secretariat of the Environment, the Brazilian Institute of Forestry Development, the Superintendency of Fisheries Development and the Superintendency of Rubber (Law No 7.735/89).
47 Constituição da Republica Federativa do Brasil, Article 24, VI.
48 For instance, the states of Amazonas, Paraná and Minas Gerais.
49 Forest Act 1093/1996.
50 'Instrução Normativa' No 7 of 1999.
51 IN 7/99, Article 1, para 1. This applies to areas less than 200ha; clear-cutting in areas greater than 1000ha requires a special licence from the appropriate state environmental agency.
52 Pentti Lähteenoja, senior adviser of the Ministry of Agriculture and Forestry, interview on 6 October 1999, Helsinki.
53 The Brazilian government collected only 27 per cent of the total amount of levied fines for the period between 1994 and 1997, and the maximum rate of collection was logged in 1997 at 34 per cent. In the Legal Amazon, the average collection rate for this period was even lower, at 20 per cent (Hirakuri, 2003, from IBAMA source).

54 The budget allocation for forest management and forest resources stewardship in all of Brazil diminished from US$1.5 million in 1995 to US$710,000 in 1997, and it took a dramatic drop in 1998, when only US$240,000 was allocated; of this, US$150,000 was allocated for Legal Amazon (Hirakuri, 2003, based on IBAMA source).
55 Law No 5.371 of 15 December 1967, Article 1, VII; and Decree No 563 of 6 August 1992, Article 2, IX.
56 Francisco Potiguara, chief of Canindé Local Unit, FUNAI/Pará, interview on 11 December 1997, Belém, Pará, Brazil.
57 Act on the Financing of Sustainable Forestry (*Laki kestävän metsätalouden rahoituksesta*) No 1094 of 12 December 1996, Chapter 1, Section 1. Financial support is available for forest regeneration, prescribed burning, young-stand management, forest fertilization, harvesting of energy wood, ditching renovation and forest road construction.
58 The Fiscal Incentive Law, enacted in 1966, provided incentives for planting fast-growing species such as eucalyptus, pines, araucaria pine, fruit trees and palms – especially in the southern states of Brazil (Kengen, 1999).
59 The National Development Plan II, introduced in 1974, included the Programme for Charcoal-Fuelled Iron Ore Smelters and the National Programme for Pulp and Paper Production, which offered a 50 per cent federal tax exemption for these industries.
60 Decree laws No 1.134/70 and No 1.478/76.
61 José D'Avila, Banco da Amazônia, interview on 22 August 1996, Belém, Pará, Brazil.
62 By March 1999, the total certified forest area under the Finnish Forest Certification System comprised 13.5 million hectares, which corresponds to over half of all Finland's forests. This was expected to rise to 22 million hectares by the end of 2000 (MTK, 2000).
63 *Grupo de Compradores* was established for the purpose of creating initiatives to further domestic consumption of certified forest products. This group of 50 companies has already increased the demand for certification timber to 1 million cubic metres per year. Their target for 2005 is to buy 100 per cent of all certified timber coming from Brazil's natural forests (*FSC Notícias*, www.fsc .org.br/menú. htm, accessed 30 October 2000).
64 In 1994, the Brazilian subsidiary of the Tropical Forest Foundation, an international non-profit organization dedicated to conserving tropical forests through sustainable forestry, began to offer technical training programmes for the conduct of low-impact logging operations in the Amazon (*Wood Technology*, 1998).
65 Ari Niiranen, head of the North Karelia Forestry Centre, interview on 30 September 1999, Joensuu, Finland.
66 Representative works include the novel *The Seven Brothers,* written in the late 19th century by Aleksis Kivi, Arto Paasilinna and Veikko Huovinen; paintings by Akseli Gallen-Kallela, Pekka Halonen and Eero Järnefelt; music by Jean Sibelius; photographs by I. K. Inha and architecture by Eliel Saarinen and Alvar Aalto (Reunala, 1999).
67 The so-called 'riverines' (or *ribeirinhos*) lived along riversides and developed an economy dependent upon rivers or streams (Hummel and Minette, 1990). They extracted logs from floodplain forests (*várzea*) using rudimentary tools and methods. But following the arrival of widespread mechanized logging activities, this culture quickly disappeared.

68   Two per cent of the population hold 57 per cent of all agricultural land in rural properties of more than 1000ha each (Colchester and Lohman, 1993).
69   Ari Ekroos, professor of law at Helsinki University of Technology, interview on 17 September 1999, Espoo, Finland.
70   The president of IBAMA, Eduardo Martins, admitted in a 1997 interview that corruption presents a serious problem for forest law enforcement (Júnior, 1997).

References

Amigos da Terra (1997) 'Políticas Públicas Para a Amazonia – Rumos, Tendências e Propostas', Documento Apresentado à Reunião dos Participantes do Programa Piloto para a Proteção das Florestas Tropicais do Brazil, Outubro 1997, Manaus, Brazil

Barreto, P. and Hirakuri, S. R. (1999) *Sugestões Para o Controle do Uso do Solo na Amazônia,* Versão Preliminar, Belém, Instituto do Homem e Meio Ambiente da Amazônia

Brevig, F. K. (1997) 'Forest extension, training and continuing education', in *People, Forests and Sustainability*, Report of the FAO/ECE/ILO Committee on Forest Technology, Management and Training, Geneva, International Labour Office

Bromley, D. (1999) 'Deforestation – institutional causes and solutions', in Palo, M. and Uusivuori, J. (eds) *World Forests, Society and Environment,* Dordrecht, Kluwer Academic Publishers

Colchester, M. and Lohman, L. (eds) (1993) *The Struggle for Land and the Fate of the Forests,* London, World Rainforest Movement and Zed Books

Ecoporé (1992) *Levantamento Sobre Exploração Madeireira através de Planos de Manejo Florestal no Estado de Rondônia no Período de 1987–1991,* Rolim de Moura/RO, Ecoporé-Ação Ecológica Vale do Guaporé

de Freitas, V. P. (1995) *Direito Administrativo e Meio Ambiente,* first edition, Curitiba, Juruá Editora

FAO (United Nations Food and Agriculture Organization) (1996) *FAO Production Yearbook 1996,* Rome, Italy, FAO

FAO (1999) *State of the World's Forests 1999,* Rome, Italy, FAO

FAO (2005) *FAOSTAT data,* www.faostat.fao.org/faostat/collections?version=ext&has bulk=0&subset=forestry, accessed 22 February 2006

FDC Tapio (Forestry Development Centre Tapio) (1997a) *From Forest Policy to Practical Forestry,* Helsinki, Finland, FDC Tapio

FDC Tapio (1997b) *Tapion Vuosikirja 1996 [Tapio Yearbook],* Helsinki, Finland, FDC Tapio

FDC Tapio (1997c) *Tapion Vuosikirja 1997 [Tapio Yearbook],* Helsinki, Finland, FDC Tapio

FDC Tapio (1999a) *Fact Sheet HN 21.9.99,* Helsinki, Finland, FDC Tapio

FDC Tapio (1999b) *Forest Planning in Finland,* Helsinki, Finland, FDC Tapio

FDC Tapio (1999c) *Tapio Fact Sheet 1999,* Helsinki, Finland, FDC Tapio

FFA (Finnish Forestry Association) (1998) *Forestry in Finland,* Finland, FFA

Forestry Centre of North Karelia (1999) *Fact Sheets,* 26 July 1999, Joensuu, Finland, Forestry Centre of North Karelia

Greeley, W. B. (1953) *Forest Policy,* New York, McGraw-Hill

GTA (Grupo de Trabalho Amazônico)/Amigos da Terra (1997) *Políticas Públicas Coerentes – Para Uma Amazônia Sustentável,* São Paulo, GTA/Amigos da Terra

Hafner, O. (2000) *The Role of Corruption in the Misappropriation Tropical Forest Resources and in Tropical Forest Destruction*, Transparency International Working Paper, Berlin, 23 October 1998, www.transparency.de/documents/work-papers/ohafner.html, accessed 23 October 2000

Hakkila, P. (1995) *Procurement of Timber for the Finnish Forest Industries*, Research Paper No 557, Vantaa, Finland, Finnish Forest Research Institute

Hannelius, S. and Kuusela, K. (1995) *Finland the Country of Evergreen Forest*, Helsinki, Forssan Printing

Hänninen, H. (1999) 'Many actors influencing forest policy', in Reunala, A., Tikkanen, I. and Åsvik, E. (eds) *The Green Kingdom*, Keuruu, Otava

Hirakuri, Sofia R. (2003) *Can Law Save the Forest? Lessons from Finland and Brazil*, Jakarta, Indonesia, Centre for International Forestry Research

Huber, R. M., Ruitenbeek, J. and da Mota, R. S. (1998) *Market-Based Instruments for Environmental Policymaking in Latin America and the Caribbean: Lessons from Eleven Countries*, World Bank Discussion Paper No 381, Washington, DC, World Bank

Hummel, A. C. (1995) *Legislação Ambiental: Aspectos Gerais do Controle da Atividade Madeireira na Amazônia Brasileira*, Monografia apresentada a Faculdade de Direito da Universidades do Amazonas, Manaus, Amazonas

Hummel, A. C. and Minette, L. J. (1990) *Aspectos do Setor Florestal do Estado do Amazonas*, Campos do Jordão, São Paulo, 6 Congresso Florestal Brasileiro

Hyde, W. (1998) *Public Support for Harvesting and Wood Products Research: Beneficial Policy for Landowners, Consumers, the Industry, and the Environment*, Tapio, Helsinki, Metsätalouden Ympaäristöfoorumi, Metsätalouden Kehittämiskeskus (Forestry Development Centre)

IBAMA (Brazilian Institute of the Environment and Renewable Natural Resources) (1997) *Vive Atolado em Dificuldades: A Província do Pará, Caderno 1*, Belém, IBAMA

IBAMA/MMA (Ministério do Meio Ambiente) (1997) *Projetos de Controle Ambiental da Amazônia Legal: Avaliação dos Planos de Manejo Florestal Sustentável da Amazônia, Fase 1 – Análise de Documentos*, Brasilia, IBAMA, MMA

INPE (National Institute of Spatial Research)/IBAMA/MMA (1998) *Amazonia: Deforestation 1995–1997*, Brasília, INPE, IBAMA, MMA

Júnior, P. (1997) 'Está Tudo Errado – Entrevista: Eduardo Martins', *Veja*, 2 July 1997, São Paulo

Kengen, S. (1999) 'Forest policies in Brazil', in Palo, M. and Uusivuori, J. (eds) *World Forests, Society and Environment*, Dordrecht, Kluwer Academic Publishers

Lele, U., Viana, V., Verissimo, A., Vosti, S., Perkins, K. and Husain, S. A. (2000) *Brazil – Forests in the Balance: Challenges of Conservation with Development*, Evaluation Country Case Study Series, Washington, DC, World Bank

Leonard, H. J. (ed) (1985) *Divesting Nature's Capital: The Political Economy of Environmental Abuse in the Third World*, New York, Holmes and Meier

MAF (Ministry of Agriculture and Forestry) (1999a) *Fact Sheet*, Helsinki, Finland, MAF

MAF (1999b) *Finland's Forests – Jobs, Income and Nature Values*, Helsinki, Finland, MAF

Magalhães, J. P. (1998) *A Evolução do Direito Ambiental Brasileiro*, first edition, São Paulo, Editora Oliveira Mendes

Mahar, D. (1989) *Government Policies and Deforestation in Brazil's Amazon Region*, Washington, DC, World Bank

Meirelles, H. L. (1995) *Direito Administrativo Brasileiro*, 20th edition, São Paulo, Malheiros Editores

METLA (Finnish Forest Research Institute) (1998a) *Finnish Statistical Yearbook of Forestry 1998,* Helsinki, Finland, METLA
METLA (1998b) Leaflet, Helsinki, Finland, METLA
Ministry of the Environment (1999) *Everyman's Right in Finland,* Helsinki, Finland, Ministry of the Environment
MMA (Ministério do Meio Ambiente) (1998) *Política de Recursos Florestais,Versão 1.0,* Brasília, Secretaria de Formulação de Políticas e Normas Ambientais
MTK (Central Union of Agricultural Producers and Forest Owners) (2000) 'PEFC logo available in forest certification this year', Press release, 9 March 2000, Helsinki, Finland
Nascimento, J. R. (1988) *O Papel do Setor Florestal no Desenvolvimento da Amazônia: Uma Perspectiva Teórica,* I Encontro Brasileiro de Economia Florestal, Curitiba
Nascimento, J. R. and Kengen, S. (1988*) Desapropriação para Reforma Agrária e o Setor Florestal na Amazônia: Análise da Legislação Recente,* I Encontro Brasileiro de Economia Florestal III, Curitiba
Nikunen, U. and Ranta, R. (1991) 'Forestry planning in privately owned non-industrial forest in Finland', in *Proceedings of the 10th World Forestry Congress,* Paris, Food and Agricultural Organization of the United Nations
Niskanen, A. and Pirkola, K. (1997) 'Economical, ecological and social sustainability in the New Forest Policy in Finland', in Tikkanen, I., Glück, P. and Solberg, B. (eds) *Review on Forest Policy Issues and Policy Processes: European Forest Institute Proceedings No 12,* Joensuu, Finland, European Forest Institute
Palo, M. (1999) 'No end to deforestation?' in Palo, M. and Uusivuori, J. (eds) *World Forests, Society and Environment,* Dordrecht, Kluwer Academic Publishers
Pandolfo, C. (1994) *Amazonia Brasileira – Ocupação, Desenvolvimento e Perspectivas Atuais e Futuras,* Coleção Amazoniana 4, Belém, Editora Cejup
Parviainen, J. (1992) 'Long-term and sustainable forestry', *Paper and Timber* vol 74, no 2, reprint Paperi Ja Puu, Helsinki, Finnish Forest Research Institute
Portin, A. (1998) 'Forestry meeting diverse needs in society', in Tikkanen, I. and Pajari, B. (eds) *Future Forest Policies in Europe – Balancing Economic and Ecological Demands,* EFI Proceedings No 22, Joensuu, Finland, European Forest Institute and International Union of Forestry Organizations
Repetto, R. (1990) 'Deforestation in the tropics', *Scientific American* vol 262, no 4, pp36–42
Reunala, A. (1999) 'Forests and Finnish culture: "There behind yonder woodland..."', in Reunala, A., Tikkanen, I. and Åsvik, E. (eds) *The Green Kingdom,* Keuruu, Otava
Ribas, L. C. (1991) *Setor Florestal: Novo Posicionamento a Partir do Contexto Municipal,* II Encontro Brasileiro de Economia Florestal, Curitiba
Rietbergen, S. (ed) (1993) *Tropical Forestry,* London, Earthscan
Ripatti, P. (1996) 'Factors affecting partitioning of private forest holdings in Finland', *Acta Forestalia Fennica* no 252, Helsinki, Finland, Finnish Society of Forest Science and the Finnish Forest Research Institute
SAE (Secretaria de Assuntos Estratégicos) (1997) *A Exploração Madeireira na Amazônia,* Brasília, Grupo de Trabalho sobre Política Florestal, Relatório
SBS (Sociedade Brasileira de Silvicultura) (1995) 'IBAMA é Piloto na Reforma Administrativa do Governo', *Silvicultura* vol 16, no 64, pp5–8
Shepherd, G., Brown, D., Richards, M. and Schreckenberg, K. (eds) (1998) *The EU Tropical Forestry Sourcebook,* London, Overseas Development Institute and European Commission

Silva, J. N. (ed) (1997) *Diagnóstico dos Projetos de Manejo Florestal no Estado do Pará – Fase Paragominas,* Belém, Pará, CPATU-EMBRAPA with IBAMA, FCAP, SECTAM, SUDAM, UFPA/NAEA and FIEPA

Viana, G. (1997) *Amazônia,* Brasília, Câmara dos Deputados

Wainer, A. H. (1991) *Legislação Ambiental Brasileira,* first edition, Rio de Janeiro, Forense

Wood Technology (1998) 'Grant to fund training for Latin foresters', *Wood Technology*, vol 125, no 2, 1 March, p15

WRI (World Resources Institute) (1985) *Tropical Forests: A Call for Action, Part I: The Plan,* Washington, DC, Report of an International Task Force convened by the World Resources Institute, the World Bank and the United Nations Development Programme

11
Verification and Certification of Forest Products and Illegal Logging in Indonesia

Luca Tacconi

Introduction

Some countries in the European Union (EU), the EU itself, and some NGOs seek to promote the import of verified and certified forest products and to increase producer countries' supply of such products in order to reduce illegal logging.[1]

The likelihood of certification leading to a reduction in illegal logging has been questioned. Richards (2004) argues that progress in tropical country certification requires a reduction in illegal logging, rather than certification being a means to reduce the latter. Forest concessions that experience illegal logging cannot be certified. Furthermore, illegal logging depresses log prices (Seneca Creek Associates and Wood Resources International, 2004), further reducing the financial benefits of certification, which are already insufficient to attract the majority of producers in tropical countries (Gullison, 2003). It is uncertain, however, whether initiatives aimed at promoting the production of verified and certified forest products and the expansion of their markets could actually overcome the limits faced by the voluntary adoption of certification and lead to a reduction in illegal logging. In particular, the EU is seeking to implement trade restrictions to exclude illegally sourced timber products from the European market through voluntary agreements with exporting countries. NGOs are lobbying European and Asian governments to introduce public procurement guidelines for timber products that direct public organizations to source only verified or, preferably, certified forest products. NGOs are also implementing programmes to support the adoption of verification and certification, and to link producers of verified and certified forest products with customers in importing countries.

This chapter focuses on Indonesia because it appears to have one of the most significant illegal logging problems globally (Tacconi et al, 2003; Seneca Creek Associates and Wood Resources International, 2004; WWF, 2005), and has seen a sustained effort by donor organizations and NGOs over the past decade to promote certification and, more recently, verification. The Nature Conservancy (TNC) and the World Wide Fund for Nature (WWF) have been implementing activities to reduce illegal logging in Indonesia through an initiative called the Alliance to Promote Certification and Combat Illegal Logging (hereafter referred to as the alliance), which received funding mainly from the US Agency for International Development and the UK Department for International Development.[2] The work of the alliance is considered not only because it is one of the major non-governmental initiatives on illegal logging in Indonesia, but also because it has adopted two key hypotheses[3] that underlie the work on verification and certification carried out by other organizations.

The *first hypothesis* states that social and international pressure will lead governments to adopt procurement policies favouring verified and certified forest products. Other NGOs that lobby governments to adopt public procurement guidelines have also implicitly adopted this hypothesis (e.g. Toyne et al, 2002; Garforth, 2004). New procurement policies would expand the market for verified and certified forest products, while trade restrictions through the voluntary agreements noted above would reduce the markets for illegal timber products.

The *second hypothesis* states that production of verified and certified forest products brings higher financial returns and market share than doing business as usual (i.e. producing illegally and/or unsustainably harvested timber). With regard to market share, this hypothesis depends partly upon the previous one, and partly upon consumers' and retailers' demand for verified and certified products. In relation to higher financial returns, it depends upon the costs of producing verified and certified forest products and the sale price. Given that production costs for these products are higher than for illegal and non-certified forest products, price premiums would have to eventuate for better financial returns to occur. Higher financial returns are relevant to the trade restrictions proposed by the EU because producers would lobby their governments to sign voluntary trade agreements in order to achieve higher financial returns.

The chapter considers whether these hypotheses hold in order to gain a better understanding of whether the voluntary agreements proposed by the EU, lobbying for public procurement policies in importing countries, and NGO initiatives on verification and certification can contribute to a (considerable) reduction in illegal logging in Indonesia and in other countries.

The chapter proceeds by first considering the global market for certified forest products and factors thought to affect their supply. It then details the strategy adopted by the Alliance to Promote Certification and Combat Illegal Logging before discussing initiatives on illegal logging in some of the most significant importing countries. This is followed by a quantitative assessment of illegal logging in Indonesia, the patterns of consumption of and trade in Indonesian forest products, and an assessment of the implications for policy. The uptake of certification in Indonesia and the impacts of the alliance are then discussed. The final section draws conclusions on the potential role of verification and certification in addressing illegal logging in Indonesia, as well as in other countries.

Supply and demand of certified forest products

Production and market share of certified timber

The production and marketing of verified timber is not yet occurring on any significant scale. Therefore, this section considers only information on certified forest products (CFPs).

Certification was initially seen as an innovative instrument to reverse the degradation of forests (Cashore et al, 2004), particularly in the tropics. However, certification has made significant progress only in temperate countries. The total area of certified forest was estimated at about 270 million hectares by mid 2006, or about 7 per cent of the global forest area. The northern hemisphere's temperate and boreal forests in the most developed countries account for most of the certified forest (58 per cent located in North America and 29 per cent in Western Europe). In Western Europe, about half of the total forest area is certified (countries such as Finland and Austria are close to reaching or have already reached 100 per cent certification of their forests), compared with one third in North America, 3 per cent in Oceania and, at most, 1 per cent in the other regions. The annual potential roundwood supply from certified forests in 2006 is estimated at about 370 million cubic feet, which corresponds to about 25 per cent of the global production of industrial roundwood.[4]

Most of this production is traded without certified status. For example, the marketing and communications director of the Forest Stewardship Council (FSC) reported that some 80 per cent of the timber from FSC-certified forests was not reaching the market as FSC-labelled products (Environmental Data Services, 2004). This relative invisibility is the result of a lack of demand for CFPs by consumers and a lack of incentives for producers to sell CFPs (due to the absence of a price premium) – hence, potential supply of CFPs exceeds actual demand in many markets (UNECE and FAO, 2006). The market share of timber products sold as CFPs in Europe was estimated at only about 5 per cent of the total timber trade volume, with the largest market share in the UK (10 per cent of total timber product trade and 1 per cent of paper), followed by The Netherlands (7 per cent of total timber product trade), Belgium and Denmark (5 per cent), and Germany (1 per cent) (UNECE and FAO, 2003). The same source estimated the market share of CFPs in Canada at 5 per cent (wood and paper), 2 per cent in the US (wood products) and 0.2 per cent in Japan (wood products). The reliability of these estimates is uncertain because the trade in CFPs is difficult to quantify as a result of the lack of official figures and trade classifications. An alternative approach to assessing trends in business-to-business interest in the trade of CFPs is the number and type of chain of custody certificates.[5] The number of these certificates increased by 20 per cent between May 2005 and May 2006 (UNECE and FAO, 2006). France is the country with the highest number of certificates, followed by Germany, the UK, the US and Poland. Outside Europe and North America, the countries with the largest number of certificates are Japan and China. In Asia, Viet Nam, Malaysia and Indonesia have also experienced growth in certificates. UNECE and FAO (2006) note, however, that companies in Asian countries, with the exception of some in Japan, normally export to North America and Europe, rather than supply their domestic markets, which do not yet demand CFPs.

Factors affecting the supply of certified products

The two main factors thought to influence producers' uptake of eco-labelling schemes are price premiums and security of access to markets (UNEP, undated). The relevance of price premiums is determined by the financial rationale of certification. Timber producers are supposed to incur the additional costs of sustainable forest management in the hopes of receiving a price premium that would offset the loss of some of the profits from unsustainable forestry (Richards, 2004). Price premiums of 5 to 65 per cent have been observed, but only for a small amount of CFPs (Eba'a Atyi and Simula, 2002). Generally, buyers do not seem to tolerate a price premium of more than 5 per cent for certified products (CCIF, 2002), a margin insufficient to pay for certification costs (CCIF, 2002; Gullison, 2003). In some cases, buyers may be unable to pay price premiums at all, as is often the case for buyers in high-volume, low-margin sectors, such as construction and paper (CCIF, 2002). Not only are price premiums unlikely to be significant, anecdotal evidence indicates that producers are least likely to capture a significant share of any premium (UNEP, undated). Furthermore, even if price premiums eventuate, they are normally not sustained because of increases in the supply of eco-labelled products that satisfy demand and thus reduce prices (UNEP, undated). The economic rationale of a price premium is therefore rooted in short-term scarcity or reduction of production costs (Sedjo and Swallow, 1999; UNEP, undated).[6] The latter does not apply to CFPs, however, as they normally involve higher production costs. It should be stressed that a price premium does not necessarily lead to increased profits since the latter is also determined by production costs, and CFPs have higher production costs than non-certified products. Short-term data seem, indeed, to support the view that price premiums are likely to be small or non-existent. In the UK, certified softwood timber products were available on the market in late 2005 at no price premium (Oliver, 2006).

Access to markets and the predictability of future access to markets seem to be more significant determinants of the uptake of certification schemes than price premiums, as indicated, for example, by the fact that some producers would be willing to sacrifice a price premium for longer-term supply contracts

(UNEP, undated). This also applies to the forest sector, where the main reason for seeking certification is market access, while the least important is price premiums (Raunetsalo et al, 2002).

Strategy and rationale of the TNC–WWF alliance in Indonesia

The TNC and WWF established the Alliance to Promote Certification and Combat Illegal Logging (initially a three-year initiative) to improve market access by those Indonesian producers who chose to produce verified and certified forest products and to support them in achieving verification and certification. The alliance specifically aims to:

- strengthen market signals to expand certification and combat illegal logging;
- increase the supply of certified Indonesian wood products;
- demonstrate practical solutions to achieve certification and differentiate legal and illegal supplies;
- reduce financing and investment in companies engaged in destructive or illegal logging in Indonesia; and
- share lessons learned from the project (TNC and WWF, undated).[7]

East Asian countries are the major importers of Indonesian forest products, as documented later, and they have largely been indifferent to the environmental impacts of their purchases. As a consequence, suppliers exploit weak governance in countries such as Indonesia to mine wood through predatory logging.[8] The alliance seeks to transform this trade by supporting an expansion of the market for verified and certified products in China and Japan. The objectives of building the capacity of Indonesian companies to produce verified and certified timber, and of improving forest management by promoting the adoption of the concept of high-conservation value forests complement the focus on importing countries. The following outcomes were to be achieved by the end of the third year (September 2005) of the alliance's activities:

- Timber imports from Indonesia by Global Forest and Trade Network members will be 30 per cent of total imports in Japan and 15 per cent in China.
- The volume of exported timber with third-party verification of legality in Indonesia will be 30 per cent of total export volume.
- The volume of third-party certified wood will be 10 per cent of total export volume from Indonesia.[9]

The alliance views the transition in international market demand from non-certified to certified forest products as an essential step to eliminate exploitative operations and to bring about a widespread adoption of sustainable management practices in Indonesia, as well as in other countries where the TNC and WWF operate. Effecting a change in unsustainable forestry practices through raising market demand for certified forest products and reducing demand for non-certified ones is viewed as more effective than simply waiting for the support and cooperation of governments in producing countries. The alliance therefore seeks to increase the size of the market for certified products by enhancing consumer awareness of the environmental benefits of certification, by convincing governments to adopt public procurement guidelines, and by convincing producers and traders to join the Global Forest and Trade Network (GFTN).

The GFTN is a WWF global initiative (broader than the alliance's activities) that seeks to support a stepwise approach to certification and to promote international trade in certified forest products as a means of achieving an improvement in forest management practices. Improving timber harvesting to meet certification criteria involves significant changes in practices that require time and investment of resources. Several organizations, including WWF and the International Timber Trade Organization, have recognized the difficulties encountered by timber producers in reaching the high harvesting standards required for certification. In a stepwise approach, producers improve forest management practices step by step, starting, for instance, with the acquisition or regularization of harvesting rights, then moving to the establishment of legal verification and ending with the certification of sustainably managed forests (e.g. Nussbaum et al, 2003).

Members of GFTN are required to develop action plans for the elimination of illegally sourced timber from their supply chains and to increase the proportion of certified forest products. GFTN acts as an umbrella organization for regional forest and trade networks. The aim of the networks is to raise private-sector awareness of the impacts of poor forestry practices and to influence policies that govern forest product procurement.[10] There are buyer networks active mainly in Europe and North America; but there is also a buyer network in Japan.[11] The WWF is paying increasing attention to the establishment of networks of producers because of a perceived shortage in the supply of verified and certified forest products.[12] Membership of a producer network supposedly provides easier access to the international market while the company prepares to achieve certification. A number of producer groups are in various stages of development in Latin America, West and Central Africa, Eastern Europe and South-East Asia, including Indonesia, as discussed later.

In summary, the alliance seeks to expand market demand for verified and certified forest products and to support the efforts of companies that aim to gain verification and certification. The alliance essentially adopts a market approach. This approach involves self-regulation (through which the industry undertakes to produce and trade only verified and certified forest products) and government public procurement policies requiring timber purchased by public organizations to be legally verified or certified.

Promoting the demand for verified and certified forest products

International NGOs such as Global Witness (e.g. Global Witness, 1999), the Environmental Investigation Agency (Environmental Investigation Agency, 1996), Greenpeace (Greenpeace, 2000), the Forest and European Research Network (e.g. Garforth, 2004) and WWF (e.g. Toyne et al, 2002) have been at the forefront of activities that raise awareness of the illegal logging problem and lobbied national governments and international organizations to address it. NGOs are still a significant factor behind the actions of governments and international organizations.[13] It needs to be recognized, however, that once NGOs had stimulated the interest in illegal logging, activities by national governments and international organizations, including the EU, have also been supported by officials in those organizations who are committed to improving forest practices and maintaining informal networking links with NGOs. Since the early 2000s, this network of environmental activists and officials in national and international organizations has generated increasing pressure on individual importing and exporting countries to adopt measures to reduce illegal logging – for instance, through several international and national conferences and workshops on illegal logging. The following analysis, therefore, is not meant to assess the impacts of lobbying by NGOs, but the outcomes of these cumulative efforts.

European Union

The European Commission launched the EU Action Plan on Forest Law Enforcement, Governance and Trade in May 2003. The measures set out by the action plan include:

- support for improved governance and capacity-building in timber-producing countries;
- development of voluntary partnership agreements with timber-producing countries to prevent illegal timber products from reaching the EU market;
- activities to reduce the consumption of illegal timber in the EU and to discourage EU institutions from investing in projects that may encourage illegal logging.[14]

Partnership agreements with producer countries would be implemented on a voluntary but binding basis.[15] The scheme aims to ensure that only legal timber from producing partner countries enters the EU. Timber products exported to the EU by producing partner countries would be identified as legal by means of certificates issued by accredited third-party organizations in the exporting partner country. The scheme would initially cover sawn wood, roundwood, plywood and veneer. The range of products covered by the scheme is particularly important, as discussed later.

With regard to the adoption of public procurement guidelines, they could have an important impact on timber markets given that the public sector accounts for about 20 per cent of total timber demand in Europe. However, only a few European countries, including Belgium, Denmark, France, Germany, The Netherlands, Sweden and the UK, have introduced or are considering the introduction of public procurement guidelines. Furthermore, no country has yet introduced mandatory procurement guidelines (SGS Trade Assurance Services, 2002), and in some cases they are considered to be weak. For example, the French one 'fails to clearly describe which timber or paper qualifies as legal or sustainable, and so allows a great deal of scope for greenwashing'.[16] Furthermore, efforts to develop public procurement policies are only at the level of national governments, which use considerably less timber than local and regional governments (UNECE and FAO, 2006).

Within the private sector, several large timber importers have adopted processes that aim to reduce or eliminate the use of timber from illegal sources, possibly in response to pressure from environmental NGOs. Some companies have therefore stopped the purchase of timber from high-risk countries such as Indonesia and source products from FSC-certified sources – for example, Brazil and, in the case of verified timber, Ghana (Jurgens, 2006). Some companies collaborate with NGOs to find verified or certified forest products in Indonesia (e.g. GFTN members, which is discussed later).

Japan

The government of Japan contributed to raising the profile of the illegal logging problem at the G8 Kyushu–Okinawa Summit in 2000. Japan has focused on three areas:

1 promotion of bilateral, regional and multilateral international cooperation;
2 technical cooperation to develop systems to identify legally/illegally harvested timber; and
3 assistance to private-sector initiatives (Tsuru, 2005).

As part of its international cooperation activities, Japan signed a memorandum of understanding to combat illegal logging and its related trade with Indonesia. The memorandum focuses on:

- the development, testing and implementation of systems for the verification of legal compliance; and
- cooperation of enforcement agencies and networks to address the trade in illegal forest products.

There is no evidence, however, that implementation of the memorandum has led to practical steps to reduce the import of illegally sourced timber products from Indonesia. Japan has announced the adoption of a public procurement policy aimed at excluding purchases of timber whose legality has not been verified.[17] The policy has been criticized, however, because 'it relies heavily on the forestry industry and sets no clear criteria by which timber and/or paper would qualify'.[18]

In relation to the private sector, the Japan Federation of Wood Industry Associations set up an investigation committee in 2001 with the objective of identifying issues on, and collecting information on, overseas illegal logging.[19] The committee carried out a survey of the wood industry in Japan. Nearly all respondents said that people in the wood industry in the country were aware of illegal logging, and 62 per cent thought that illegally logged timbers were imported into Japan. Yet only 3 per cent indicated that they themselves had dealt with such timber. Regarding the issue of legality, 54 per cent of respondents stated that the legality of their timber had been confirmed in some way, and 20 per cent of respondents mentioned the 'FSC, the Programme for the Endorsement of Forest Certification Schemes, etc.'. These statistics are probably misleading, however, as recognized in the report itself, given that there are not yet secure tracking methods to establish legality. About 37 per cent of the respondents noted that there was no alternative to dealing in timber of suspicious origin. It is notable, also, that only 10 per cent of dealers have timber purchasing guidelines, and 19 per cent have corporate social responsibility rules. With regard to the degree of their customers' awareness of the legality issue, they responded that 38 per cent of their customers were, to some extent, aware of it. In addition, 61 per cent of the customers were interested in the place of origin of the timber. Customers were therefore more interested in the place of origin than in the legality of the timber.

We noted earlier that the Alliance to Promote Certification and Combat Illegal Logging sought to increase the market share of verified and certified forest products in Japan. The alliance works closely with buyers of Indonesian pulp and paper (particularly those sourcing from Riau Province) to exert pressure on the producers to cease the conversion of natural forest for pulp and paper plantations. One major pulp and paper buyer in Japan – Ricoh – signalled that it would take steps to ensure that the products it buys originate from responsible pulp and paper operations (Tacconi et al, 2004a). With regard to increasing the awareness of certification through the establishment of GFTN networks, 25 Japanese companies have joined the network.[20] In spite of these developments, there appears to be little demand for certified wood. According to WWF's Japanese Forest and Trade Network, by the end of 2005 there had not been any requests from Japanese importers for certified forest products from Indonesia (Jurgens, 2006). In summary, the market for verified and certified forest products in Japan does not seem to have expanded to any significant extent during recent years.

China

China and Indonesia signed a Memorandum of Understanding Concerning Cooperation in Combating Illegal Trade of Forest Products in 2002. To our best knowledge, this memorandum has yet to yield practical outcomes, as exemplified by an interview with Indonesian Minister of Forestry Kaban in the *Jakarta Post* (the main national English language newspaper) on 18 July 2006, in which he said:

> Take China, for example. At the International Timber Trade Organization, they expressed a commitment to combat illegal logging. But we still don't know which institution regulates timber trading there and why illegal logs [from Indonesia] are still entering the country.[21]

The Alliance to Promote Certification and Combat Illegal Logging hopes that the example set by major companies will be followed by other timber trading firms. In China, the leaders in the purchase and use of certified timber are IKEA, Kingfisher (a trader that purchases forest products for B&Q UK) and Carrefour, which together account for about 1 per cent of the processing and trading of furniture products in China (Tacconi et al, 2004a). By working with these and other target companies in China, the alliance seeks to develop consumer awareness of the importance of purchasing verified or certified forest products. Consumers, however, use only a small share of traded timber products (Oliver, 2006), thus significantly limiting the potential impacts of the alliance's approach.

Some Chinese companies are becoming aware of illegal logging issues. The WWF China forest trade network has three members that are importers of Indonesian timber products and that have exposure to UK and North American markets. These include Kingfisher and Yingbin, the latter of which works with WWF to identify verified timber in Indonesia (Jurgens, 2006). As noted earlier in relation to companies holding chain-of-custody certificates, the companies that are aware of illegal logging and take steps to source verified or certified timber do so because they have links to European and North American buyers.

US

The US was among the initial promoters of the East Asia Forest Law Enforcement and Governance process, which led to the first conference on the issue of illegal logging in Bali in 2001. Unlike the other countries considered above, it has not engaged in bilateral agreements with Indonesia, however, and does not have a public procurement policy concerning timber (Garforth, 2004).

Summary

It is evident from the above discussion that some six years since the issue of illegal logging was raised at the Kyushu–Okinawa G8 Summit, only a few European governments are considering the introduction of public procurement guidelines for timber products, and in the cases of France and Japan, the proposed drafts are seen to be too weak. China, one of the largest importers of Indonesian timber, has not introduced procurement guidelines. Therefore, social and international pressures have not resulted in (at least until the time of writing) many governments adopting procurement policies favouring verified and certified forest products, as hypothesized by the Alliance to Promote Certification and Combat Illegal Logging, as well as other organizations. There has also been limited change in market access to Asian and European markets for verified and certified forest products with regard to business-to-business activities.

It is likely that concerns about the economic impacts of promoting trade in verified and certified products are contributing to the very limited progress made in this area by most governments, particularly those in Asia (which, furthermore, face limited or inexistent pressures from domestic environmental movements). By some estimates, the world's forestry output would decrease between 0.3 and 5.1 per cent with the adoption of certification at a global level, while the world price would rise between 1.6 and 34.6 per cent, following increases of between 5 and 25 per cent in timber production costs resulting from certification. The major net importers of forest products (many of them East Asian countries) would suffer the most significant economic impacts from global certification (Gan, 2005).

We now turn to assessing the extent of illegal logging in Indonesia, the markets for Indonesian forest products and the implications for the initiatives that seek to reduce illegal logging by restricting market access for illegally logged timber.

Illegal logging in Indonesia and the timber trade

The concern about illegal logging in Indonesia expressed by some countries and NGOs stems from an apparently high rate of illegal harvest. The rate of illegal logging appears to have increased between 2000 and 2003 (see Table 11.1), despite the national and international attention given to illegal logging in Indonesia, which resulted in the establishment of the East Asia Forest Law and Governance process and governments' commitment to address illegal logging in the region (expressed formally at a conference on the issue in Bali in 2001).

Let us now consider timber use by sector and timber sales by market because they have implications for policies aimed at reducing illegal logging.

Pulp and paper production account for the largest share of timber consumption followed by plywood and sawnwood (Table 11.1). These three sectors account for about 90 per cent of the total consumption of roundwood equivalent. In terms of priorities for intervention, the pulp and paper and the plywood sectors should receive particular attention. Given the increasing share of timber used by the pulp and paper industry (from 31 per cent in 2000 to 42 per cent in 2003), policies need to focus particularly on this sector to have a significant impact on illegal logging in Indonesia.[22]

International demand for wood products is larger than domestic consumption, the former accounting for 59 per cent of roundwood equivalent

Table 11.1 Estimated use of roundwood by sector, illegal harvest and export of timber products from Indonesia[*]

		Year 2000			Year 2003		
	Roundwood equivalent (RWE) (m³)	Sectoral use of RWE[a] (percentage)	Sectoral share exported[b] (percentage)	RWE exported[c] (percentage)	Sectoral use of RWE[a] (percentage)	Sectoral share exported[b] (percentage)	RWE exported[c] (percentage)
Log export	1,606,600	3	100	5	–	–	–
Illegal log export	3,000,000	6	100	9	5	100	9
Sawn wood	11,700,000	22	29	10	24	24	10
Fibreboard	768,600	1	77	2	1	77	2
Particleboard	280,000	1	90	1	1	61	1
Plywood	18,860,000	36	62	35	25	83	35
Veneer	131,100	0	25	0	1	6	0
Pulp	16,394,400	20	34	31	18	43	31
Paper	–	11	37	7	24	28	12
Log import	−186,700						
Total harvest (a)	52,554,000	100	63	100	100	59	100
Official harvest (b)	33,946,500						
Illegal harvest (a−b)	19,057,500						
Proportion of illegal harvest (percentage)	57			74			

Notes: [*] See the Appendix at the end of this chapter for the method used in deriving estimates.
[a] 'Sectoral use of RWE' represents the share of total roundwood equivalent that each sector consumes.
[b] 'Sectoral share exported' represents the percentage of the output of a sector that is exported. The total percentage of the column is the ratio of RWE export to total RWE.
[c] 'RWE exported' is the contribution of each sector to total RWE export.

Source: FAOSTAT database for production and trade statistics (accessed August 2006), with the exclusion of illegal log export; the volume of illegal log export is a conservative estimate that applies to Kalimantan (Tacconi et al, 2004b), and it is possible that the illegal export of logs is over 5 million cubic metres per year; the

in 2003. The pulp and paper and the plywood sectors export between two-thirds and three-quarters of their roundwood equivalent production. Plywood and pulp and paper exports account for over one third of the use of total roundwood equivalent and about 70 per cent of the total export of roundwood equivalent. It is therefore clear that policies need to target exports of plywood and pulp and paper to have a significant impact on illegal logging.

Regarding the export of plywood and pulp and paper by country of destination (Figure 11.1), the largest importers of Indonesian products are China (which in 2003 imported 44 per cent of Indonesian pulp and 40 per cent of paper),[23] Japan (40 per cent of plywood), South Korea (24 per cent of pulp) and the EU (19 per cent of pulp). The share of paper exports to 'other' countries (about 35 per cent) is distributed among many countries, which individually account for relatively small shares of total exports. The conclusion from these statistics is that unless China, Japan, South Korea and the EU target their respective imports of plywood and pulp and paper from Indonesia, the impact of trade measures on illegal logging in Indonesia is bound to be rather limited. The voluntary partnership scheme of the EU does not include pulp and paper, which represent the EU's largest timber products imports from Indonesia.

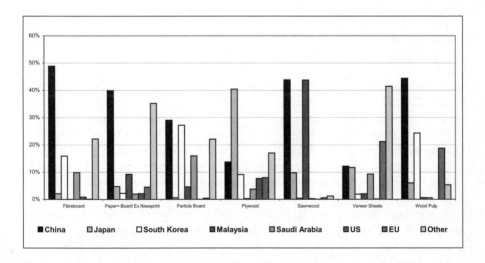

Figure 11.1 *Destination of timber product exports from Indonesia in 2003 (volume)*

Source: FAOSTAT, http://faostat.fao.org, accessed August 2006

Trade measures such as voluntary partnership agreements could have an indirect effect on Indonesia if they resulted in a reduction of non-verified timber product imports from third countries that, in turn, import from Indonesia. The most obvious case is China, one of Indonesia's main timber export markets. The quantity of timber processed and exported by China is equivalent, in terms of volume, to over 70 per cent of its timber imports.[24] The main forest products exported by China in 2005 were furniture (12.7 million cubic metres of roundwood equivalent) and plywood (some 10 million cubic metres of roundwood equivalent). The US, Japan and Hong Kong are the main markets for Chinese timber products, while EU countries import smaller amounts. Currently, the US, Japan and Hong Kong do not plan to introduce trade measures (e.g. similar to the EU's voluntary partnership scheme) that could be applied to China. Therefore, there will be no indirect impacts on the export of timber products from Indonesia to China. Whether there will be impacts on exports from Indonesia to China from changes in trade between the EU and China depends upon whether the latter two trading partners sign a voluntary partnership agreement. It should be noted that a voluntary partnership agreement between China and the EU would not necessarily lead to a reduction in China's import of illegally sourced timber products from Indonesia. Chinese companies could use Indonesian illegal timber products to supply the domestic market or export to markets other than the EU, while sourcing verified timber products from non-Indonesian producers for processing and re-export to the EU.

Let us now consider domestic demand for timber products in Indonesia (see Table 11.1). Estimated domestic consumption (about 19.4 and 23.3 million cubic metres of roundwood equivalent, respectively, in 2000 and 2003) is large but below the official harvest. This finding has relevant implications for policy. Consider the event that all countries importing timber products from Indonesia required proof of legal origin. If Indonesian producers attained verification to access export markets and were able to do so, for example, at the 2003 level of export (about 33.2 million cubic metres of roundwood equivalent), the official harvest would be required to supply the export market. Therefore, the equivalent of the demand for timber products in the domestic market could be met through imports or timber illegally harvested in Indonesia. This assumes that domestic supply of verified timber could not be increased above the level of the 2003 official harvest, a rather realistic assumption given that the Ministry of Forestry had set the annual allowable harvest from production forests at 6.89 million cubic metres in 2003 and 5.74 million cubic metres in 2004. It needs to be stressed that the annual allowable harvests for 2003 and 2004 did not seem to be based on scientific assessments of sustainable harvest; but it is also relevant to note that there is a lack of documentation concerning the sources of the official production that exceeds the annual allowable harvest. Therefore, ensuring that all timber products exported from Indonesia are verified would not necessarily solve the illegal logging problem in Indonesia unless it started to import most of the timber products needed to meet domestic demand.

Let us now turn to considering the progress made in the development of verification and certification in Indonesia.

Verification and certification in Indonesia

The progress of certification in Indonesia

Certification has experienced minimal progress since its introduction in Indonesia, as in most tropical countries. There are two certification systems in Indonesia: the FSC system and that developed by the Indonesian Eco-labelling Institute (known locally as LEI).

LEI was established to develop an Indonesian certification system and to act as an accreditation institution for certifying bodies in the country. The establishment of LEI was promoted in 1994 by a working group that involved stakeholders from academia, government, NGOs and the private sector under the leadership of Emil Salim, a former Indonesian minister for the environment. The initial goal of the working group was to design a forest standard for the Indonesian context. In 1998, it introduced certification guidelines for natural forests, followed by the chain-of-custody system in 2002. Over the following two years, LEI piloted certification schemes for plantation forests and community-managed forests.

In 1999, the FSC and LEI signed a joint certification protocol for natural forests to certify enterprises that meet the requirements of both certification systems. The LEI has accredited only one natural forest management unit, totalling about 90,000ha.[25] That forest management unit also has FSC certification. Following certification, the Ministry of Forestry granted this company the privilege to self-approve the annual work plan and to harvest ramin (*Gonystylus bancanus*), which is listed as a protected species in Appendix 3 of the Convention on International Trade in Endangered Species of Wild Flora and Fauna (CITES) (Tacconi et al, 2004a).

The establishment of an Indonesian certification system has not led to the certification of a significant share of the forest area. The FSC system covers about 739,000ha,[26] or about 2.7 per cent of the 27.43 million hectares of natural forests included in forest concessions.[27] Thirteen forest concession companies, covering a total area of 2.43 million hectares, have initiated the process towards certification with LEI.[28] Three of these concessions, amounting to about 649,000ha, are already FSC certified. Although the number of companies applying for certification may seem to indicate an increase of interest in certification, this does not suggest that they will all achieve certification if the past success rate is any indication of possible outcomes. Between 1998 and 2003, LEI-accredited certification bodies assessed 14 natural forest management units covering 2.5 million hectares, and only one achieved certification (Muhtaman and Prasetyo, 2004).

The activities of the Alliance to Promote Certification and Combat Illegal Logging

Given the relative lack of uptake of certification, the alliance introduced the stepwise approach to certification in Indonesia. Criteria of legality were needed for implementing verification. The alliance therefore developed a draft definition of legality which, however, has experienced rather slow progress. The initial draft was criticized by local NGOs, which thought that it did not protect the rights to land of local communities. The initial draft was revised and the Legality Standard for Tracking Timber to the Source is now being considered by a group led by the LEI and including the TNC and WWF.[29]

To support producers in achieving certification in Indonesia, WWF also launched the producer group *Nusa Hijau* (Green Archipelago) in 2003 as part of the development of its GFTN network. The goal of this group is to increase the supply of verified and certified timber in Indonesia. In return for membership in *Nusa Hijau*, companies are required to commit to a stepwise process that eventually leads to full certification of forests in the case of a forest participant, and chain of custody in the case of a trade participant. The following are the stated benefits of joining *Nusa Hijau* for producers:

- support for the certification process by providing information and training on verification and certification;
- easier access to markets through the provision of information on the forest management status of its members, thus allowing for the early recognition of efforts to become certified – for example, buyers may source timber from members of a producer group since this should guarantee that the product is legal;
- publicity for forestry firms through recognition of their efforts to move towards sustainable forest management by a well-known environmental NGO such as the WWF;
- potential for collective bargaining with the government of Indonesia for the easement of logging quotas and other assistance provided by the Ministry of Forestry for certification.[30]

Nusa Hijau's recruitment success has been limited. By August 2006, it had three members that produce timber from natural forests, one of which is already certified at FSC standards.[31] Some companies that already had market linkages for the sale of certified timber reportedly preferred to delay consideration of membership in *Nusa Hijau* until they had achieved certification (Jurgens, 2006). This seems to indicate that some companies do not find the stepwise approach to certification proposed by *Nusa Hijau* a worthwhile approach. A possible explanation which needs testing is that the costs to become verified are not too dissimilar from those of certification, and companies prefer to become certified because the market for certified forest products is more established than that for verified timber. The fact that some companies prefer to join the producer group after certification points to the possibility that the support it provides for achieving verification and certification may not be a significant benefit for companies. Market access and higher financial returns are probably more important factors in driving membership of the producer network than the attainment of verification and certification. These factors have already been considered in the foregoing analysis, particularly market access. Let us consider some details of financial returns specific to the Indonesian case.

Jurgens (2006) reports that a forest concession in West Papua expressed interest in joining *Nusa Hijau*; but it would do so only if the producer group could guarantee a price premium of about 150 per cent of the price of illegally sourced *Intsia bijuga* (a high value timber locally known as merbau).[32] *Nusa Hijau* was reported to be seeking expressions of interest through the GFTN; but it had yet to find buyers (Jurgens, 2006). Further evidence indicates that the financial benefits derived from illegal logging are currently significantly higher than those from verified or certified logging. In Indonesia, the estimated cost to a mill to have a large forest concessionaire deliver legal timber to its door has been estimated at US$85 per cubic metre (including bribes of 20 per cent),[33] while a small concessionaire can deliver it for US$46 per cubic metre (URS Forestry, 2002). The cost of illegal timber has instead been estimated for small and medium operations at between US$5 per cubic metre (at the roadside) and US$32 per cubic metre (at the mill door) (URS Forestry, 2002).

The evidence presented above indicates that the hypothesis that the production of verified timber brings better financial returns than doing business as usual does not appear to hold in Indonesia. This is similar to conclusions already reached at the global level with regard to the certification of timber (e.g. Gullison, 2003) and which also apply to certification in Indonesia, as stated by the Tropical Forest Trust in a news release on its website in September 2005:

> Despite the UK trades' commitment to stay engaged in Indonesia and despite the UK government's new procurement rules favouring FSC-certified products, Tjipta can't find a UK buyer. Why not? ... The simple answer, of course, is price. Two years on, the UK market has almost completely moved away from Indonesia to cheaper suppliers elsewhere, most notably China, but also Malaysia. Brazilian FSC plywood has gained ground; but even that product is now sitting in warehouses across the UK, struggling to compete with Chinese product.[34]

Conclusion

Lasting price premiums have not eventuated for certified forest products and there is no obvious reason for their occurrence for verified timber products, except for possible short-term gaps between supply and demand. We have shown that the cost of producing illegal timber is considerably lower than that of legal timber, and verification and chain-of-custody costs would further increase the cost of verified timber. In the absence of a price premium at least equal to the difference between legal and illegal timber (i.e. one that leads at least to unchanged profits), the majority of producers can be expected to take up the production of verified timber products only if they otherwise face restricted market access, which we have considered in the form of public procurement guidelines and trade restrictions through voluntary partnership agreements in the EU.

Pressure on importing countries to introduce public procurement guidelines for timber products has been largely unsuccessful to date. In Asia, China has yet to consider this issue, while Japan has introduced guidelines that were considered too weak and unlikely to significantly affect the import of illegal timber. Only a handful of countries in Europe have begun to give serious consideration to the introduction of public procurement guidelines. Even in their case, however, if the procurement guidelines do not apply to local-level governments, as currently appears to be the case, they are unlikely to have significant impacts on timber markets given that local-level governments account for a large share of the public sector's timber demand.

A voluntary partnership agreement between the EU and Indonesia is unlikely to have a significant impact on illegal logging. First, as currently specified by the EU, it would exclude pulp, the main timber product imported by the EU from Indonesia. Furthermore, even if all timber product exports from Indonesia to the EU were to be included in a voluntary partnership agreement, it would not necessarily lead to a reduction of illegal logging in Indonesia. Until there is market demand for legal timber, it is likely that illegal timber production will continue. In other words, there will be production of verified and certified forest products for markets that require them (if it is profitable to do so), as well as illegal production for the markets that do not require proof of legality or certification (if it is profitable to do so).

This argument seems to lead to the conclusion that unless global trade restrictions are implemented through which all countries restrict the import of non-verified and non-certified forest products, illegal logging will continue. Unfortunately, global adoption of trade restrictions seems to be unlikely in the near future. We have noted that there are economic costs associated with the global adoption of certification, which are also likely to apply to the global adoption of verification. Therefore, it can be expected that some countries will resist imposing restrictions on the import of illegal timber products. This outcome seems, indeed, to be taking place, as demonstrated by the slow pace of adopting public procurement guidelines and other trade restrictions similar to those being proposed by the EU.

Indonesia has a large domestic market for timber products. The global adoption of restrictions on the import of illegal timber products would not necessarily stop illegal logging given that producers could supply the domestic market with illegal timber, unless it also demanded verified or certified forest products (which seems unlikely to occur at a significant scale in the near term). Therefore, unless the government acted to control the sale of illegal timber or to reduce illegal logging in the field, international trade restrictions are unlikely on their own to stop illegal logging in Indonesia. Previous chapters have highlighted some of the political economic factors that seem to rule out, unfortunately, that the government of Indonesia will be able to stop illegal logging in the near future.

Acknowledgement

I would like to thank Rowena Humphreys for her assistance in collating and processing data on timber production and trade.

Notes

1 Certification involves the establishment of forest management processes in accordance with specified sustainable forest management criteria, such as those of the Forest Stewardship Council, which are 'certified' by an independent third party (Upton and Bass, 1995). *Verification of legality* refers to a process that ensures that timber is harvested according to specified legal criteria. To demonstrate that a timber product is of *legal origin* or certified, a *chain of custody* needs to be established, which traces the timber used in its production from harvest through the market chain, to the transformation into the product and to the end user. Verified timber is not necessarily harvested sustainably, while certified timber could be expected to be of legal origin. The latter, however, is the case only if the criteria used for certification include appropriate legality criteria. For this reason, the Forest Stewardship Council is reviewing its criteria to ensure that this is the case (Ryder and Amariei, 2003). In the remainder of this chapter it is assumed that certified forest products are of legal origin.
2 For information on this initiative, see, for example, www.assets.panda.org/downloads/tncwwfalliancenews7.pdf, accessed 8 August 2006.
3 The hypotheses were stated in the logical framework for project activities, which is a project management tool that specifies project objectives, outputs, activities, inputs and underlying assumptions.
4 The data presented in this paragraph are derived from UNECE and FAO (2006).
5 The number of chain-of-custody certificates is not necessarily a good indicator of the volume and/or value of products traded because, for instance, there could be a large number of small producers and traders in a country.
6 Sedjo and Swallow (1999, p15) find that 'if the amount of new demand created by certification is modest, the market is less likely to generate a price premium for the certified product... However, to the extent that the costs of certification are small and certification creates significant new demand, the two-price alternative [the occurrence of a price premium] is increasingly likely to be generated by voluntary market activities.'
7 The current analysis focuses on the activities of the alliance concerned with the verification and certification of forest products. This chapter is not concerned with assessing whether the alliance has been successful. Rather, it focuses on one of its activities, the promotion of forest products certification, in order to understand its potential contribution to the reduction of illegal logging. An earlier and more comprehensive analysis of the alliance's activities is presented in Tacconi et al (2004a).
8 Derived from the project design document submitted by the alliance to USAID.
9 They were indicative and not contractually binding indicators, given that they were included in the project proposal but not in the contract awarding the funds (Nigel Sizer, TNC, pers comm, April 2003).

10 At the time of writing, the main web page of GFTN was reporting that members traded an annual timber volume of about 179 million cubic metres (about 50 per cent of the estimated global production of certified forest products), with a value of about US$33 billion: see www.panda.org/about_wwf/what_we_do/forests/our_solutions/responsible_forestry/certification/gftn/index.cfm, accessed 27 August 2006. However, the 'GFTN participants' section of the website reported a total trade of about 33.6 million cubic metres: see www.panda.org/about_wwf/what_we_do/forests/our_solutions/responsible_forestry/certification/gftn/gftn_participants/index.cfm, accessed 27 August 2006.
11 See www.panda.org/about_wwf/what_we_do/forests/our_solutions/responsible_forestry/certification/gftn/gftn_participants/index.cfm, accessed 27 August 2006.
12 Members of GFTN manage about 15 million hectares of forest, about 5.7 million of which are FSC certified. See www.panda.org/about_wwf/what_we_do/forests/our_solutions/responsible_forestry/certification/gftn/gftn_participants/index.cfm, accessed 27 August 2006.
13 This view was expressed, for example, by a representative of a trade federation at a workshop on illegal logging organized in Indonesia (August 2004) by the Association of South-East Asian Nations and the Swedish International Development Agency.
14 FLEGT Briefing Notes, www.europa.eu.int.
15 The voluntary basis is needed in order to ensure compliance with the rules of the World Trade Organization.
16 *EU Forest Watch* (2006) Issue 107, July/August, www.fern.org.
17 See www.rinya.maff.go.jp/policy2/ihou/eiyaku.pdf, accessed 28 August 2006.
18 *EU Forest Watch* (2006) Issue 107, July/August, www.fern.org.
19 The information reported in this paragraph is derived from Tsuru (2005).
20 See www.panda.org/about_wwf/what_we_do/forests/our_solutions/responsible_forestry/certification/gftn/gftn_participants/index.cfm, accessed 27 August 2006.
21 See www.illegal-logging.info/news.php?newsId=157, accessed 8 August 2006.
22 This statement assumes that a significant number of illegally harvested logs are used in the production of pulp. We do not have any information indicating the contrary.
23 Paper imports by China saw a huge increase as they had accounted for 25 per cent of Indonesia's paper exports in 2001.
24 The data on China's timber production and trade presented in this paragraph are derived from White et al (2006).
25 See www.lei.or.id/english/akreditasi.php?cat=19, accessed 15 August 2006.
26 See www.fsc.org, accessed 8 August 2006.
27 Area of forest concessions sourced at www.dephut.go.id/INFORMASI/PH/HPH_per_Prop.htm, accessed 8 August 2006.
28 See www.lei.or.id/english/akreditasi.php?cat=20, accessed 8 August 2006.
29 See www.assets.panda.org/downloads/tncwwfalliancenews7.pdf, accessed 15 August 2006.
30 See www.forestandtradeasia.org/guidance2/Indonesia/English/98/237/3/, accessed 8 August 2006.
31 See www.forestandtradeasia.org/guidance2/Indonesia/English/98/237/6/, accessed 15 August 2006.
32 Merbau timber is used in flooring and joinery products. The illegal trade in merbau has been the subject of a campaign by the NGOs Environmental Investigation Agency and Telapak. See www.eia-international.org/cgi/news/news.cgi?t=template&a=320, accessed 16 August 2006.

33 Payment of bribes could be interpreted to mean that timber is necessarily illegal; but in many situations, even companies that have legal rights to harvest timber are required to pay bribes in order to obtain permits (e.g. timber travel authorizations) that should be issued to them.
34 See www.tropicalforesttrust.com/archives/2005/fscplywood.htm, accessed 17 August 2006. The company in question, Tjipta, was seeking to sell 500 cubic metres of certified plywood.

References

Cashore, B., Auld, G. and Newsom, D. (2004) *Governing through Markets: Forest Certification and the Emergence of Non-state Authority*, New Haven, Yale University Press

CCIF (Conservation and Community Investment Forum) (2002) *Analysis of the Status of Current Certification Schemes in Promoting Conservation*, San Francisco, CCIF

Eba'a Atyi, R. and Simula, M. (2002) *Forest Certification: Pending Challenges for Tropical Timber*, ITTO Technical Series No 19, Yokohama, International Tropical Timber Organization

Environmental Data Services (2004) 'Forest certification schemes are "undermining sustainability" ', *ENDS Report* 349 (February), p36

Environmental Investigation Agency (1996) *Corporate Power, Corruption and the Destruction of the World's Forests: The Case for a New Global Forest Agreement*, London, Environmental Investigation Agency

EU Forest Watch (2006) Issue 107, July/August, www.fern.org.

Gan, J. (2005) 'Forest certification costs and global forest product markets and trade: A general equilibrium analysis', *Canadian Journal of Forest Research*, vol 35, no 7, pp1731–1743

Garforth, M. (2004) *To Buy or Not to Buy: Timber Procurement Policies in the EU*, Moreton in Marsh, Forests and the European Union Resource Network

Global Witness (1999) *The Untouchables: Forest Crimes and the Concessionaires – Can Cambodia Afford to Keep Them*, London, Global Witness

Greenpeace (2000) *Against the Law: The G8 and the Illegal Timber Trade*, Greenpeace, www.greenpeace.org/raw/content/usa/press/reports/against-the-law-the-g8-and-th.pdf

Gullison, R. E. (2003) 'Does forest certification conserve biodiversity?', *Oryx*, vol 37, no 2, pp153–165

Jurgens, E. (2006) *Linking Demand for Certified and Legality Verified Forest Products to Supply in Indonesia*, Bogor, Centre for International Forestry Research

Muhtaman, D. R. and Prasetyo, F. A. (2004) 'Forest certification in Indonesia', Paper presented at the Symposium Forest Certification in Developing and Transitioning Societies: Social, Economic, and Ecological Effects, Yale School of Forestry and Environmental Studies, New Haven, Connecticut, US, 10–11 June

Nussbaum, R., Gray, I. and Higman, S. (2003) *Modular Implementation and Verification (MIV): A Toolkit for the Phased Application of Forest Management Standards and Certification*, Oxford, ProForest

Oliver, R. (2006) *Price Premiums for Verified Legal and Sustainable Timber*, Timber Trade Federation and UK Department for International Development, www.illegal-logging.info/papers/FII_Price_Premiums_Feb06.pdf, accessed 8 August 2006

Raunetsalo, J., Juslin, H., Hansen, E. and Forsyth, K. (2002) *Forest Certification Update* for the UNECE Region – Summer 2002, Geneva Timber and Forest Discussion Papers, ECE/TIM/DP/25, Geneva, United Nations Commission for Europe and United Nations Food and Agriculture Organization

Richards, M. (2004) *Certification in Complex Socio-political Settings: Looking Forward to the Next Decade*, Washington, DC, Forest Trends

Ryder, S. and Amariei, L. (2003) 'FSC certification and strengthening legal compliance in the forest products trade', Paper presented at the seminar on Strategies for the Sound Use of Wood, Poiana Brasov, Romania, 24–27 March

Sedjo, R. A. and Swallow, S. K. (1999) *Eco-labelling and the Price Premium*, Washington, DC, Resources for the Future

Seneca Creek Associates and Wood Resources International (2004) *'Illegal' Logging and Global Wood Markets: The Competitive Impacts on the US Wood Product Industry*, Poolsville, MD, American Forest and Paper Association

SGS Trade Assurance Services (2002) *Forest Law Assessment in Selected African Countries*, Draft report, Washington, DC, World Bank/WWF Alliance

Tacconi, L., Boscolo, M. and Brack, D. (2003) *National and International Policies to Control Illegal Forest Activities*, Bogor, Centre for International Forestry Research

Tacconi, L., Obidzinski, K. and Prasetyo, A. (2004a) *Learning Lessons to Promote Forest Certification and Control Illegal Logging in Indonesia*, Bogor, Centre for International Forestry Research

Tacconi, L., Obidzinski, K., Smith, J., Subarudi, and Suramenggala, I. (2004b) 'Can "legalization" of illegal forest activities reduce illegal logging? Lessons from East Kalimantan', *Journal of Sustainable Forestry*, vol 19, no 1/2/3, pp137–151

TNC (The Nature Conservancy) and WWF (World Wide Fund for Nature) (undated) *Promoting Forest Certification and Combating Illegal Logging in Indonesia*, Project proposal for USAID/Indonesia, Jakarta, TNC and WWF

Toyne, P., O'Brien, C. and Nelson, R. (2002) *The Timber Footprint of the G8 and China: Making the Case for Green Procurement by Government*, Gland, Switzerland, WWF-International

Tsuru, S. (2005) *Current Activities to Combat Illegal Logging and Associated Trade in Illegally Sourced Wood Products in Japan*, Ibaraki, Japan, Forestry and Forest Products Research Institute

UNECE (United Nations Commission for Europe) and FAO (United Nations Food and Agriculture Organization) (2003) *Forest Products Annual Market Analysis 2002–2004*, Timber Bellini LVI (2003)-3, Geneva, UNECE and FAO

UNECE and FAO (2006) *Forest Products Annual Market Analysis 2005–2006*, Geneva Timber and Forest Study Paper 21, Geneva, UNECE and FAO

UNEP (United Nations Environment Programme) (undated) *The Trade and Environmental Effects of Ecolabels: Assessment and Response*, Nairobi, UNEP

Upton, C. and Bass, S. (1995) *The Forest Certification Handbook*, London, Earthscan

URS Forestry (2002) *Review of Formal and Informal Costs and Revenues Related to Timber Harvesting, Transporting and Trading in Indonesia*, Draft report, Jakarta, World Bank

White, A., Sun, X., Canby, K., Xu, J., Barr, C. and Katsigris E. (2006) *China and the Global Market for Forest Products: Transforming Trade to Benefit Forests and Livelihoods*, Washington, DC, Forest Trends

WWF (World Wide Fund for Nature) (2005) *Failing the Forests: Europe's Illegal Timber Trade*, Godalming, WWF-UK

Appendix: Method for assessing illegal harvest, consumption and export of timber

The total log harvest (QL_t) is equal to the official (i.e. legal) log harvest (QL_l) plus the illegal log harvest (QL_i):

$$QL_t = QL_l + QL_i \qquad [11.1]$$

Therefore, illegal logging is derived as:

$$QL_i = QL_t - QL_l \qquad [11.2]$$

If data on consumption, export and import of wood products was available, the total log harvest could be assessed as following:

$$QL_t = CRWE_t + EXPRWE_t - IMPRWE_t \qquad [11.3]$$

where $CRWE_t$ is the total roundwood equivalent of total domestic consumption, $EXPRWE_t$ is total exports and $IMPRWE_t$ is total imports of wood products. Consumption data, however, are not available or are highly unreliable.

A conservative estimate of illegal logging can be derived by calculating the roundwood equivalent of total domestic production of the different sectors ($PRORWE_t$) and subtracting from this the total official harvest (QL_l):

$$QL_i = PRORWE_t - QL_l \qquad [11.4]$$

$PRORWE_t$ is defined as follows:

$$PRORWE_t = LOGEXP_t + ILLOGEXP_t - LOGIMP_t + SAWRWE_t \\ + FIBRWE_t + PARRWE_t + PLYRWE_t + VENRWE_t + PULRWE_t \qquad [11.5]$$

where $LOGEXP_t$ is total legal round logs export; $ILLOGEXP_t$ is illegal log export; $LOGIMP_t$ is log import; $SAWRWE_t$ is production of sawn wood roundwood equivalent; $FIBRWE_t$ is production of fibreboard roundwood equivalent; $PARRWE_t$ is production of particleboard roundwood equivalent; $PLYRWE_t$ is production of plywood roundwood equivalent; $VENRWE_t$ is production of veneer roundwood equivalent; and $PULRWE_t$ is production of pulp roundwood equivalent.

The total roundwood equivalent of paper production is not included in Equation 11.5. Paper is produced from pulp, whose wood requirements have already been accounted for. The inclusion of paper would result in double counting.

The relative contribution of domestic and international demand for wood products to the problem of illegal logging may be calculated on the basis of the assumption that each wood product contributes to illegal logging proportionally to its total volume:

$$QL_i = QL_i \times EXPPRORWE_t/PRORWE_t + QL_i \times DOMPRORWE_t/PRORWE_t \quad [11.6]$$

where $EXPPRORWE_t$ is the roundwood equivalent produced domestically and exported, and $DOMPRORWE_t$ is the roundwood equivalent produced domestically and consumed domestically. $EXPPRORWE_t/PRORWE_t$ and $DOMPRORWE_t/PRORWE_t$ are the shares of exports and domestic consumption in domestic production.

Total export of roundwood equivalent is:

$$EXPPRORWE_t = LOGEXP_t + ILLOGEXP_t + EXPSAWRWE_t + EXPFIBRWE_t + EXPPARRWE_t + EXPPLYRWE_t + EXPVENRWE_t + EXPPULRWE_t + EXPPAPRWE_t \quad [11.7]$$

where $EXPSAWRWE_t$ is the export of sawn wood roundwood equivalent; $EXPFIBRWE_t$ is the export of fibreboard roundwood equivalent; $EXPPARRWE_t$ is the export of particle board roundwood equivalent; $EXPPLYRWE_t$ is the export of plywood roundwood equivalent; $EXPVENRWE_t$ is the export of veneer roundwood equivalent; $EXPPULRWE_t$ is the export of pulp roundwood equivalent; and $EXPPAPRWE_t$ is the export of paper roundwood equivalent. The export of pulp roundwood equivalent, $EXPPULRWE_t$, represents the pulp exported directly, whereas $EXPPAPRWE_t$ represents pulp exported in the form of paper. To calculate the amount of pulp exported in the form of paper, the following approach was adopted. It is assumed that all of the pulp not exported directly is used to produce paper. The domestic pulp content of paper produced and exported is calculated on the basis of 'domestic pulp requirement factors'. The ratios of 'pulp for exported paper' to total pulp for paper and of 'pulp for paper consumed domestically' to total 'pulp for paper used domestically' are used to allocate proportionally the pulp produced domestically to domestic consumption and export.

The volume of roundwood equivalent consumed by the domestic market is calculated by subtracting the export from the domestic production of the product. This implies that inaccurate values for domestic production result in inaccurate estimates of domestic consumption.

Conversion factors used to calculate roundwood equivalent are presented in Table 11.2.

Table 11.2 *Conversion factors used to calculate roundwood equivalent*

Fibreboard	1.8
Particleboard	1.4
Plywood	2.3
Pulp	4.4
Sawn wood	1.8
Veneer	1.9

12

Illegal Logging and the Future of the Forest

Luca Tacconi

Introduction

Previous chapters have considered the causes of illegal logging, its impacts, policy options to address illegal logging, and gaps in knowledge that need to be filled to deal with illegal logging in an effective and equitable way. This chapter draws together key findings from previous chapters and from other literature on illegal logging and forest management, and considers their implications for policies aimed at addressing illegal logging. Detailed policy options to address illegal logging have been considered in previous chapters and in several other reports (e.g. Contreras-Hermosilla, 2002; Tacconi et al, 2003; FAO, 2005; Colchester et al, 2006; World Bank, 2006). The chapter focuses, therefore, on the broad implications for policy of the analysis of the causes of illegal logging and some key policy options raised in previous chapters. The chapter also points out key gaps in knowledge about illegal logging that remain to be addressed.

The chapter is structured according to the key dimensions of illegal logging that were outlined in Chapter 1:

1 types and extent of illegal logging;
2 social, economic and environmental impacts;
3 causes of illegal logging; and
4 policies to address illegal logging and its impacts.

Types and extent of illegal logging

Illegal logging appears to occur in many countries around the world, and estimates point to some 70 countries that may be affected (WWF, 2002). Saying that so many countries are affected does not clarify the nature of the problem. Illegal logging needs to be broken down into more detailed types of illegality to better understand its nature.

Significant emphasis has been devoted to the quantity of logs harvested illegally, and it would seem that many countries experience high rates of illegal harvest (WWF, 2002; Seneca Creek Associates and Wood Resources International, 2004; SGS Trade Assurance Services, 2002). The uncertain nature of those estimates becomes obvious, however, when one looks more closely at the issues involved and the relevant statistics. Let us consider the cases of Cameroon and Indonesia.

Cameroon has often been said to have up to 50 per cent of its timber harvest derived from illegal logging (e.g. Cerutti and Tacconi, 2006). A closer analysis of the various reports on illegal logging in Cameroon reveals that the supposed 50 per cent level is incorrect. Several reports have, in fact, misquoted previous reports that referred to illegalities of various forms rather than to the volume of harvested timber (Cerutti and Tacconi, 2006). This is a significant issue because, for example, a company may have the permits to harvest timber, but might fail to comply with a regulation guiding timber harvest. This clearly is an infringement that may need to be corrected or penalized.[1] There is, however, a significant difference in terms of the impacts on the environment arising from illegal harvest outside the authorized boundary, or even in protected areas, and the breaking of some aspects of a harvesting regulation that may lead to limited or no environmental impact. Infringements of regulations can certainly lead to negative environmental impacts; but the point is that those impacts cannot be presumed. They need to be assessed, and, unfortunately, the information to do so is unavailable in most countries. The analysis of statistics on timber harvest and trade from Cameroon shows illegal harvest in only four years over the period 1990–2004. The rates of illegal harvest ranged from 48.5 per cent in 1999 to 9.3 per cent in 2004 (Cerutti and Tacconi, 2006). The activities of small-scale logging operators that had been suspended by the Ministry of Forestry (see Chapter 8), which overstepped its authority (Cerutti and Tacconi, 2006), also contributed to illegal harvest. Small-scale logging operations are mainly active in the non-permanent forest estate, which means that the forest has been allocated to conversion to other uses. Therefore, illegal logging in those areas does not necessarily involve environmental harm.[2]

Chapter 11 presents estimates of illegal logging in Indonesia. Chapters 3 and 4 show that district governments promoted logging that was considered illegal by the national government. Timber harvested with permits issued by local-level governments would result as illegal timber because it was not accounted for in national statistics. It could be argued that the districts were not supposed to issue timber harvesting permits and therefore the timber harvested with those permits is, indeed, illegal. Problems inherent to the centralized forest management system of the Suharto regime and its contested nature are well documented (e.g. McCarthy, 2000). To a certain extent, this system endures in Indonesia partly as a result of the failure to decentralize forest management. It may be legally correct, therefore, to state that the logging permits issued by district governments are illegal. This argument disregards, however, the socially and politically contested nature of the control over forest resources. Addressing how communities (and their local governments) view logging is an important aspect of the definition of the problem (see Chapter 2). The issue also affects its quantitative extent – that is, what is the amount of non-contested versus contested illegal harvest? If communities and local governments consider logging a legitimate activity, even if banned or restricted by the national government, a significantly different approach would be required to address the problem compared to a situation in which there was a shared view of the illegal nature of the activity. Let us consider in more detail the nature of illegal logging.

The illegal harvest rate may indicate how widespread the problem is. It does not provide, however, sufficient details about the nature of the problem. A further aspect to be considered is whether illegal logging is mainly carried out by large commercial enterprises or by small-scale activities. These two types of logging operators may, at times, log in separate areas. For example, large companies normally operate in large concessions, whereas small-scale operations are often active outside concession areas (Cerutti and Tacconi, 2006), although this is not always the case – for instance, in Indonesia, small-scale logging operations have been found to operate in concession areas, sometimes illegally, at other times subcontracted by the concessionaire. If the different types of operators are active in separate areas, estimating the size of their illegal harvest could provide information on their respective impacts on the environment. Knowledge of the types of operators involved in illegal logging is also needed to assess the social impacts of illegal logging and law enforcement, as we will see below. This type of analysis is required because the cases of Cameroon, Honduras and Nicaragua (Chapters 7 and 8) represent the conditions of many other countries that experience illegal logging as a result of regulations that prohibit the exploitation of forests by small-scale operators who sell the timber domestically (e.g. SGS Trade Assurance Services, 2002).

Another type of illegality is illegal deforestation, which may be defined as the clearing of forest in areas that are supposed to be maintained under forest cover. Illegal deforestation is normally carried out to change the land use from forest to other uses, especially agriculture. There is a considerable lack of data on the extent of illegal deforestation (see Chapter 9). Illegal deforestation needs to receive more attention because it is one of the types of illegality with the greatest potential for environmental impacts. It also has significant social and economic implications because illegal deforestation is often carried out by farmers who need land for their livelihood activities. The significance of illegal deforestation is exemplified by the fact that some 56 per cent of protected forest in the Indonesian part of Kalimantan was deforested between 1985 and 2001 (Curran et al, 2004).³ The potential impacts of illegal deforestation and other illegalities are discussed below. It is important to point out here that national- and global-level deforestation assessments should consider providing information on illegal deforestation.

It is clear from this discussion that there is a considerable lack of knowledge about the extent of the various illegalities. Filling this gap should be considered a priority by initiatives at the country level so that the actual social, economic and environmental impacts of illegal logging can be assessed more accurately to inform the development of policies.

Social, economic and environmental impacts

Many reports on illegal logging point out its negative environmental, social and economic impacts without considering its potential positive impacts (Tacconi et al, 2003). Several chapters in this book note that illegal logging can, in fact, have positive social and economic implications. The most obvious cases are those in which rural people carry out activities that are illegal but are essential or important to their livelihoods, such as timber harvesting, collection of non-timber forest products and agricultural development (see Chapters 3 and 6–9), although there may be instances when the poorest people benefit less than the elites from illegal logging (Chapter 7; Tacconi et al, 2004). The contribution to community cohesion is a more subtle (but, nonetheless, important) social aspect of illegal logging (Chapter 2). When illegal logging has positive social and economic impacts, law enforcement is bound to have some negative impacts (Chapters 2, 6 and 8). In the short term, the negative impacts of law enforcement may be attenuated by non-enforcement or selective implementation of the law (Chapters 2, 6 and 8). Forestry regulations should be reviewed, however, to evaluate how they can be changed to accommodate the livelihood needs of the rural population. This review of legislation is particularly important in countries that have low per capita incomes and where illegal logging takes place predominantly in small-scale activities carried out by rural people.

The World Bank (2002) suggested that some US$10 billion of government revenues may be lost yearly because of illegal logging, although this estimate was not based on detailed research.⁴ Illegal logging has also been estimated to depress global prices of timber products by an average of 7–16 per cent, depending upon the product (Seneca Creek Associates and Wood Resources International, 2004).⁵ This is probably the best available estimate of global impacts on prices arising from illegal logging. That assessment relied, however, on existing rough estimates of illegal harvests, which are rather unreliable, as noted above. At the country level, previous chapters noted that illegal logging might have positive economic implications in some cases; but there is a considerable lack of knowledge about these impacts. The few estimates currently available do not account for economy-wide impacts. They do not rely on rigorous economic modelling to take into account, for example, revenue losses from unpaid taxes, as well as the positive economic impacts of greater economic activity resulting from higher harvest levels. For instance, such modelling for Indonesia would need to account for the short- to medium-term effects of reducing the output of forestry products by some 50 per cent below current levels (see Chapter 11). The effects of changes in harvest level would be felt in the forestry sector, in the labour market, in the agricultural sector and in the industrial and services sectors. Economic assessments of economy-wide impacts of changes in illegal logging should also take into account the environmental costs that it might cause. Research on the economic impacts of illegal logging is needed to understand the factors that may be leading producer countries to avoid taking serious action on illegal logging and to assess the policies that could be introduced to limit the negative economic impacts originating from a reduction in illegal logging.

The environmental impacts of illegal logging have been noted in many reports (Tacconi et al, 2003). There is still only scant knowledge of their actual magnitude, however, as a result of the lack of information on the extent of illegal logging and a significant lack of knowledge about the environmental impacts arising from deforestation and forest degradation. Deforestation has well-established climate change effects. Greenhouse gas emissions from land-use change, including deforestation, have been estimated at about 18 per cent of global emissions (Houghton, 2003). Illegal deforestation accounts for part of these emissions, whereas forest degradation from illegal logging does not result in climate impacts because there is no net emission of greenhouse gases if forests regenerate after illegal logging.[6] Deforestation and forest degradation may also have negative impacts on biodiversity. These impacts are less straightforward than those on climate change, however, because deforestation and forest degradation do not necessarily lead to irreversible losses of biodiversity. Extinctions of populations are influenced by changes in size and isolation of vegetation patches in the landscape, with the critical extinction threshold estimated at 30 per cent of remaining habitat cover (Andrén, 1994). The impacts of deforestation on biodiversity also depend upon, among other factors:

- the types of forests affected (e.g. secondary versus old growth) and whether species in the areas affected are adapted to human pressure (Laurance, 2007); and
- whether the protected areas system is representative of the diversity of habitats and species found in the relevant region (e.g. Margules and Pressey, 2000).

The local environmental impacts of illegal forest degradation and deforestation are similarly uncertain. The most significant local environmental benefits of forests are thought to be watershed services and natural hazard regulation. Mounting evidence shows, however, that the removal of forest cover does not necessarily result in increased soil erosion or in a reduction of the quantity and quality of freshwater (Bruijnzeel, 2004). The vegetation that replaces forests is normally sufficient to protect the soil, according to Bruijnzeel (2004), and removal of forest cover actually results in an increased total annual water yield, the bulk of the increase being in low flows if surface disturbance is limited. With regard to protection from natural hazards, mangrove forests have been shown to reduce the impacts of storm surges, including tsunamis on coastal areas (Danielsen et al, 2005); but forests generally do not reduce the occurrence of floods (FAO and CIFOR, 2005).

The biophysical impacts discussed above were considered from a scientific perspective. Communities near or far from the forests affected may hold other, more intangible, environmental values, such as existence and option values (e.g. Pearce and Moran, 1994). These values are an important factor in decision-making and partly determine whether illegal logging and associated forest degradation and deforestation come to be seen as unacceptable acts, thus leading to community demands for action against illegality (see Chapter 2).

I have outlined some significant gaps in knowledge about the social, economic and environmental impacts of illegal logging. Some of these gaps, such as those related to the biophysical impacts of forest degradation and deforestation, cannot be filled solely by research on illegal logging. Other gaps are more specific to the issue of illegal logging, however, and focused research is needed to fill these gaps and to contribute to improved understanding of their implications for choosing the most suited policies to address the impacts of illegal logging and its causes. Governments that face significant trade-offs need increased knowledge about the actual impacts of illegal logging, as discussed further on in this chapter.

The causes of illegal logging

Previous chapters have identified several causes of illegal logging, which may be thus summarized: community perceptions that illegal logging is not a criminal and/or harmful activity and that it can be accommodated; a biased, inconsistent or over-complex regulatory framework that leads forestry companies and rural people to infringe it; lack of government willingness and/or capacity to enforce the law; conflicting interests over forest management between central- and lower-level governments; and corruption. This section considers these causes with particular attention to the fact that illegal logging, like legal logging and any other economic activity, is driven primarily by the benefits that it generates, whether they are financial benefits from the sale of harvested products or benefits derived from the direct use of the products.

Community views are likely to be affected by moral values, as well as economic considerations (Tacconi, 2000; see also Chapter 2). Poorer communities are more likely to focus on the extractive livelihood values of forests (i.e. logging and conversion to agriculture) than other values of natural resources, such as existence values, as demonstrated by the fact that agricultural development is the leading cause of deforestation (Geist and Lambin, 2002), and rural people clearly stress their interest in agricultural development when asked directly (Tacconi, forthcoming). Hence, in low-income countries it can be expected that communities are more tolerant, or approve, of illegal logging than communities in higher-income countries. If this hypothesis were true, overall economic development leading to rising per capita income would lead communities to increasingly see illegal logging as a reproachable, if not criminal, activity. This point is consistent with the fact that higher income levels are associated with better governance (Kaufmann and Kraay, 2003; Sachs, 2005). Can this hypothesis be reconciled with the fact that even in a high-income country such as the US illegal logging persists and is accommodated by communities and enforcement agencies (see Chapter 2)? The fact that illegal harvesting in the US is a sporadic rather than a widespread activity would seem consistent with the hypothesis.

Forestry regulatory frameworks at times marginalize rural people, who are thus forced to infringe the regulations to make a living (see Chapters 6–8). Reform of the regulatory framework would therefore be required. The question is whether reform would resolve the problem of illegal logging. The answer depends upon whether the new regulatory framework allowed livelihood activities to be carried out without impediments. If livelihood activities could be carried out without environmental impacts, there would be no need for the regulatory framework to restrict livelihood activities. Unfortunately, this win–win scenario is rather uncommon (e.g. Wunder, 2001). The trade-offs between environmental conservation and maintenance of livelihoods imply that the regulatory framework either does not limit livelihood activities and allows environmental harm, or seeks to limit the latter by restricting livelihood activities, with the implication that illegal logging would continue. These trade-offs are at the root of the conflicts between rural people and protected areas that experience illegal logging (Cernea and Schmidt-Soltau, 2006). This common scenario needs to be addressed with appropriate instruments aimed at solving the conflicts, rather than just relying on law enforcement, as discussed later.

Trade-offs between sustainable forest management and returns from logging are also at the core of unsustainable logging because the latter yield higher returns (Rice et al, 1997; Pearce et al, 2003). Similarly, illegal logging yields higher returns than legal and sustainable logging because it reduces costs (see Chapter 11). Commercial logging companies therefore have an incentive to carry out illegal logging no matter how simplified the regulatory framework is because illegality results in higher profits. There are risks associated with carrying out illegal activities, such as fines, jail terms and possible loss of market share. Law enforcement activities contribute to raising the costs of carrying out illegal activities, which, in turn, influences profits and whether companies become involved in illegal logging (Tacconi et al, 2004).

Law enforcement requires resources. These resources may not be forthcoming because the relevant government agencies do not have them or do not consider enforcement a priority (see Chapter 2). The former reason relates to lack of government capacity. Government officials may not consider enforcement a priority because they may prefer to maintain good relations with the community – particularly when forest crimes account for a small share of the total output – and seek to enforce the law only when a crime surpasses a certain threshold (see Chapter 2). In this case, it is obvious that institutional and community views of what constitutes a crime, and considerations associated with the costs and benefits of law enforcement influence whether and how the law is enforced. Lack of, or limited, law enforcement may also be due to perceived negative impacts of the law on livelihoods, and law enforcement officers may avoid enforcing it (see Chapter 8). In these cases, lack of law enforcement in itself is not a cause of illegal logging.

Lack of government capacity to enforce the law is best seen in the overall context of the determinants of government effectiveness. This aspect is complicated by the fact that we lack a satisfactory and complete theory of government behaviour in managing natural resources. Ascher (1999) proposes that governments may not build the capacity of specific agencies because by leaving them weak the government can achieve objectives that may be against those agencies' goals, such as sustainable forest management.[7] These objectives include financing controversial development programmes (with revenues diverted from natural resource exploitation); creating rent-seeking opportunities to gain private actor cooperation in pursing other objectives; capturing natural resource rents for the central treasury; and evading accountability through reliance on low-visibility resource management arrangements. These objectives do not imply that government officials seek to achieve them for personal gain – that is, corruption. Ascher (1999) essentially asserts that government agencies can be effective when governments decide that they should be so. His case studies support the hypothesis that the objectives listed above are, indeed, followed, at times, by government agencies and officials. The case studies do not demonstrate, however, that lack of capacity is not a constraint in some situations. Ascher's study also does not include corruption as one of the possible explanatory variables, thus leaving open the possibility that corruption was also a relevant factor in his case studies. A further variable to be considered is the influence of business on government (e.g. Dauvergne, 2001). Ascher (1999) notes that some states that had been said to be weak (and thus liable to be considerably influenced by business) were, indeed, rather strong and actually dealt with business in a way that furthered their interests. This does not exclude, obviously, that there are weak states where business can significantly influence governments. For example, weak states with forestry sectors include the Democratic Republic of Congo, Papua New Guinea and the Solomon Islands.[8] We also need to stress that business may also influence policy-making in strong states. The implication of the above discussion is that there are several motives that may be driving government decisions and they need to be considered in each specific situation.

In relation to government commitment, we should note that timber producer country members of the International Tropical Timber Organization had committed to implementing sustainable forest management by the year 2000. This objective has not been achieved, given that only about 7 per cent of the natural forest in the permanent forest estate in those countries was managed sustainably as of 2005 (ITTO, 2006). This outcome was influenced to a considerable extent by the fact that governments were not truly committed to sustainable forest management because it yields lower economic benefits than unsustainable management, as with the conversion of forests to other land uses (Norton-Griffiths and Southey, 1995; Kishor and Constantino, 1993; Tomich et al, 2001) and because of the use of forests for patronage purposes, noted above. The significant influence of the economics of forest and land-use management on government decision-making is clearly demonstrated by tropical forest countries' statements that they need financial compensation to implement sustainable forest management.[9]

Given the economics of forest and land-use management, it should not come as a surprise that governments may not allocate sufficient resources to stop illegal logging. Governments may consider supporting the production of forest products and their contribution to the economy as a higher priority than the reduction of illegal logging, particularly because some of the areas logged illegally will be converted to other land uses. This is demonstrated by the case of Indonesia, which seems to have experienced an increase in the level of illegal harvest between 2000 and 2003 despite the government's stated intention to address illegal logging (see Chapter 11). Several governments have explicitly stated their commitment to addressing illegal logging or even to implement sustainable forest management. This does not mean, however, that they actually intend to do that. International relations theory shows that governments do make commitments, but often do so in the knowledge that they will not necessarily be forced to implement them (Hempel, 1996) as a result of a lack of compliance mechanisms, as in the case of the forest law enforcement and governance (FLEG) declarations prepared by several ministerial conferences on illegal logging.

Corruption has been identified as problematic for forest management. Chapter 5 stressed the importance of distinguishing between collusive and non-collusive corruption. It is helpful to further note that the existence of various types of corruption results in uncertainty about its actual contribution to the illegal harvest. If corruption allows loggers to avoid law enforcement and to harvest illegally, then corruption clearly contributes to the illegal harvest. The presence of corruption does not necessarily imply, however, that there is illegal harvesting or that corruption is necessarily contributing to the illegal harvest. If corruption involves the payment of bribes to government officials for services that should be provided by those officials (i.e. 'grease payments'), such as issuing timber transport permits, then corruption does not contribute to the illegal harvest. The types of corruption affecting a specific country need, therefore, to be considered by initiatives aimed at reducing illegal logging. It is worth noting that the empirical evidence on the impact of corruption on forest management is still inconclusive. One study finds no statistically significant relationship between corruption and the management of natural forest in Africa (Smith et al, 2003), whereas a global cross-country econometric study finds that corruption explains between 11 and 30 per cent of deforestation in tropical developing countries (Barbier et al, 2005).

Various causes discussed above – government effectiveness, the economics of forest and land-use management, and corruption – could be at the root of the conflicts between central governments and local governments over the management of forests.[10] Previous chapters have noted that while corruption may be affecting local governments, they also seek to appropriate revenues from forest management for their budgets and they are willing, at times, to allow forest activities considered illegal by the central government (Chapters 3–5). Local governments also seek the conversion of forests to agriculture (Chapter 3) and, at times, authorize development in areas demarcated by the central government as permanent forest estate. These projects are often supported by rural people who benefit from their development (Tacconi, forthcoming). Therefore, while the resolution of disputes between central and local governments over their respective authority over resource management would be beneficial, it should not be expected in itself to reduce forest degradation and deforestation if there are no significant changes in the economics of forest and land management. Financial incentives would need to be provided to local-level governments to actually promote the conservation of forests (Tacconi, forthcoming).

I have hypothesized that two main common threads link the various apparent causes of illegal logging: government objectives and economic and financial benefits from forest management and other land uses. I am not suggesting that these two variables are the only ones that fully explain illegal logging. Community views of the problem are partly independent from economic development, and corruption is not necessarily fully supported by governments; but it is influenced by economic development (Kaufmann and Kraay, 2003; Sachs, 2005). The hypothesis proposed above is based on the analysis of several case studies that had looked at only some of the potential causes of illegal logging. The hypothesis should be further tested by a study that includes at least all of the relevant explanatory causes of illegal logging summarized above, including the influence of business on government policy. Such a study would also contribute to improving our understanding of the drivers of forest management. It would also contribute to the development of a more general theoretical framework of natural resource management than is currently available.

Implications for policy and the future of the forest

Government objectives determine whether forest agencies are supported. This point implies that if the lack of support by the government is intentional, external attempts at strengthening those agencies will not lead to tangible outcomes. This is a familiar picture to many development projects in the forest sector. In other words, capacity-building projects to stop illegal logging will not lead to substantial change unless the recipient government is committed to the overall objective of reducing illegal logging. I am sceptical about the commitment of governments of timber producer countries to stopping illegal logging. The evidence about lack of commitment to sustainable forest management has already been detailed above. My scepticism was further strengthened by the lack of direct participation in the FLEG processes, particularly the one for Eastern Asia. Despite long lists of policy reform issues drawn up at the related ministerial meetings (and approved by acclamation, *not signed, not ratified*), there has been very limited, if any, action on those reforms. Even the negotiation process has struggled to continue.[11]

Detailed lists of policy options to reduce illegal logging are, therefore, useful as long as a government is committed to them. The case study on Bolivia (see Chapter 9) shows that governments can, indeed, implement reforms to allow the poorer and marginalized groups in society to access forest resources legally. The key issue to be understood is if and how commitment to policy reform may be generated and followed through from an initial situation in which government commitment is lacking. This type of analysis should precede and underlie capacity-building activities aimed at supporting the reform of specific aspects of forest management.

Social movements can be expected to have a role in influencing government commitment to reform. There is a need to understand, therefore, how illegal logging and, more generally, logging are perceived among rural communities and how it affects them. There has been a tendency to emphasize the negative effects of logging. There are also positive effects, not only from small-scale (legal or illegal) logging (Chapters 7–8), but also from large-scale logging, as demonstrated by the fact that when people control the forest and have the opportunity to choose whether to have a logging project or not, they often choose logging (McCallum and Sekhran, 1997; Tacconi, 2000; Chapter 10). Increased understanding of rural communities' views of logging and illegal logging would be relevant to the understanding of whether there are ways of influencing the perceived thresholds of environmental harm linked to logging (Chapter 2).

Once there is government commitment to address illegal logging, the reform of the forestry regulatory framework should take precedence in order to remove biases against rural livelihoods (Chapter 6) and to reduce the regulatory costs faced by both small-scale and large-scale forestry operations (Chapters 6–8). The issue of the extent to which a forestry regulatory framework aimed at ensuring sustainable forest management can be simplified remains to be addressed. Sustainable forest management is a complex task, and it may be that only a complex regulatory framework can properly regulate it.

Government commitment is also needed to address corruption. Tackling corruption is a task that needs a whole-of-government approach, and it is unlikely to be achieved on a sectoral basis. Some of the policies required to combat corruption, such as increased economic and political competition (Chapter 5), are clearly beyond the scope of activities in the forest sector. Initiatives to increase accountability and transparency in the forest sector could contribute, however, to the fight against corruption.

Government commitment is influenced by the economics of forest management.[12] Market demand for verified and certified forest products is limited, and even if demand increased, possibly through trade mechanisms such as those being developed by the European Union, it is conceivable that illegal logging would continue to supply domestic markets in producer countries and in non-environmentally sensitive international markets (Chapter 11). The economics of sustainable forest management would need to see significant changes for governments to increase their commitment to sustainable management and to promote its implementation throughout the forest estate, rather than just in those forest management areas that produce for environmentally sensitive markets. Such a change could be brought about by financial transfers from developed to developing countries for the conservation of forests in order to avoid emissions of greenhouse gases (Former et al, 2006). The fact that these payments can make sustainable timber harvesting competitive with non-sustainable harvesting has already been established (Pearce et al, 2003). It remains to be seen, however, whether financial incentives that originate from carbon markets could make sustainable timber harvesting competitive with the conversion of forest to alternative land uses. The future of the forest and illegal logging is, therefore, partly linked to the politics and economics of climate change.

The politics and economics of biodiversity conservation is another factor with a bearing on the future of illegal logging and the forest. Illegal logging in protected areas may seem to be the most obvious target of law enforcement. This point is exemplified by the high-profile case of Tanjung Puting National Park in Indonesia, where the operators of a sawmill that used illegal timber from the park held hostage, assaulted and injured environmental activists who were investigating illegal logging in the area. In this case, it is obvious that the law should be enforced, at least in relation to the events concerning the environmental activists – although it is instructive to note in relation to government commitment that the owner of the sawmill, who is a member of the national parliament, was not prosecuted. In the case of many protected areas, however, the use of law enforcement as the main means of addressing illegal logging is probably not the best option. There are many examples around the world of rural people being dispossessed of their land to establish protected areas or being denied access to them for their livelihood activities (see Chapter 6; Cernea and Schmidt-Soltau, 2006). The relationship between livelihoods and poverty outside and inside protected areas, and how to address it, is the focus of ongoing debate (Sanderson and Redford, 2004; Sanderson, 2005; Brockington et al, 2006). The outcomes of this debate and the practical implementation of measures aimed at conserving biodiversity while benefiting rural people will influence the future of illegal logging and the forest.

Finally, macro-economic factors influence forest management and illegal logging. In the forest transition model, countries first experience a decline in forest cover, normally due to the expansion of agricultural activities, which is then followed by an increase in forested area.[13] As countries develop, the industrial and service sectors provide opportunities for off-farm employment, and marginal lands are abandoned and eventually revert to forests. Illegal logging can be expected to decrease with economic development. Higher salaries and more employment opportunities in other sectors will draw people away from the forest and from working with (illegal) logging operations – jobs that are considered dangerous and that tend to attract mostly the poorest people (Tacconi, forthcoming). Economic development also leads, as noted above, to better governance and to increased awareness of environmental problems and community demands for improved environmental management by forestry companies. Evidence on forest transitions, however, shows that some countries experience it only at rather low levels of forest cover – for example, below 10 per cent. Thus, even if the forest transition occurred, deforestation may have resulted in irreversible losses of biodiversity and considerable emissions of greenhouse gases. The likelihood of a possible forest transition should not be used, therefore, as a reason for doing nothing about illegal logging.

The stage of economic development of a country and its position along the forest transition path is likely to influence the policy options that need to be considered to reduce illegal logging and to enforce the law. More cross-country comparisons, such as the one presented in Chapter 10, are needed. They should consider which government policies were successful in reducing illegal logging and deforestation in countries that have now completed the deforestation phase in the forest transition path. The key question is the extent to which economic development drives forest management and, in turn, determines the future of illegal logging.

Notes

1 Forestry regulations are often too complex (see Chapter 6). This complexity implies that, in some cases, the regulation should be changed instead of punishing those who infringe it. This is also at the root of 'lenient' law enforcement discussed in Chapter 8.
2 Environmental harm may occur if forest conversion takes place in areas unsuitable to conversion. It is useful to recall that even guidelines by the World Conservation Union (Poore, 2003) recognize that forest conversion may be appropriate under certain conditions.
3 The study adopted the ground cover threshold of 80 per cent to define forest. A significantly lower level of deforestation could be expected if the study had adopted a threshold to define forest similar to international definitions, such as, for instance, the 10 per cent level adopted by the FAO.
4 Jim Douglas, former head of the Forestry Team at the World Bank, pers comm, May 2005.
5 This estimate was derived using rigorous economic modelling with the Global Forest Products Model.
6 Illegal logging could facilitate forest conversion to other land uses by opening up roads to farmers and ranchers. In this case, illegal logging could be said to contribute to climate change. This would depend, however, upon whether or not land-use change would have happened even without illegal logging.
7 Ascher (1999) uses a comparative case study approach that looks at several resource sectors, including forestry, in various countries.
8 An index of *failed* states, which does not perfectly overlap with the concept of weak states, is provided at www.foreignpolicy.com/story/cms.php?story_id=3098, accessed 1 September 2006.

9 Papua New Guinea and Costa Rica made a request to this effect at the meeting of the Conference of the Parties to the Climate Convention and the Kyoto Protocol in Montreal during November–December 2005.
10 These causes could affect one or both government levels simultaneously.
11 As an example, the author was nominated to coordinate the activities of the group on timber demand and supply of the FLEG task force in East Asia. The respective governments did not even nominate contact points in their countries, an undertaking agreed upon at the relevant meeting.
12 A predatory regime or one interested in other objectives, as specified by Ascher (1999), should not be expected to support the implementation of sustainable forest management even if the latter represented the most efficient land-use option.
13 The discussion of the forest transition hypothesis is based on Rudel et al (2005).

References

Andrén, H. (1994) 'Effects of habitat fragmentation on birds and mammals in landscapes with different proportions of suitable habitats: A review', *Oikos*, vol 71, pp 355–366

Ascher, W. (1999) *Why Governments Waste Natural Resources: Policy Failures in Developing Countries*, Baltimore, Johns Hopkins University Press

Barbier, E. B., Damania, R. and Leonard, D. (2005) 'Corruption, trade and resource conversion', *Journal of Environmental Economics and Management*, vol 50, no 2, pp276–299

Brockington, D., Igoe, J. and Schmidt-Soltau, K. (2006) 'Conservation, human rights, and poverty reduction', *Conservation Biology*, vol 20, no 1, pp250–252

Bruijnzeel, L. (2004) 'Hydrological functions of tropical forests: Not seeing the soil for the trees?', *Agriculture, Ecosystems and Environment*, vol 104, no 1, pp185–228

Cernea, M. M. and Schmidt-Soltau, K. (2006) 'Poverty risks and national parks: Policy issues in conservation and resettlement', *World Development*, vol 34, no 10, pp1808–1830

Cerutti, P. O. and Tacconi, L. (2006) *Forests, Illegality and Livelihoods in Cameroon*, Working Paper No 35, Bogor, Centre for International Forestry Research

Colchester, M., Boscolo, M., Contreras-Hermosilla, A., del Gatto, F., Dempsey, J., Lescuyer, G., Obidzinski, K., Pommier, D., Richards, M., Sembiring, S. N., Tacconi, L., Vargas Rios, M. T. and Wells, A. (2006) *Justice in the Forest: Rural Livelihoods and Forest Law Enforcement*, Bogor, Centre for International Forestry Research

Contreras-Hermosilla, A. (2002) *Law Compliance in the Forestry Sector: An Overview*, WBI Working Papers, Washington, DC, World Bank

Curran, L. M., Trigg, S. N., McDonald, A. K., Astiani, D. Y., Hardiono, M., Siregar, P., Caniago, I., and Kasischke, E. (2004) 'Lowland forest loss in protected areas of Indonesian Borneo', *Science*, vol 303, pp1000–1003

Danielsen, F., Sorensen, M. K., Olwig, M. F., Selvam, V., Parish, F., Burgess, N. D., Hiraishi, T., Karunagaran, V. M., Rasmussen, M. S., Hansen, L. B., Quarto, A. and Suryadiputra, N. (2005) 'The Asian tsunami: A protective role for Coastal vegetation', *Science*, vol 310, no 5748, p643

Dauvergne, P. (2001) *Loggers and Degradation in the Asia-Pacific: Corporations and Environmental Management*, Melbourne, Cambridge University Press

FAO (United Nations Food and Agriculture Organization) (2005) *Best Practices for Improving Law Compliance in the Forestry Sector*, Forestry Paper No 145, Rome, FAO

FAO and CIFOR (Centre for International Forestry Research) (2005) *Forests and Floods: Drowning in Fiction or Thriving on Facts?*, RAP Publication 2005/03, Forest Perspectives 2, Bangkok and Bogor, FAO and CIFOR

Former, C., Blaser, J., Jotzo, F. and Robledo, C. (2006) 'Keeping the forest for the climate's sake: Avoiding deforestation in developing countries under the UNFCC', *Climate Policy*, vol 6, no 3, pp275–294

Geist, H. J. and Lambin, E. F. (2002) 'Proximate causes and underlying driving forces of tropical deforestation', *BioScience*, vol 52, no 2, pp143–150

Hempel, L. (1996) *Environmental Governance: The Global Challenge*, Washington, DC, Island Press

Houghton, R. A. (2003) *Emissions (and Sinks) of Carbon from Land-Use Change (Estimates of National Sources and Sinks of Carbon Resulting from Changes in Land Use, 1950 to 2000)*, Report to the World Resources Institute, Falmouth, MA, Woods Hole Research Center

ITTO (International Tropical Timber Organization) (2006) *Status of Tropical Forest Management 2005*, Yokohama, ITTO

Kaufmann, D. and Kraay, A. (2003) *Governance and Growth: Causality Which Way? – Evidence for the World*, Washington, DC, World Bank

Kishor, N. and Constantino, L. F. (1993) *Forest Management and Competing Land Uses: An Economic Analysis for Costa Rica*, Washington, DC, World Bank

Laurance, W. F. (2007) 'Have we overstated the typical biodiversity crisis', *Trends in Ecology and Evolution*, vol 22, no 2, pp65–70

Margules, C. R., and Pressey, R. L. (2000) 'Systematic conservation planning', *Nature*, vol 405, no 6783, pp243–253

McCallum, R. and Sekhran, N. (1997) *Race for the Rainforest: Evaluating Lessons from an Integrated Conservation and Development 'Experiment' in New Ireland, Papua New Guinea*, Port Moresby, PNG Biodiversity Conservation and Resource Management Programme

McCarthy, J. (2000) 'The changing regime: Forest property and *reformasi* in Indonesia', *Development and Change*, vol 31, no 1, pp91–129

Norton-Griffiths, M. and Southey, C. (1995) 'The opportunity costs of biodiversity conservation in Kenya', *Ecological Economics*, vol 12, no 2, pp125–139

Pearce, D. W. and Moran, D. (1994) *The Economic Value of Biodiversity*, London, Earthscan

Pearce, D., Putz, F. E. and Vanclay, J. K. (2003) 'Sustainable forestry in the tropics: Panacea or folly?', *Forest Ecology and Management*, vol 172, pp29–247

Poore, D. (2003) *Changing Landscapes: The Development of the International Tropical Timber Organization and its Influence on Tropical Forest Management*, London, Earthscan

Rice, R. E., Gullison, R. E. and Reid, J. W. (1997) 'Can sustainable management save tropical forest?', *Scientific American* (April), pp44–49

Rudel, T. K., Coomes, O. T., Moran, E., Achard, F., Angelsen, A., Xu, J. and Lambin, E. (2005) 'Forest transitions: Towards a global understanding of land use change', *Global Environmental Change, Part A*, vol 15, no 1, pp23–31

Sachs, J. D. (2005) *The End of Poverty: How We Can Make It Happen in Our Lifetime*, London, Penguin

Sanderson, S. (2005) 'Poverty and conservation: The new century's "peasant question"?', *World Development*, vol 33, no 2, pp323–332

Sanderson, S. E. and Redford, K. H. (2004) 'The defence of conservation is not an attack on the poor', *Oryx*, vol 8, no 2, pp146–147

Seneca Creek Associates and Wood Resources International (2004) *'Illegal' Logging and Global Wood Markets: The Competitive Impacts on the US Wood Product Industry*, Poolsville, MD, American Forest and Paper Association

SGS Trade Assurance Services (2002) *Forest Law Assessment in Selected African Countries*, Draft report, Washington, DC, World Bank/WWF Alliance

Smith, R. J., Muir, R. D. J., Walpole, M. J., Balmford, A. and Leader-Williams, N. (2003) 'Governance and the loss of biodiversity', *Nature*, vol 426, pp67–70

Tacconi, L. (2000) *Biodiversity and Ecological Economics: Participation, Values and Resource Management*, London, Earthscan

Tacconi, L., Boscolo, M. and Brack, D. (2003) *National and International Policies to Control Illegal Forest Activities*, Bogor, Centre for International Forestry Research

Tacconi, L., Obidzinski, K., Smith, J., Subarudi and Suramenggala, I. (2004) 'Can "legalization" of illegal forest activities reduce illegal logging? Lessons from East Kalimantan', *Journal of Sustainable Forestry*, vol 19, no 1/2/3, pp137–151

Tacconi, L. (forthcoming) 'Decentralization, forests and livelihoods: Theory and narrative', *Global Environmental Change*

Tomich, T., van Noordwijk, M., Budidarsono, S., Gillison, A., Kusumanto, T., Murdyiarso, D., Stolle, F. and Fagi, A. M. (2001) 'Agricultural intensification, deforestation, and the environment: Assessing tradeoffs in Sumatra, Indonesia', in Lee, D. R. and Barrett, C. B. (eds) *Tradeoffs or Synergies? Agricultural Intensification, Economic Development and the Environment*, Wallingford, CAB International

World Bank (2002) *A Revised Forestry Strategy for the World Bank Group*, Washington, DC, World Bank

World Bank (2006) *Guidelines for Formulating and Implementing National Action Plans to Combat Illegal Logging and Other Forest Crime*, Draft report, Washington, DC, World Bank

Wunder, S. (2001) 'Poverty alleviation and forests – what scope for synergies?', *World Development*, vol 29, no 11, pp1817–1833

WWF (World Wide Fund for Nature) (2002) *Forests for Life – Working to Protect, Manage and Restore the World's Forests*, Washington, DC, WWF

Index

Page numbers in *italic* refer to Figures, Tables and Boxes.

AAFs (annual area fees) 178–81, *178*
access rights 147–50, *148*, 193–4, *194*, *195*, 221
   loss of 114–15, 117, *119*, 124, *130*, 208, 214
accommodation, institutional 31–9
accountability 11, 63, 83, 85, 107, 157
   Bolivia 13, 198, 204, 208, 214
   of institutions 127, 129
adat authorities 75–9, 82
administration 160, 185, 188, 222–3, 226–7
   failures 38, 154–7, *155*, *156*
AFE-COHDEFOR (State Forestry Administration-Honduran Forestry Development Corporation) 144, *145*, 147, *148*, 150, *151*
affiliated timber theft 21–2
Alliance to Promote Certification and Combat Illegal Logging 14, 252, 255–6, 259, 260, 261, 266–7
annual allowable cut 12, 120, 125, 160
annual area fees *see* AAFs
anticipatory accommodation 35–7
ASLs (Asociaciones Sociales del Lugar, Bolivia) 194, *195*, 196, *196*, 201, 203, 206

assimilatory accommodation 33–5
atrophic accommodation 37–8

barriers to legality 12, 140, 144–52, *145*, *148*, *149*, *151*, 154
   public administration failures 154–7, *155*, *156*
   reducing 160–1
benefits from illegal logging 8, 14, 60, 83, 158, 280, 283
Berau (East Kalimantan) 43, 48, *50*, *52*, 60
   illegal logging 46, 49, *50*, 51–5, 61, 62
biodiversity 28, 63, 286
Brazil 224
Finland 221, 222, 238, 239
   loss of 5, 125, 279, 287
Bolivia 12–13, 83, 191, 198–9, 204–8, 213–14, 285
   access rights 193–4, *194*, *195*
   accountability 204, 208, 214
   forest management plans *196*, 197, 214
   forest users 198–204, *200*
   governance 194–7, *196*, 208, 214
   illegal logging 208–10, *209*
   indigenous peoples 193, *193*, 194, 199, *200*, 202–3, 204, 205, 207, *207*, 214
   institutions 194–7, *196*
   land tenure 192–3, *193*

livelihoods 208–11, *209*
policy options 211–13
regulatory framework 192–8, *193, 194, 195, 196*
taxation 197–8, 202
boundaries 6, 20, 160, 172
delineating 38, 48, 55
transgressions 36, 98, 276
Brazil 224, *225*, 240–2, 267
comparison with Finland 228–42, *230–1*
corruption 228, *231*, 238, 240, 242
deforestation 218, 224, *225*, 226, 238, 241
forest administration 224–7
illegal logging 39, 218–19
law enforcement 219, 226–7, 228, *230–1, 234*
non-compliance 13, 219, 228, 240
possible reforms 228–30, *230–1*
regulatory approach 219, 229–37, *230, 234, 235*
Brazilian Institute of the Environment and Renewable Natural Resources *see* IBAMA
bribes 7, 92–4, 105, 118, 126, *131*, 283
causing loss of income 114, 116, *119, 130*
elites' use of 110, 115
Indonesia 95, 101
to speed up processes 122, 123, 232
British Columbia (Canada) 26–31, 41
brokers 62, 76, 77, 78, 95
bureaucracy 121, 123, 150–1, 205, 212, 232
see also paperwork
bush meat 115, 125, 174

Cameroon 4, 12, 167–8, 169, *169*
commercial logging 175–81, *176, 178, 180*
community forests 183–5, *186*
hunting and gathering 173–5, *175*
illegal logging 4, 276
law enforcement 168, 172–3, 187–8
regulations 168–71, 173–5, *175*, 277
small-scale logging 181–3, *182, 183*, 277
zoning and classifying forests 170–3, *171, 173*
capacity, lack of 6–7, 14, 160, 281, 282–3
capacity building 6–7, 212, 257, 284, 285
carbon markets 14, 286
causes of illegal logging 5–8, 8–9, 14, 280–4
Central America, timber trade 142–3
Central Kalimantan 48, *51, 56, 58*, 62
illegal logging 43, 46, 50–1, *51*, 55–9
see also Kotawaringin Timur
certification 9, 14, *231*, 238–9, 251–3, 253, 256, 265–7
CFPs (certified forest products) 14, 252, 253–61, 268, 285–6
CFs (community forests) 183–5, *186*
chainsaws 79, 181
prohibition of 13, 116, 120–1, 128, 197, 205
China 123, 127–8, 268
certification 9, 254, 255
exports 104, 264, 267
trade with Indonesia 259–60, 263, *263*
chronic tree theft 32–3
civil society 81, 83, 106, 129, *130*, 161
clandestine production 142–3, *142*
classifying forests 170–3, *171, 173*
clear-cutting 27, 28, 29–30, 221, 232, *236*
climate change 279, 286, 287
COATLAHL (Cooperativa Agroforestal Colón, Atlántida

Honduras Limitada) 153, *153*, 158–9
collusive corruption 92–3, 93–4, 104, 105, 283
   Honduras and Nicaragua 144, *149*, 161
   Indonesia 72, 81, 95, 101, 104, 107
commercial logging 13, 125–6
   Cameroon 175–81, *176*, *178*, *180*
   see also logging companies
communal forests 171
communities 63–4, 75–9, 124, 133, 144, *145*, 184–5
   property rights 12, 78, 79, 82–3
   stability 10, 21–2, 26, 32, 39
   and timber theft 21–2
community, shared meaning of 18–19
community forestry 118, 126, 191, 198, 205, 214
community forests *see* CFs
community views 10–11, 14, 277, 280–1, 282, 284, 285
community-based law enforcement 40, 132–3
competition 104–6
compliance 13, 205, 240
   Brazil 13, 219, 228, 240
   Finland 13, 219, 228, 233
   see also barriers to legality
concessions 175–6, *176*, 193–4, *194*, *195*, *196*
   see also large-scale concessions; small-scale concessions
conflicts 5, 71, 102, 124, 198, 281
   between central and local governments 14, 280, 284
consensus-oriented approach 13, *231*, 239–40
conservation 79–80, 127, 286
conservation areas 124, 125
control strategies 103–6, 107
Cooperativa Agroforestal Colón, Atlántida Honduras Limitada *see* COATLAHL

corruption 3, 7, 8, 11, 12, 14, 110, 283
   Bolivia 214
   Brazil 228, *231*, 238, 240, 242
   controlling 63, 103–7, 129, 285
   Honduras and Nicaragua 152, 154–5, 155, 160, 161
   and illegal logging 280, 282, 284
   Indonesia 11, 91–2, 94–5, 96, 102–3, 104, 107
   and non-enforcement 31–2
   South Aceh 71, 72, 75
   types 92–3
   see also collusive corruption; non-collusive corruption
crime 8, 19, 25, 26, 282
   organized 155–6, 157, 160
criminalization 26–31, 40, 117–18, 140, 145
cultures, local 113, 118, *119*, 125, 128, *131*
customary rights 47–8, 96, 142, 172

'deadwood' licences 144, *145*
death threshold 28–9
decentralization 83, 94, 105, 161, 188, 198, 227
   and illegal logging in Indonesia 11, 44, 48, 84–5, 101, 277
   Indonesia 46, 48, 58, 60, 64, 69, 70, 83–5, 95–6, 102, 107
deforestation 5, 115, 117, 123, 160, 283, 287
   Brazil 218, 224, *225*, 226, 238, 241
   Cameroon 167–8
   environmental impacts 279–80, 287
   Honduras and Nicaragua 139
   illegal 13, 210–11, 277–8, 279
   Indonesia 82, 96, 277–8
   permits for 203–4
   by small farmers 202
democracy 63, 94, 105–6, 107, 115, *119*, *131*

Department for International Development *see* DFID
development fees 76, 77
deviance 18–19, 32
  illegal logging as 18–19, 21, 23–5, 26–31, 40
  spectrum of 25–6, *25*
DFID (Department for International Development) 6, 112, 113, 252
disenfranchisement 115, 117–18
domestic demand 143
  Indonesia 14, 261, *262*, 264, 268
donor support 118, 129, 161

East Kalimantan 43, 48, 49, *50*, 51–5, *52*, 62
  see also Berau
economic crisis 46–7, 80–1, 95
economic development 1, 286–7
economic growth *119*, *131*
economic impacts of illegal logging 60–1, 278–9
elites 83, 94, 111, 115, *119*, 132–3, 158, 185
  Indonesia 61, 84, 85, 127
emissions 279, 286, 287
employment 60, 79, 114, 116–17, 123, 158
enforcement *see* law enforcement
entrepreneurs 45, 53, 54, 55, 94–5, 96, 100–1
  South Aceh 71–2, 78, 79, 80, 82, 85
environmental impacts of illegal logging 2, 62–3, 110, 176, 210, 276, 279–80
environmental services 115, *119*, *131*
environmental values 280
EU (European Union) *263*
  FLEGT 1, 6, 8, 257–8
  voluntary partnership agreements 14, 251, 257, *263*, 264, 268, 286
extent of illegal logging 3–4, 10, 276–8

external interventions 79–80, 82

fees 76, 77, 122, 203, 210–11
  AAFs 178–9, *178*, 180–1
Finland 220–1, *231*
  certification *231*, 238–9
  comparison with Brazil 228–42, *230–1*
  compliance 13, 219, 228, 233
  forest administration 220–4
  forest culture *231*, 240–1
  law enforcement 222–4, 228, *230–1*
  regulatory approach 229–37, *230*, *234*, *235*
FLEG (forest law enforcement and governance) processes 8–9, 260, 261, 285
FLEGT (forest law enforcement, governance and trade, EU) 1, 6, 8, 257–8
forest administration 220–7
forest community 38–9, 39–40
forest culture 38–9, *231*, 240–1
forest damage *119*, *131*
forest degradation 85, 115, 117, 123, 160, 253, 279–80
forest ecosystems 120
forest extension 13, 219, 222–3, 228, *231*, 236, 239–40
forest law enforcement *see* law enforcement
forest law enforcement and governance *see* FLEG
forest law enforcement, governance and trade *see* FLEGT
forest management 8, 40, 121, 124, *131*, 133–4, 284
forest management plans 120, 160, 203
  Bolivia 13, *196*, 197, 203–4, 205, 214
  Brazil 228, 229, *230*
  Cameroon 171, 172, 175, 176–7, 179, 183–4, 187
  Finland 228, 229, *230*, 238

Honduras 150–1, *152*
Nicaragua 149, 150, 150–1
Forest Service (Canada) 30, 31
Forest Service (US) 17, 20, 24–5, 25, 32–9, 37
Forest Stewardship Council *see* FSC
forest transition model 286–7
forest users, Bolivia 198–204, *200*
forestry law reform 128
forestry officials 120, 129, 212
  corruption 7, 79, 81, 82, 111, 115, 161, 283
forests
  future of 41, 284–7
  regeneration 63, 120
  zoning and classifying 170–3, *171*, *173*
formalization of illegal logging 44, 48, 55
FSC (Forest Stewardship Council) 253, 258, 259, 265
fuelwood 116, 126, 133, 147, 202
future of the forest 41, 284–7

gathering 173–5, *175*
  see also NTFPs
GFTN (Global Forest and Trade Network) 255, 258, 259, 267
GLNP (Gunung Leuser National Park, Indonesia) 70, 81
governance 63, 64, 85, 107, 151–2, 257, 287
  Bolivia 194–7, *196*, 208, 214
  weak 1, 255
governments
  commitment 14, 282–3, 284–5
  conflicts with local governments 14, 280, 284
  lack of capacity 6–7, 14, 160, 281, 282–3
  legitimacy 82, 106, *231*, 241–2
  transitions 81, 93–4, 105–6, 107
  weak 6, 93–4, 101, 102–3, 107
greenhouse gases 279, 286, 287
Gunung Leuser National Park *see* GLNP

Hak Pemungutan Hasil Hutan see HPHHs
harvest limits 150, *151*
harvesting rotations 120, 136
Honduras 139, 277
  access rights 147, *148*, 150
  barriers to legality 140, 144–52, *145*, *148*, *149*, *151*, *152*
  corruption 154–5
  forests and livelihoods 140–1
  illegal logging 139–40, 142–3, 153–4, *153*, 157–60, *159*
  land tenure 146–7
  public administration failures 154–7, *155*, *156*
HPHHs (*Hak Pemungutan Hasil Hutan*) 48, 63–4
human rights 111, 124
hunting 115, 125, 171, 173–5, *175*, 188
hunting and gathering 173–5, *175*
  see also NTFPs

IBAMA (Brazilian Institute of the Environment and Renewable Natural Resources) 219, 225, 227, 228, *230*, 234, 237
illegal deforestation 13, 210–11, 277–8, 279
illegal exports 98, 99–101, *100*
illegal forest production, definition 141–2, *142*
illegal logging
  activities involved in 2–3, *4*
  awareness of 1–2
illegal timber trade, Central America 142–4
impacts of illegal logging 3, 4–5, 9, 278–80, 285
  economic 60–1, 113–14, *119*, 278–9
  environmental 2, 62–3, 115, *119*, 176, 276, 279–80
  negative 5, 60–1, 62, 113–15, *119*
  positive 5, 60, 61–2

social 61–2, 114–15, 115, *119*, 277, 278
INAFOR (National Forestry Institute, Nicaragua) 144, 149–50, 157
incentives 13, 64, 198, 211, *230–1*, 237–8, 284
  for illegal logging 153–4, *153*, 210
  for sustainable forest management 161, 198
incomes 5, 112, 140–1, 278
  loss of 114, 116, *119*, 123, 124–5, *130*
India 126, 127–8, 132
indigenous peoples 13, 124, 140, 241
  Bolivia 193, *193*, 194, 199, *200*, 202–3, 204, 205, 207, *207*, 214
  land rights 147, *148*
  ownership by 193, *193*, 202–3
  property rights 121, 193, *193*
  rights *4*, 118, 121, 128, 132, 214
Indonesia 1, 63, 106, 127, 283
  adat authorities 75–9
  after fall of Suharto 94, 95–6
  certification 14, 251–2, 254, 255–6, 258, 265–8
  corruption 11, 91–2, 94–5, 96, 102–3, 104, 107
  decentralization and illegal logging 11, 44, 48, 69–70, 84–5, 101, 277
  decentralization process 46, 48, 58, 60, 64, 69, 70, 83–5, 95–6, 102, 107
  deforestation 82, 96, 277–8
  domestic markets 14, 261, *262*, 264, 268
  economic crisis 46–7, 80–1, 95
  exports 98, 99–101, *100*, 261–4, *262*, 263
  formalization of illegal logging 44, 48, 55
  illegal logging 43, 60–3, 127, 261, *262*, 276–7

illegal logging after fall of Suharto 11, 43, 46–8, 49–51, *50*, *51*, 53–5, 57–9, 81–2, 85, 96, 104
illegal logging during Suharto era 44–6, 71–2, 85
large-scale concessions 54–5, 71, 94, 277
law enforcement 64, 106, 127
law and order decline 46, 47
local governments 44, 48, 276–7
the military 6, 94, 95, 101, 106
political transition 81, 93–4, 105–6, 107
small-scale concessions 45, 48, 53–4, 54–5, 55, 60, 61–2
Suharto era (New Order) 6, 45, 70–1, 82, 84, 277
trade 258–60, 263, *263*, 264, 267, 268
verification 9, 264, 265–7
weak government 101, 102–3, 107
see also HPHHs; IPPKs; Kalimantan; Sumatra
Indonesian Eco-labelling Institute *see* LEI
informal activities 116, 126, 127
informal payments 54–5, 61–2, 71, 79, 101, 129
INPE (National Institute of Spatial Research, Brazil) 224
INRA (Instituto Nacional de Reforma Agraria, Bolivia) 192, *193*, *195*, 203, 213
inspections 123, 223–4, 230, 232–3, *234*, 236
institutional accommodation 31–9
institutionalization of illegal logging 63
institutions 83, 127, 129, *131*, 159, 213–14, 236–7
  Bolivia 194–7, *196*, 213–14
  Indonesia 69–70, 106, 107
  problems 6, 206, 228, *230*, 234–7
Instituto Nacional de Reforma Agraria *see* INRA
intellectual property rights 175

intermediaries 144, *145*, *152*, *156*, 204, 210
IPPKs (*Izin Pemungutan dan Pemanfataan Kayu*) 48, 53–4, 71, 71–2, 96, 96–101, *97*, 102–3
operators 98, 102, 106
ITTO (International Tropical Timber Organization) 1, 256, 282–3

Japan 1, 104, 256, 258–9, 263, *263*, 264, 268
verification and certification 9, 254, 255
judiciary 101, 106, 107, 152

Kalimantan (Indonesia) 10–11, 63–4, 77, 98, 99–101, *100*, 277–8
North-East 96–101, *97*
see also Berau; Central Kalimantan; East Kalimantan; Kotawaringin Timur
knowledge gaps 9, 14, 276, 278, 280
Kotawaringin Timur (Central Kalimantan) 43, 48, 60, 62–3
illegal logging 46, 50–1, 55–9, *56*, *58*, 61, 62

land rights 3, 12, 147, *148*, 172, 194, 266
see also ownership; property rights; titling
land tenure 146–7, *231*, 241
Bolivia 192–3, *193*
land titling 146–7, 160, *193*, *195*, 202–3
large-scale concessions 54–5, 71, 94, 202, 277
law enforcement 10, 11, 105, 277, 278, 281–2, 286
Bolivia 13, 197, 213
Brazil 219, 226–7, 228, *230 1*, *234*
Cameroon 172–3, 187–8
community-based 40, 132–3

Finland 222–4, 228, *230–1*
Honduras 156–7, *156*
impacts on rural livelihoods 11–12, 110–12, 113, 115–18, *119*, 122–8, 172–3, 187–8, 213
Indonesia 64, 106, 127
lack of 7, 31–9
Nicaragua 152, 156–7, *157*
North America 31–9, 40
and sustainable forest management 118, 120–2
weak 7, 82, 118, 156–7, *156*, *157*, 160, 236, 280, 282
law enforcement officers *see* LEOs
laws 128
commercial logging 125–6, 179–81, *180*
and rural livelihoods 134, 179–81, *180*, 202–8, *207*
see also law enforcement; legislation; regulations
legal framework 221–2, 225–6
legalization of illegal logging 44, 63, 84, 96, 98, 142, 144, 154
legalized production 142, *142*, 143
legislation 110, 118–20, 127
see also laws
legitimacy 82, 83, 106, 126, 141–2, *231*, 241–2
legitimacy threshold 30–1
LEI (Indonesian Eco-labelling Institute) 265, 266
LEOs (law enforcement officers) 24, 25, 31–9
Leuser Ecosystem 70, 81
livelihoods *see* rural livelihoods
local cultures 113, 118, *119*, 125, 128, *131*
local governments 14, 83–4, 196, 258, 280, 284
Indonesia 44, 48, 63, 64, 83–4, 85, 276–7
local taxes 48, 53–4, 57–9, 60, 63, 72–5, *73*, 78
logging bans 117, 123

logging companies 7, 46, 111,
    125–6, 187, 210, 277
  and communities 132, 133, 184
logging networks 70, 81, 82–3, 85,
    144
looking the other way 31–9
loss of life 28–9

macro-economic factors 43, 286–7
magnitude threshold 27
management plans *see* forest
    management plans
market access 261, 267, 267–8
market premiums 14, 254–5, 267,
    267–8
market share 252, 253–4
market-oriented approaches 13, 14,
    230–1, 237–9
markets 5, 8, 143, 286
Menggamat (Sumatra) 75–80
the military 5, 6, 94, 95, 106, 101
minimum diameters 120, 125
monitoring 6–7, 13, 49, 128, 132–3,
    161, 213, 223–4

National Environmental
    Management System *see*
    SISNAMA
National Forestry Institute *see*
    INAFOR
National Institute of Spatial
    Research *see* INPE
national parks 101, 286
The Nature Conservancy *see* TNC
negotiation *231*, 240
NGOs (non-governmental
    organizations) 1–2, 7, 129, 161,
    212, 251–2, 266
  and community forestry 184, 185
  environmental 1–2, 44, 47, 258
  in South Aceh 79–80, 81
Nicaragua 139, 140–1, 277
  access rights 148–50
  barriers to legality 140, 144–52
  illegal logging 139–40, 143, 153–4
  impacts of illegal logging on
    livelihoods 139, 157–60, *159*
  public administration failures
    154–7, *157*
nilam 78, 79, 80
non-collusive corruption 92, 93, 104,
    105, 283
  Indonesia 95, 101, 104, 107
non-compliance 13, 219, 228
non-enforcement 31–9, 174
North-East Kalimantan 96–101, *97*
NTFPs (non-timber forest products)
    2, 47, 79, 171, 173–5, *175*, 278
  extractors 199–201, *200*, 203, *207*
Nusa Hijau 266–7

Olancho (Honduras) 147, *149*
old-growth trees 28–9
open access 13, 210, 212–13
ownership *4*, 6, 47, 241
  Bolivia 192–3, *193*, *195*, 202–3,
    212–13
  see also property rights; titling

palm plantations 52–3, 53, 55, 60
paperwork 110, 115, 116, 121–2,
    150–1
  costs 115, 118, 123, 232
patchouli 78, 79, 80
patrons 114, *119*, *130*
penalties 7, 8, 13, 226, *230*, 233–4
  weak 156–7, *156*, *157*
physical security 112, *119*, *130*, 208
  threats to 114, 117, 124, 125, 126,
    158
plantations 71, 117, 224, 238
  oil palm 52–3, 53, 55, 60
policy implications 284–7
policy options 128–34, *130–1*,
    211–13, 287
political transition 81, 93–4, 105–6,
    107
Port Renfrew (British Columbia) 41
poverty 5, 139–40, 187–8, 278, 286
  alleviation 113
  Bolivia 191, 214
  Cameroon 167

Honduras and Nicaragua 140
power relations 69–70, 75–9, 81, 154–5, *155*, 161
price premiums, for CFPs 14, 254–5, 267, 267–8
prices 5, 251, 252, 261, 278
private landowners 202, 208, 211
private sector 258, 259, 260
procurement, public policies 252, 256, 258–9, 260–1, 268
property rights 6, 127, 146–7, 241
　of communities 78, 79, 82–3
　customary 172
　indigenous peoples' 121, 193, *193*
　intellectual 175
　see also land rights; ownership; titling
prosecution threshold 36
protected areas 124, 127, 276, 281, 286
　inhabitants 110, 116, 117, 124
public administration failures 154–7, *155*, *156*
public institutions 106
public procurement policies 252, 256, 258–9, 260–1, 268

reformasi 46, 47–8
regional autonomy *see* decentralization
regional co-ordination 161
regulations *4*, 7, 49, 153, 277
　Bolivia 192–8, *193*, *194*, *195*, *196*, 197, 206
　complex 12, 14, 150–1, 153, 280
　and rural livelihoods 12, 13, 168–70, *170*, 173–5, *175*, 278, 281
　see also laws
regulatory approach
　Brazil 229–37, *230*, *234*, *235*, *236*
　Finland 229–37, *230*, *234*, *235*
regulatory framework 7, 13, 281, 285
　Bolivia 192–8, *193*, *194*, *195*, *196*
　Brazil 225–6
　Finland 221–2

over-complex 12, 14, 150–1, 153, 280
remoteness 6–7, 7, 29–30
retribusi 72–5, *73*
revenues 112, *119*, 158, *159*, *182*, 214
　from taxes 5, 73–4, 124, *130*
　loss of 5, 114, *119*, 209, 278–9
rights 28, 121
　customary 47–8, 96, 142, 172
　human rights 111, 124
　indigenous peoples' *4*, 118, 121, 128, 132, 147, *148*, 193, *193*, 214
　intellectual property 175
　land rights 3, 12, 147, *148*, 172, 266
　to forest resources 64, 128
　see also access rights; property rights
Rio Platano Biosphere Reserve (Honduras) *156*
rule of law 46, 47, 102, 106, 110, 114, 115, *119*
rural livelihoods 112–13, *119*, 120, *130–1*, 286
　Bolivia 208–11, *209*
　Cameroon 167–8, 171–3, *173*, 182–3, *182*, *183*, 185, *186*
　Honduras 140–1
　illegal logging and 139–40, 157–60, *159*, 182–3, *182*, *183*, 208–11, *209*
　impacts of illegal logging on 9, 113–15, *119*, 157–60, *159*
　impacts of law enforcement on 11–12, 110–12, 113, 115–18, *119*, 122–8, 172–3, 187–8, 213
　laws and 134, 179–81, *180*, 202–8, 207
　Nicaragua 140–1
　regulations and 12, 13, 168–70, *170*, 173–5, *175*, 278, 281

Sabah (Malaysia) 98, 99–101, *100*

sanctions 40, 226, *230*, 233–4
SF (Superintendencia Forestal, Bolivia) 194–7, *196*, 204, 209, 210, 211, 212
  perceptions of 204–5, 213, 214
shared meaning
  of community 18–19
  of illegal logging 19–26
shifting cultivation 118, 121, 124–5, 128, 171, *200*, 220
Sico-Paulaya Valley (Honduras) 154–5, *155*, 156, 158
SISNAMA (National Environmental Management System, Brazil) 226–7
slash-and-burn cultivation 118, 121, 124–5, 128, 171, *200*, 220
small farmers, Bolivia *200*, 201–2, 203–4, *207*, 208
small-scale concessions, Indonesia 45, 48, 53–4, 54–5, 55, 60, 61–2
small-scale logging, Cameroon 181–3, *182*, *183*, 276, 277
small-scale timber producers 277
  Bolivia *200*, 201, 203, 205, 207, *207*
social capital 113, 115, 117–18, *119*, 125, *131*
social control approach *231*, 239–40
Social Forestry System *see* SSF
social functions of illegal logging 39–40
social impacts of illegal logging 61–2, 277, 278
South Aceh (Sumatra) 70, 71–5, *73*, 81–3, 84, 85
  logging in Menggamat 75–81
SSF (Social Forestry System, Honduras) 144, *145*, 147, *148*, *151*, *152*
State Forestry Administration-Honduran Forestry Development Corporation *see* AFE-COHDEFOR
subsidies 104, *230–1*, *231*, 237–8
Sumatra (Indonesia) 11

  see also South Aceh
Superintendencia Forestal *see* SF
sustainability 60, 111, 120, *131*, 160, 238–9
sustainable forest management 13, 113, 191, 238–9, 256, 281, 282–3, 285
  Bolivia 198, 204, 214
  economics of 286
  Finland 220, 223, 224, 238
  incentives for 161, 198, 238
  law enforcement and 118, 120–2
swidden cultivation 118, 121, 124–5, 128, 171, *200*, 220

tax revenues 5, 73–4, 124, *130*
taxation
  Bolivia 197–8, 202, 206
  Cameroon 178–9, *178*
taxes 112, 122, 176, *230–1*, 237–8
  evasion 98–9, 101, 114, *119*, *130*
  local 48, 53–4, 57–9, 60, 63, 72–5, *73*, 78
  unpaid 210–11, 279
timber prices 5, 46, 154, 251, 252, 261, 278
timber theft 19, 20–1
timber trade 7–8, 142–3, 257–65, *262*
timber trespass 19, 20
titling 146–7, 160, *193*, *195*, 202–3
TNC (The Nature Conservancy) 14, 252, 255–6, 266
traditions 113, 118, *119*, 125, *131*
transparency 63, 85, 127, 128, 129, 161, 176
  Bolivia 13, 198, 204, 208, 214
  lack of 75, 114, 157
tree poaching 23–5
tree theft 10, 19–26, *25*, 31–9
types of illegal logging 276–8

UK (United Kingdom) 253, 254, 258, 260, 267
  see also DFID
unaffiliated timber theft 22–3

US (United States) 260, 264, 281
    Forest Service 17, 32–9, 37

verification 9, 14, 251–3, 256, 264, 265–7
verified products, demand for 14, 252, 253–4, 255–61, 268, 285–6
village heads 75–9, 82–3
violators, biggest 129–32
visibility threshold 29–30

volume constraints 150, *151*, 160
voluntary partnership agreements (EU) 14, 257, 263, 264, 268

wages 61, *130*, 158, 182
WWF (World Wide Fund for Nature) 2, 79, 257
    and certification 14, 252, 255–6, 259, 260, 266

zoning of forests 170–3, *171*, *173*

Join our
online community
and help us save paper and postage!

www.earthscan.co.uk

By joining the Earthscan website, our readers can benefit from a range of exciting new services and exclusive offers. You can also receive e-alerts and e-newsletters packed with information about our new books, forthcoming events, special offers, invitations to book launches, discussion forums and membership news. Help us to reduce our environmental impact by joining the Earthscan online community!

How? – Become a member in seconds!

>> Simply visit **www.earthscan.co.uk** and add your name and email address to the sign-up box in the top left of the screen – You're now a member!

>> With your new member's page, you can subscribe to our monthly **e-newsletter** and/or choose **e-alerts** in your chosen subjects of interest – you control the amount of mail you receive and can unsubscribe yourself

Why? – Membership benefits

- ✔ Membership is free!
- ✔ 10% discount on all books online
- ✔ Receive invitations to high-profile book launch events at the BT Tower, London Review of Books Bookshop, the Africa Centre and other exciting venues
- ✔ Receive e-newsletters and e-alerts delivered directly to your inbox, keeping you informed but not costing the Earth – you can also forward to friends and colleagues
- ✔ Create your own discussion topics and get engaged in online debates taking place in our new online Forum
- ✔ Receive special offers on our books as well as on products and services from our partners such as *The Ecologist*, *The Civic Trust* and more
- ✔ Academics – request inspection copies
- ✔ Journalists – subscribe to advance information e-alerts on upcoming titles and reply to receive a press copy upon publication – write to info@earthscan.co.uk for more information about this service
- ✔ Authors – keep up to date with the latest publications in your field
- ✔ NGOs – open an NGO Account with us and qualify for special discounts

Join now?
Join Earthscan now!

name

surname

email address

Earthscan Member
[Your name]

Click to Change

My profile
My forum
My bookmarks
All my pages

www.earthscan.co.uk

Your *essential* resource for responsible global travel

The Ethical Travel Guide
Introduction by *Guardian* journalist Polly Pattullo for Tourism Concern

Published by EARTHSCAN April 2006

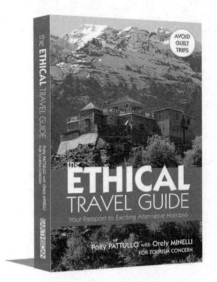

"As ever, Tourism Concern is at the forefront of efforts to ensure that the benefits of travel are shared much more equitably."

JONATHON PORRITT
Forum for the Future

Want to have exciting holidays that have a positive impact on local people *and* the environment?

- **300 holidays in 60 countries that benefit local people** *directly*
- **Find hundreds of amazing places not listed in other guide books – from simple local style holidays to luxury retreats**
- **Read *Guardian* journalist Polly Pattullo's expert 'state-of-the-industry' introduction to tourism and ethical travel**
- **Tourism is the world's largest and fastest growing industry – here's how be part of the solution, not part of the problem.**

Available from all good bookshops or direct from Earthscan:
Visit www.earthscan.co.uk or phone 01256 302699 and provide this ISBN: 1-84407-321-1

RRP: £12.99 – 10% online discount: £11.69

www.earthscan.co.uk